LBS
1060107
AF749698

Econometric Analysis of Count Data

Springer
Berlin
Heidelberg
New York
Hong Kong
London
Milan
Paris
Tokyo

Rainer Winkelmann

Econometric Analysis of Count Data

Fourth Edition

With 19 Figures and 20 Tables

Springer

Professor Dr. Rainer Winkelmann
University of Zürich
Department of Economics
Zürichbergstraße 14
8032 Zürich
Switzerland
winkelmann@sts.unizh.ch

ISBN 3-540-40404-X Springer-Verlag Berlin Heidelberg New York
ISBN 3-540-67340-7 3rd edition Springer-Verlag Berlin Heidelberg New York

Cataloging-in-Publication Data applied for
A catalog record for this book is available from the Library of Congress.
Bibliographic information published by Die Deutsche Bibliothek
Die Deutsche Bibliothek lists this publication in the Deutsche Nationalbibliografie; detailed bibliographic data is available in the Internet at <http://dnb.ddb.de>.

This work is subject to copyright. All rights are reserved, whether the whole or part of the material is concerned, specifically the rights of translation, reprinting, reuse of illustrations, recitation, broadcasting, reproduction on microfilm or in any other way, and storage in data banks. Duplication of this publication or parts thereof is permitted only under the provisions of the German Copyright Law of September 9, 1965, in its current version, and permission for use must always be obtained from Springer-Verlag. Violations are liable for prosecution under the German Copyright Law.

Springer-Verlag Berlin Heidelberg New York
a member of BertelsmannSpringer Science+Business Media GmbH

http://www.springer.de

© Springer-Verlag Berlin · Heidelberg 1994, 1997, 2000, 2003
Printed in Germany

The use of general descriptive names, registered names, trademarks, etc. in this publication does not imply, even in the absence of a specific statement, that such names are exempt from the relevant protective laws and regulations and therefore free for general use.

Hardcover-Design: Erich Kirchner, Heidelberg

SPIN 10937716 42/3130-5 4 3 2 1 0 – Printed on acid-free paper

To Gerhard

Preface

When asked why he wanted to climb Mount Everest, George Mallory famously replied, "Because it is there." Similarly, a great deal of the recent interest in count data modelling is explicable by the simple fact that such discrete data that take on only non-negative integer values are frequently encountered in empirical practice. Examples are as diverse as the number of congressional budget vetoes, the number of children in a household, the number of trips to a recreational site, the number of visits to a physician, or the number of mechanical defects in a production line.

The frequent collection of statistical information by counting, in itself, would not be a sufficient reason for developing a whole class of specialised models tailored to such data. After all, standard methods such as linear regression might work just fine even if applied to count data. However, as will be shown throughout this book, the standard linear model framework is not nearly as rich and complex as is necessary to understand count data. For example, in the case of count data any explanatory variable will affect not only the conditional mean but the conditional variance and each probability as well. Moreover, much can be learned by interpreting count data as resulting from an underlying count process, where issues such as unobserved heterogeneity, duration dependence, and varying intensity become immediate concerns. All these aspects become apparent once one recognizes the discrete structure of the data and analyses them within the framework of full probability models for count data.

The primary objective of this book, now in its fourth edition, is to introduce these models at a level suitable for graduate students, researchers or interested practitioners, building on the latest developments in the literature. While the models presented in this book are in general motivated by substantive issue and examples from applied social sciences, economics in particular, the book should also include much relevant material for other empirical disciplines such as biometrics and technometrics. The unifying theme is a focus on conditional distribution models where the probability distribution of the dependent count variable is modeled as a function of explanatory variables.

The main objects of interest are estimates of how changes in an explanatory variable affect various aspects of the outcome distribution, and predictions of the outcome distributions for given values of the explanatory variables.

This latest edition has many new features. One important aspect is that I place now more emphasis on the interpretation of parameters and models. Indeed, given the progress in IT capabilities, and the inclusion of many specialised count data modules in standard statistical software, estimation of these models, all of them highly nonlinear, has become quite straightforward. What remains a challenge, however, is the proper interpretation of the parameter estimates. Also, count data models are inherently more complex than simple linear regression models, and offer a richer set of interesting inferences. For example, "marginal probability effects" can be computed in order to trace the response of the entire count distribution to small changes in an explanatory variables. To what extent has a variable different effects in different parts of the outcome distribution? Seen from this perspective all newly developed, generalized, count data models have to pass a simple test: do they offer additional interesting interpretations of the data generating process that were not possible with simpler models?

There are many other changes to this fourth edition. The literature has been updated to include innovations up to the time of printing in 2003. A new chapter on quantile regression for count data has been added, as has been a chapter on measurement error. The chapters on endogenous regressors, simultanous equations and hurdle models have been completely rewritten. They offer now much more information and detail. The chapters on selectivity have been reorganized. New application sections provide information on the use of count data models in crime and in trip frequency studies. Finally, a software review section and an author's index have been added.

My work on this book, in its various stages, has greatly benefited from comments and suggestions by students and colleagues. In particular, I would like to thank Kurt Brännäs, Malcolm Faddy, William Greene, James Heckman, Nikolai Kolev, Jochen Mayer, Daniel Miles, Andreas Million, Hans van Ophem, Pravin Trivedi, Frank Windmeijer, and in particular Joao Santos Silva for very helpful comments. Over the years, I also had the good fortune of taking part in a number of joint research projects with Tim Barmby, Siddharta Chib, Edward Greenberg, Robert Jung, and Gary King, respectively. I have learned a lot from these collaborations and I am grateful to my co-authors for sharing their knowledge and interest with me. A final thank you goes to Klaus F. Zimmermann who, as Ph.D. supervisor and co-author, initiated my interest in count data models and provided continuing advice and support.

Zürich, May 2003 *Rainer Winkelmann*

Contents

List of Figures

List of Tables

1

Introduction

In this book I consider regression models for count dependent variables, i.e., dependent variables that take the values $y = 0, 1, 2, \ldots$ without explicit upper limit. Regression analysis, narrowly defined, attempts to explain variations in the conditional expectation of y with the help of variation in explanatory variables x. Regression analysis, more broadly defined and as used in this book, refers to the estimation of conditional distribution functions of y given x.

Regression models are tools for the applied researcher who wants to explore data or weigh the evidence in data for or against a specific hypothesis. In the first case, their function is to reduce the dimensionality of complex information and to help discover structures in data, thereby motivating the direction of future theorizing. In the second case, they are used to test economic theories, quantify theoretical relations, and predict or forecast.

In either case, specialized tools are usually preferred to more blunt all-purpose tools. The econometric all-purpose regression tool is the normal linear regression model. However, for discrete data in general, and for count data in particular, normal linear regression has a number of shortcomings that render its use dubious and logically unsatisfactory. Hence, more suitable tools are required. Examples for discrete regression models are the binomial and multinomial probit and logit regression models for categorical data, ordered probit and logit models for ordinal data, and the Poisson regression model (PRM) for count data.

While the PRM is an improvement over the normal linear model as it explicitly recognizes the non-negative integer character of the dependent variable, it can be inappropriate in other respects. In fact, it is a common finding in applied work using economic count data that certain assumptions of the PRM are systematically rejected by the data. Much of this book is concerned with a unified presentation of the whole variety of count data models that have been developed to date in response to these restrictive features of the PRM. Nevertheless, we start with a brief characterization of the PRM, both in order to motivate the following presentation, and because it remains the bench-mark model for count data in much the same way as the normal linear model is the bench-mark model for continuous data.

1.1 Poisson Regression Model

The Poisson regression model has two components, first a distributional assumption, and second a specification of the mean parameter as a function of explanatory variables. The Poisson distribution is a one parameter distribution. The parameter, λ, is equal to the mean and the variance, and it must be positive. It is convenient to specify λ as a log-linear function of the explanatory variables x_i that account for observed sample heterogeneity: $\lambda_i = \exp(\beta_0 + \sum_j x_{ij}\beta_j)$ or, in vector notation, $\lambda_i = \exp(x_i'\beta)$. In this way, λ_i remains positive for all possible combinations of parameters and explanatory variables. Moreover, the systematic effects interact in a multiplicative way, and the coefficients β_j have the interpretation of a partial elasticity of $\mathrm{E}(y|x)$ with respect to (the level of) x_j if the logarithm of x_j is included among the regressors.

Assuming an independent sample, the parameters of the model can be estimated by maximum likelihood. Although the first-order condition is nonlinear and thus not solvable in closed form, iterative algorithms can be used to find the maximum which is unique as the log-likelihood function is globally concave. Under correct specification, the estimator has all the desirable properties of maximum likelihood estimators, in particular asymptotic efficiency and normality.

The lack of a mean-independent determination of the variance for the Poisson distribution contrast with the flexibility of the two-parameter normal distribution where the variance of the distribution can be adjusted independently of the mean. This feature of the PRM is likely too restrictive. However, Poisson regression is robust: the estimator for β remains consistent even if the variance does not equal the mean (and the true distribution therefore cannot be Poisson) as long as the mean function λ is correctly specified. This robustness mirrors the result for the linear model where OLS is unbiased independently of the second-order moments of the error distribution.

A different issue is that of efficient estimation. A model needs to be correctly specified in order to apply the efficiency results for maximum likelihood estimators. This requirement is more binding for count data models than for the normal linear model. In the latter model, OLS or GLS are efficient in the class of linear estimators, no matter what the distribution is, and efficiency among all estimators does not require a particular variance function as long as normality holds.

1.2 Some Thoughts on Methodology

The book emphasizes the classical function of econometric modeling: testing the predictions of economic theories. Confronting economic theory with data typically consists of a three step sequence: specification, estimation, and evaluation. The specification step defines a stochastic model that is deemed to have generated the observed data. A distinctive characteristic of econometric models is the introduction of exogenous explanatory variables. It is the role of economic theory to suggest variables that affect the endogenous variable and thus explain the observed variation in the endogenous count variable.

In parametric modeling, the distribution of the endogenous variable conditional on explanatory variables x is determined up to an unknown vector of parameters. The model parameters, which are either parameters of primary interest or nuisance parameters, are estimated as a function of the data. The estimation technique most widely used throughout this book is maximum likelihood. The basics of Bayesian posterior analysis are given as well.

In semiparametric modeling, the distribution of the endogenous variable is left unspecified. Identification of the model parameters rests on weaker assumptions, such as the form of the mean function, or mean and variance functions. Such models are estimated by pseudo-maximum likelihood, nonlinear least squares, or (generalized) method of moments (GMM).

The two considerations in choosing an estimator, robustness and efficiency, were mentioned before. There is a tradeoff as strong assumptions give strong results when correct but possibly very poor results when false. Hence, with efficient estimation being a potentially elusive goal, robust estimation has much to speak for itself. Indeed, a substantial part of the literature has dealt with semiparametric modeling, including recent applications based on GMM.

Against these trends, the book puts a disproportionate emphasis on the introduction of a rich class of parametric models for count data. A couple of points can be made in support of this approach. First, by adopting a semiparametric model in order to estimate the regression coefficients consistently, the researcher foregoes the possibility to predict probabilities of single outcomes or conditional distribution functions. If this is a requirement of the analysis, parametric models are needed. In a sense, semiparametric modeling ignores the non-negative integer nature of the dependent variable that generated the interest in a suitable class of specialized tools in the first place.

Secondly, parametric modeling gives the researcher a framework in which to formulate and study the underlying stochastic data generating process. The relationship between duration analysis and count data, sample selection, hurdle models, discrete and continuous mixture models, to name but a few, are intrinsically related to statistical distribution theory. Such developments offer important insights and re-interpretations of existing models, and suggest avenues for improved ones. Although it is the case that estimation of the resulting models can often be cast in a semiparametric framework, the parametric approach provides conceptual benefits at the design/specification level, and it is hoped that these benefits become clear throughout the remainder of the book.

1.3 Examples

The count model of choice very much depends on the type of available data. In particular, the following questions have to be answered at the outset:

- What is the nature of the count data? Are they univariate or multivariate, are they grouped or censored, what is known about the stochastic process underlying the generation of the data?
- What was the sampling method? Are the data representative of the population, or have they been sampled selectively?

Note that we assume here that the researcher is not involved personally in sampling design but rather has to work with data that have been generated elsewhere and without consultation. This should be a realistic assumption for most research undertaken by applied economists.

A crude frequency count can be helpful in selecting an initial model framework. Consider, for instance, the following examples taken from the applied count data literature:

- Kennan (1985) gives the monthly number of contract strikes in U.S. manufacturing. In his analysis, Kennan concentrates on the duration of strikes, rather than on their number per se.
- McCullagh and Nelder (1989) look at the incidence of certain ship damages caused by waves using the data provided by an insurance company. They model the number of incidents regardless of the damage level.
- Zimmermann and Schwalbach (1991) use a data set on the number of patents (stock) of German companies registered at the German Patent Office in 1982. They merge information from the annual reports of the respective companies as well as industry variables.
- Davutyan (1989) studies how the number of failed banks per year in the U.S. for 1947 - 1981 relates to explanatory variables such as a measure of the absolute profitability of the economy, the relative profitability of the banking sector, as well as aggregate borrowing from the Federal Reserve.
- Dionne, Gagné, Gagnon and Vanasse (1997) study the frequency of airline accidents (and incidents) by carrier in Canada on a quarterly basis between 1974 and 1988. Their sample includes approximately 100 Canadian carriers, resulting in around 4000 panel entries. The total number of accidents during the period was 530.
- Winkelmann and Zimmermann (1994) model completed fertility measured by the number of children. Using the German Socio-Economic Panel, they select women aged between 40 and 65 who live in their first marriage. The number of children varies from 0 to 10, the mean is 2.06, and the mode is 2.

The respective empirical frequency distributions of the dependent count variable are given in the following table:

Table 1.1. Count Data Frequency Distributions

Counts	Strikes	Ships	Patents	Banks	Airplane	Children
0	-	9	30	-	3498	61
1	12	5	6	-	411	167
2	14	2	7	2	51	297
3	11	1	2	7	3	117
4	9	2	0	4	2	52
5	14	1	3	4	-	14
6	9	2	1	4	-	12
7	4	2	2	1	-	2
8	7	0	0	3	-	1
9	10	0	1	5	-	-
10	6	0	0	3	-	1
> 10	7	11	19	2	-	-
> 100	-	-	20	-	-	-
Observations	103	34	91	35	3965	724
Maximum	18	58	9805	17	4	10
Mean	5.5	10.2	304.6	6.3	0.013	2.1
Variance	13.4	236.5	$1.6{*}10^6$	11.8	0.015	1.7

The six frequency distributions are very indicative of the type of data encountered in applied research. They differ in a number of ways. First, the realized range of observations differs significantly. In two cases, no zeros are observed, while in other cases, zero is the modal value. Some of the empirical distributions are uni-modal, while others display multiple modes. In most cases, the variance clearly exceeds the mean, while in one case (airlines) it is roughly the same, and in one case (children), the mean is greater than the variance. Second, the structure of the data differs. The three observed types of data are a cross section of individuals, a panel, and a time series. Models for all three types of data are covered in this book.

It should be noted that Tab. 1.1 shows marginal frequency distributions whereas the focus of this book is on conditional models. Such models account for the influence of covariates in a regression framework. For instance, if the conditional distribution of y given (a non-constant) x is Poisson, the marginal distribution of y cannot be Poisson as well.

1.4 Organization of the Book

Chap. 2 presents probability models for count data. The basic distributions are introduced. They are characterized both through the underlying stochastic process, and through their relationships amongst each other. Most generalizations rely on the tools of mixing and compounding – these techniques are described in some detail. A discussion of hyper-distributions reveals the differences and commonalities between the models. This chapter also draws extensive analogies between probabilistic models for duration data and probabilistic models for count data.

Chap. 3 starts with a detailed exposition of the Poisson regression model, including a comparison with the linear model. Two issues that are of particular relevance for the practitioner are the correct interpretation of the regression coefficients, and how to deal with varying time of exposure. The basic estimation techniques are discussed, and the properties of the estimators are derived. The second part of the chapter is devoted to possible misspecification of the Poisson regression model: its origins, consequences, and how to detect misspecification through appropriate testing procedures.

Chap. 4 deals with alternative models that each address a specific shortcoming of the Poisson regression model, such as unobserved heterogeneity, dependent count processes, selective sampling and endogenous regressors. Many of these models are discussed from a parametric point of view, except for the last two sections where generalized variance models using semiparametric estimation and quantile regression is discussed.

Chap. 5 is concerned with count data models for multivariate, panel and time series data. This is an area of intensive current research effort, and many of the referred papers are still at a working paper stage. However, a rich class of models is beginning to emerge and the issues are well established: the need for a flexible correlation structure in the multivariate context, and the lack of strictly exogenous regressors in the case of panel data.

Chap. 6 provides an introduction to Bayesian posterior analysis of count data. Again, many of the developments in this area are quite recent. They partly mirror the general revival of applied Bayesian analysis that was triggered by the combined effect of increasing computing power and the development of powerful algorithms for Markov chain Monte Carlo simulation. The potential of this approach is demonstrated, among other things, in an model for highly dimensional panel count data models with correlated random effects.

The final Chap. 7 illustrates the practical use of count data models in a number of applications. Apart from a literature review for applications such as accidents, health economics, demography and marketing, the chapter contains an extended study of the determinants of labor mobility using data from the German Socio-Economic Panel.

2

Probability Models for Count Data

2.1 Introduction

The introduction emphasized the need for a rich class of probability distributions when modeling count data. Since probability distributions for counts are not yet standard in the econometric literature, their properties are explored in some detail in this chapter. Special attention is paid to flexible, or 'generalized', count data distributions since they serve as building blocks for improved count data regression models. Furthermore, the properties of the underlying data generating process are studied. Count data frequently may be interpreted as outcomes of an underlying *count process* in continuous time. The classical example for a count process is the number of incoming telephone calls at a switchboard during a fixed time interval. Let the random variable $N(t)$, $t > 0$, describe the number of occurrences during the interval $(0, t)$. *Duration analysis* studies the waiting times τ_i, $i = 1, 2, \ldots$, between the $(i-1)$-th and the i-th event. *Count data models*, by contrast, model $N(T)$ for a given T. By studying the relation between the underlying count process, the most prominent being the Poisson process, and the resulting probability models for event counts N, one can acquire a better understanding of the conditions under which a given count distribution is appropriate. For instance, the Poisson process, resulting in the Poisson distribution for the number of counts during a fixed time interval, requires independence and constant probabilities for the occurrence of successive events, an assumption that appears to be quite restrictive in most applications to social sciences or elsewhere. Further results are derived in this chapter.

2.2 Poisson Distribution

2.2.1 Definitions and Properties

Let X be a random variable with a discrete distribution that is defined over $\mathbb{N} \cup \{0\} = \{0, 1, 2, \ldots\}$. X has a Poisson distribution with parameter λ, written $X \sim \text{Poisson}(\lambda)$ if and only if the probability function is as follows:

$$P(X = k) = \frac{e^{-\lambda}\lambda^k}{k!}, \qquad \lambda \in \mathbb{R}^+ ,\ k = 0, 1, 2, \ldots \tag{2.1}$$

The probability generating function of the Poisson distribution is given by

$$\begin{aligned}
\mathcal{P}(s) &= \sum_{k=0}^{\infty} s^k P(X = k) \\
&= \sum_{k=0}^{\infty} s^k \frac{e^{-\lambda}\lambda^k}{k!} \\
&= e^{-\lambda} \sum_{k=0}^{\infty} \frac{(\lambda s)^k}{k!} \\
&= e^{-\lambda + \lambda s}
\end{aligned} \tag{2.2}$$

(See Appendix A for definition and properties of the probability generating function). Conversely, the Poisson probability function is obtained as

$$P(X = k) = (k!)^{-1} \left. \frac{\mathrm{d}^k \mathcal{P}}{\mathrm{d}s^k} \right|_{s=0} \tag{2.3}$$

$$= \frac{e^{-\lambda}\lambda^k}{k!} \tag{2.4}$$

The Poisson distribution has expected value

$$\mathrm{E}(X) = \mathcal{P}'(1) = \lambda \tag{2.5}$$

and variance

$$\begin{aligned}
\mathrm{Var}(X) &= \mathcal{P}''(1) + \mathcal{P}'(1) - [\mathcal{P}'(1)]^2 \\
&= \lambda^2 + \lambda - \lambda^2 \\
&= \lambda
\end{aligned} \tag{2.6}$$

Alternatively, the expected value can be derived directly using the probability function:

$$\begin{aligned}
\mathrm{E}(X) &= \sum_{k=0}^{\infty} k \frac{e^{-\lambda}\lambda^k}{k!} \\
&= \sum_{k=1}^{\infty} k \frac{e^{-\lambda}\lambda^k}{k!} \\
&= \lambda \sum_{k=1}^{\infty} \frac{e^{-\lambda}\lambda^{(k-1)}}{(k-1)!} \\
&= \lambda
\end{aligned} \tag{2.7}$$

The equality of mean and variance is characteristic of the Poisson distribution. It plays a crucial role in the further discussion and will be referred to as *equidispersion*. Departures from equidispersion can be either *overdispersion* (variance is greater than the mean) or *underdispersion* (variance is smaller than the mean). In contrast to other multi-parameter distributions, such as the normal distribution, a violation of the variance assumption is sufficient for a violation of the Poisson assumption.

Some further properties of the Poisson distribution

1. The ratio of recursive probabilities can be written as:

   $$\frac{p_k}{p_{k-1}} = \frac{\lambda}{k} \, . \tag{2.8}$$

   Thus, probabilities are strictly decreasing for $0 < \lambda < 1$ and the mode is 0; for $\lambda > 1$, the probabilities are increasing for $k \leq \text{int}[\lambda]$ and then decreasing. The distribution is uni-modal if λ is not an integer and the mode is given by $\text{int}[\lambda]$. If λ is an integer, the distribution is bi-modal with modes at λ and $\lambda - 1$.
2. Taking the first derivative of the Poisson probability function with respect to the parameter λ, we obtain

   $$\frac{dp_k}{d\lambda} = \frac{e^{-\lambda}\cdot(-1)\lambda^k}{k!} + \frac{e^{-\lambda}k\lambda^{k-1}}{k!} = p_k\left(\frac{k}{\lambda} - 1\right) \tag{2.9}$$

   Therefore, the probabilities p_k decrease with an increase in λ (i.e., with an increase in the expected value) for $k < \lambda$. Thereafter, for $k > \lambda$, the probabilities p_k increase with an increase in λ.
3. Consider the dichotomous outcomes $P(X = 0)$ and $P(X > 0)$. The probabilities are given by

   $$p_0 = e^{-\lambda}$$

   and

   $$p_+ = 1 - e^{-\lambda} \, ,$$

respectively. These expressions coincide with the cumulative and complementary cumulative density functions of the exponential distribution. The intrinsic relation between the Poisson distribution and the exponential distribution is explored in section (2.2.6).

Proposition 1 (Sums of Poisson Random Variables). *Assume that* $X \sim \text{Poisson}(\lambda)$ *and* $Y \sim \text{Poisson}(\mu)$, $\lambda,\ \mu \in I\!R^+$. *The random variable* $Z = X + Y$ *is Poisson distributed* $Po(\lambda + \mu)$ *if and only if* X *and* Y *are independent.*

Proof: ($\Longleftarrow$) Let X and Y be independent. Then $\text{E}(s^{X+Y}) = \text{E}(s^X)\text{E}(s^Y)$. Further,

$$\begin{aligned}\mathcal{P}^{(Z)} &= \text{E}(s^{X+Y}) \\ &= e^{-(\lambda+\mu)+(\lambda+\mu)s} \end{aligned} \tag{2.10}$$

which is exactly the probability generating function of a Poisson distributed random variable with parameter $(\lambda + \mu)$. Hence, $Z \sim \text{Poisson}(\lambda + \mu)$.

Alternatively, from first principles,

$$\begin{aligned}P(Z = k) &= \sum_{i=0}^{k} P(X = k - i)P(Y = i) \\ &= \sum_{i=0}^{k} \frac{e^{-\lambda}\lambda^{k-i}}{(k-i)!}\frac{e^{-\mu}\mu^i}{i!} \\ &= \frac{e^{-\lambda-\mu}}{k!}\sum_{i=0}^{k}\frac{k!}{(k-i)!i!}\lambda^{k-i}\mu^i \\ &= \frac{e^{-\lambda-\mu}(\lambda+\mu)^k}{k!} \end{aligned} \tag{2.11}$$

where the last equality follows from the definition of binomial coefficients.

($\Longrightarrow$) The issue is whether two Poisson variables, the sum of which is again Poisson distributed, must be independent. In other words, it is to be shown that the sum of two dependent Poisson variables cannot be Poisson distributed. A later chapter gives a proof based on a more sophisticated technique using properties of the bivariate Poisson distribution. Here, an example is given showing that the sum of two dependent variables is not necessarily Poisson distributed.

Let the joint distribution of X and Y be

$$P(X = x, Y = y) = \frac{\lambda^x}{x!}\frac{(\mu-\lambda)^{y-x}}{(y-x)!}e^{-\mu}\ ,\ x, y \in \{0, 1, 2, \ldots\}$$

where $y \geq x$. Clearly, X and Y are not independent since the support of Y depends on the realization x. However, both marginal distributions are Poisson:

$$
\begin{aligned}
P(X=x) &= \sum_{y=x}^{\infty} \frac{\lambda^x}{x!} \frac{(\mu-\lambda)^{y-x}}{(y-x)!} e^{-\mu} \\
&= \sum_{y=0}^{\infty} \frac{\lambda^x}{x!} \frac{(\mu-\lambda)^{y}}{y!} e^{-\mu} \\
&= \frac{\lambda^x}{x!} e^{-\lambda} \sum_{y=0}^{\infty} \frac{(\mu-\lambda)^{y}}{y!} e^{-(\mu-\lambda)} \\
&= \frac{\lambda^x}{x!} e^{-\lambda}
\end{aligned}
$$

$$
\begin{aligned}
P(Y=y) &= \sum_{x=0}^{y} \frac{\lambda^x}{x!} \frac{(\mu-\lambda)^{y-x}}{(y-x)!} e^{-\mu} \\
&= \frac{e^{-\mu}}{y!} \sum_{x=0}^{y} \frac{y!}{x!(y-x)!} \lambda^x (\mu-\lambda)^{y-x} \\
&= \frac{e^{-\mu}\mu^y}{y!}
\end{aligned}
$$

Now, consider $Z = X + Y$ (a sum of two Poisson distributed variables):

$$
\begin{aligned}
P(Z=0) &= P\{X=0, Y=0\} \\
&= e^{-\mu}
\end{aligned}
$$

Similarly

$$
\begin{aligned}
P(Z=1) &= P\{X=0, Y=1\} \\
&= (\mu-\lambda)e^{-\mu}
\end{aligned}
$$

Assume that $Z \sim \text{Poisson}(\nu)$, $\nu \in I\!R^+$. From $P\{Z=0\} = e^{-\mu}$ follows $\nu = \mu$. Likewise, from $P\{Z=1\} = (\mu-\lambda)e^{-\mu}$ and hence $\nu = \mu - \lambda$. Thus, it must be the case that $\lambda = 0$, which contradicts the starting assumption that $\lambda > 0$. Hence, Z cannot be Poisson distributed for $\mu, \lambda > 0$. If two variables are not independent their sum is not (necessarily) Poisson distributed, which was to be shown. □

Proposition 2 (Linear Transformations). *The Poisson distribution is not closed under linear transformations, in the sense that linear transformations on the sample space do not generate again a Poisson distribution with a different value of the parameter* λ.

Proof: Let $Y = a + bX$ with $X \sim \text{Poisson}(\lambda)$ and a, b arbitrary constants. For Y to be Poisson distributed, it must be true that $\text{E}(Y) = a + b\lambda = \text{Var}(Y) = b^2\lambda$ for any $\lambda > 0$. But the equality holds if and only if $a = 0$ and

$b = 0$ or $b = 1$. Thus, Y does not have a Poisson distribution for arbitrary values of a and b. □

The distribution of $Y = a + bX$ for $b = 1$ is sometimes referred to as "shifted" or "displaced" Poisson distribution with probability function

$$P(X = k) = \frac{e^{-\lambda}\lambda^{(k-a)}}{(k-a)!}, \qquad k = a, a+1, a+2, \ldots \tag{2.12}$$

where a generally is taken to be an integer variable, although this is not necessary. For $a > 0$, such a distribution is characterized by underdispersion (see also Chap. 4.4.1).

It can be shown that within a large class of distributions, only the normal distribution is preserved under both location and scale transformation (see Hinkley and Reid, 1991).

2.2.2 Genesis of the Poisson Distribution

In most applications the Poisson distribution is used to model the number of events that occur over a specific time period (such as the number of telephone calls arriving at a switchboard operator during a given hour, the annual number of visits to a doctor, etc.). It is thus of interest to study how the Poisson distribution is related to the intertemporal distribution of events. The next section introduces the general concept needed for the analysis of this issue, the *stochastic process*. The subsequent sections present a number of underlying stochastic models that each give rise to a Poisson distribution for the number of events during the fixed time interval.

The first model is the *Poisson process* in continuous time. The second model introduces the Poisson distribution as a limiting form of a discrete time stochastic process. Finally, the Poisson distribution arises from independently and identically exponentially distributed interarrival times between events. All three derivations require as their main assumption that events occur completely randomly over time. The underlying randomness is the hallmark of the Poisson distribution.

2.2.3 Poisson Process

The Poisson process is a special case of a count process which, in turn, is a special case of a stochastic process. Hence, some general definitions will be introduced first, before the properties of the Poisson process are presented.

Definition 1. *A stochastic process* $\{X(t), t \in T\}$ *is a collection of random variables (on some probability space) indexed by time.*

$X(t)$ is a random variable that marks the occurrence of an event at time t. The underlying experiment itself remains unformalized and the definitions and arguments are framed exclusively in terms of the $X(t)$. If the index set T is an interval on the real line, the stochastic process is said to be a *continuous time* stochastic process. If the cardinal number of T is equal to the cardinal number of $\mathbb{N}$, it is called a *discrete time* stochastic process.

Definition 2. *A stochastic process $\{N(t), t \geq 0\}$ is said to be a count process if $N(t)$ represents the total number of events that have occurred before t.*

The following properties hold:

1. $N(t) \geq 0$
2. $N(t)$ is integer valued
3. For $s < t$, $N(s) \leq N(t)$
4. For $s < t$, $N(t) - N(s)$ gives the number of events that occurred in the interval (s, t)

A count process is called stationary if the distribution of the number of events in any time interval depends only on the length of the interval:

$$(\forall s > 0) \qquad N(t_2 + s) - N(t_1 + s) \overset{i.d.}{\sim} N(t_2) - N(t_1)$$

A count process has independent increments if the numbers of events which occur in disjoint time intervals are independent.

The Poisson process is a continuous time count process with stationary and independent increments. In other words, it assumes that the occurrence of a random event at a particular moment is independent of time and of the number of events that have already taken place. Let $N(t, t + \Delta)$ be the number of events that occurred between t and $t + \Delta$, $t > 0, \Delta > 0$. The two basic assumptions of the Poisson process can be formalized as follows:

a) The probability that an event will occur during the interval $(t, t + \Delta)$ is stochastically independent of the number of events occurring before t.
b) The probabilities of one and zero occurrences, respectively, during the interval $(t, t + \Delta)$ are given by:

$$P\{N(t, t + \Delta) = 1\} = \lambda\Delta + o(\Delta) \tag{2.13}$$

$$P\{N(t, t + \Delta) = 0\} = 1 - \lambda\Delta + o(\Delta) \tag{2.14}$$

where $o(\Delta)$ represents any function of Δ which tends to 0 faster than Δ, i.e., any function such that $[o(\Delta)/\Delta] \to 0$ as $\Delta \to 0$.

It follows that the probability of an occurrence is proportional to the length of the interval and the proportionality factor is a constant independent of t. Further,

$$\begin{aligned} P\{N(t,t+\Delta) > 1\} &= 1 - P\{N(t,t+\Delta) = 0\} \\ &\quad - P\{N(t,t+\Delta) = 1\} \\ &= o(\Delta) \ . \end{aligned} \tag{2.15}$$

In a sufficiently short interval, the probability of two or more events occuring approaches zero.

Assumptions a) and b) can be restated by saying that the increments of a Poisson process are *independent* and *stationary*: $N(t, t+\Delta)$ and $N(s, s+\Delta)$ are independent for disjoint intervals $(t, t+\Delta)$ and $(s, s+\Delta)$, and $P\{N(t, t+\Delta) = k\}$ is independent of t.

Let $p_k(t+\Delta) = P\{N(0, t+\Delta) = k\}$ denote the probability that k events occurred before $(t+\Delta)$. The outcome $\{N(0, t+\Delta) = k\}$ can be obtained in $k+1$ mutually exclusive ways:

$$\begin{aligned} &\{N(0,t) = k\} \text{ and } \{N(t,t+\Delta) = 0\} \text{ , or} \\ &\{N(0,t) = k-1\} \text{ and } \{N(t,t+\Delta) = 1\} \text{ , or} \\ &\qquad\vdots \\ &\{N(0,t) = 0\} \text{ and } \{N(t,t+\Delta) = k\} \ . \end{aligned}$$

By assumption of independence, the probability of each of the above outcomes equals the product of the single probabilities of its two constituent parts. For example,

$$P[\{N(0,t) = k\} \text{ and } \{N(t,t+\Delta) = 0\}] = p_k(t)(1 - \lambda\Delta) \tag{2.16}$$

Similarly,

$$P[\{N(0,t) = k-1\} \text{ and } \{N(t,t+\Delta) = 1\}] = p_{k-1}(t)\lambda\Delta \tag{2.17}$$

Furthermore, since the outcome "two or more events" has probability zero we get that

$$P[\{N(0,t) = k-j\} \text{ and } \{N(t,t+\Delta) = j\}] = 0$$

for $j \geq 2$. Finally, the two outcomes (2.16) and (2.17) are disjoint, and the probability of their union is therefore given by the sum of their probabilities. Putting everything together, we obtain:

$$p_k(t+\Delta) = p_k(t)(1 - \lambda\Delta) + p_{k-1}(t)\lambda\Delta + o(\Delta) \tag{2.18}$$

i.e.

$$\frac{p_k(t+\Delta) - p_k(t)}{\Delta} = -\lambda(p_k(t) - p_{k-1}(t)) + o(\Delta) \ . \tag{2.19}$$

Taking limits for $\Delta \to 0$:

$$\frac{dp_k(t)}{dt} = -\lambda(p_k(t) - p_{k-1}(t)) \tag{2.20}$$

and similarly that

$$\frac{dp_0(t)}{dt} = -\lambda p_0(t) \tag{2.21}$$

The differential equation (2.21) can be solved using the initial condition $p_0(0) = 1$ to obtain

$$p_0(t) = \exp(-\lambda t)$$

Setting $k = 1$ in (2.20) and multiplying through by $\exp(\lambda t)$, we obtain

$$\exp(\lambda t)\frac{dp_1(t)}{dt} + \lambda \exp(\lambda t)p_1(t) = \frac{d}{dt}[\exp(\lambda t)p_1(t)] = \lambda$$

with solution

$$p_1(t) = \lambda t \exp(-\lambda t)$$

Repeated applications of the same procedure for $k = 2, 3, \ldots$ yields the Poisson probability distribution. Alternatively, one can derive directly the probability generating function of the Poisson distribution:

$$\frac{\mathrm{d}\mathcal{P}(s;t)}{\mathrm{d}t} = \frac{\mathrm{d}\sum_{k=0}^{\infty} p_k(t)s^k}{\mathrm{d}t}$$

$$= \sum_{k=0}^{\infty}[-\lambda p_k(t) + \lambda p_{k-1}(t)]s^k \tag{2.22}$$

$$= -\lambda \sum_{k=0}^{\infty} p_k(t)s^k + \lambda s \sum_{k=1}^{\infty} p_{k-1}(t)s^{k-1} \tag{2.23}$$

$$= (-\lambda + \lambda s)\mathcal{P}(s;t)$$

where it is understood that $p_{-1} = 0$. This first order differential equation has solution

$$\mathcal{P}(s;t) = \exp[(-\lambda + \lambda s)t] \tag{2.24}$$

The length of the interval can be normalized to unity, which gives the probability generating function of the standard Poisson distribution.

2.2.4 Generalizations of the Poisson Process

Non-stationarity

A first generalization is to replace the constant λ in (2.13) by a time-dependent variable $\lambda(t)$:

$$P\{N(t, t + \Delta) = 1\} = \lambda(t)\Delta + o(\Delta)\ . \tag{2.25}$$

Define the integrated intensity $\Lambda(t) = \int_0^t \lambda(s)ds$. It can be shown that

$$P\{N(t) = k\} = \frac{e^{-\Lambda(t)}\Lambda(t)^k}{k!} . \tag{2.26}$$

$N(t)$ has a Poisson distribution function with mean $\Lambda(t)$. Hence, this generalization does not affect the form of the distribution.

Dependence

In order to explicitly introduce path dependence, it is helpful to rewrite the basic equation defining the Poisson process (2.13) in terms of the conditional probability

$$P\{N(0, t + \Delta) = k + 1|N(0, t) = k\} = \lambda\Delta + o(\Delta)$$

One generalization is to allow the rate λ to depend on the current number of events, in which case we can write

$$P\{N(0, t + \Delta) = k + 1|N(0, t) = k\} = \lambda_k\Delta + o(\Delta)$$

A process of this kind is known in the literature on stochastic processes as a *pure birth* process. The current intensity now depends on the history of the process in a way that, in econometric terminology, is referred to as "occurrence dependence". In this case, N is not Poisson distributed.

There is a vast literature on birth processes. However, much of it is barely integrated into the count data literature. An exception is Faddy (1997), who uses properties of the pure birth process in order to develop generalized count data distributions. This framework can also be used to give a simple re-interpretation of over- and underdispersion. For instance, if $\lambda_0 < \lambda_1 < \lambda_2 < \ldots$ ("positive occurence dependence") the count N can be shown to be overdispersed relative to the Poisson distribution. Similarly, if $\lambda_0 > \lambda_1 > \lambda_2 > \ldots$ ("negative occurence dependence") the count N is underdispersed relative to the Poisson distribution. In order to derive parametric distributions based on birth processes, one needs to specify a functional relationship between λ_k and k. For instance, it can be shown that a pure birth process gives rise to a negative binomial distribution if this function is linear, i.e., for $\lambda_k = \alpha + \beta k$. These results and extensions are presented in greater detail in Chap. 2.5.3.

2.2.5 Poisson Distribution as a Binomial Limit

Consider an experiment all outcomes of which can be unambiguously classified as either success (S) or failure (F). For example, in tossing a coin, we may call *head* a success and *tail* a failure. Alternatively, drawing from an urn that contains only red and blue balls, we may call *red* a success and *blue* a failure. In general, the occurrence of an event is a success and the non-occurrence is a failure. Let the probability of a success be denoted by p. Then $0 < p < 1$ and the probability of a failure is given by $q = 1 - p$.

Now suppose that the experiment is repeated a certain number of times, say n times. Since each experiment results in either an F or an S, repeating the experiment produces a series of S's and F's. Thus, in three drawings from an urn, the result red, blue, red, in that order, may be denoted by SFS. The order may represent discrete time. Thus, the first experiment is made at time $t = 1$, the second at time $t = 2$, and the third at time $t = 3$. Thereby, the sequence of outcomes can be interpreted as a discrete time stochastic process. The urn drawing sequence with replacement is the classical example of an independent and stationary discrete time process: The outcomes of experiments at different points in time are independent, and the probability p of a success is constant over time and equal to the proportion of red balls in the urn. In this situation, all permutations of the sequence have the same probability.

Define a variable X as the total number of successes obtained in n repetitions of the experiment. X is called a count variable and n constitutes an upper bound for the number of counts. Under the assumptions of independence and stationarity, X has a binomial distribution function with probability generating function

$$\mathcal{P}(s) = [q + ps]^n \tag{2.27}$$

The binomial distribution and its properties are discussed in Chap. 2.3.2 in greater detail.

Up to this point, n was interpreted as the number of repetitions of a given experiment. To explicitly introduce a time dimension, consider a fixed time interval $(0, T)$ and divide it into n intervals of equal length. p is now the probability of success within the interval. What happens if the number of intervals increases beyond any bound while T is kept constant? A possible assumption is that the probability of a success is proportional to the length of the interval. The length of the interval is given by T/n, where T can be normalized without loss of generality to 1. Denote the proportionality factor by λ. Then $p_n = \lambda/n$, i.e., $p_n n = \lambda$, a given constant. Moreover, let $q_n = 1 - \lambda/n$. Substituting these expressions for P_n and q_n into (2.27) and taking limits, we obtain

$$\begin{aligned} \lim_{n\to\infty} \mathcal{P}(s) &= \lim_{n\to\infty} \left[1 - \frac{\lambda}{n} + \frac{\lambda}{n} s\right]^n \\ &= \lim_{n\to\infty} \left[1 + \frac{\lambda(s-1)}{n}\right]^n \\ &= e^{\lambda(s-1)} \end{aligned} \tag{2.28}$$

But (2.28) is precisely the probability generating function of the Poisson distribution. Dividing the fixed time period into increasingly shorter intervals, the binomial distribution converges to the Poisson distribution. This result is known in the literature as 'Poisson's theorem' (See Feller, 1968, Johnson and Kotz, 1969). The upper limit for the number of counts implicit in a binomial distribution disappears, and the sample space for the event counts approaches $\mathbb{N}_0$. Also note that in the limit the variance and expectation of the binomial (if they exist) are identical:

$$\begin{aligned}\lim_{n\to\infty} \text{Var}(X) &= \lim_{n\to\infty} [np(1-p)] \\ &= np \\ &= \text{E}(X)\end{aligned} \tag{2.29}$$

As for the Poisson process, this discrete time stochastic process assumed independence and stationarity (i.e., "randomness") of the successive Bernoulli trials.

2.2.6 Exponential Interarrival Times

The durations separating the arrival dates of events are called *waiting times* or *interarrival times*. Let τ_i be the waiting time between the $(i-1)$-th and the i-th event. It follows that the arrival date of the k-th event is given by $\vartheta_k = \sum_{i=1}^k \tau_i$, $k = 1, 2, \ldots$ Let $N(T)$ represent the total number of events that have occurred between 0 and T. Following the definitions of Chap. 2.2.3, $\{N(T), T > 0\}$ is a count process, while for fixed T, $N(T)$ is a count variable. The stochastic properties of the count process (and thus of the count) are fully determined once the joint distribution function of the waiting times τ_i, $i \geq 1$, is known. In particular it holds that the probability that at most $k-1$ events occurred before T equals the probability that the arrival time of the k-th event is greater than T:

$$P(N(T) < k) = P(\vartheta_k > T) \tag{2.30}$$

Moreover

$$\begin{aligned}P(N(T) = k) &= P(N(T) < k+1) - P(N(T) < k) \\ &= P(\vartheta_{k+1} > T) - P(\vartheta_k > T) \\ &= F_k(T) - F_{k+1}(T)\end{aligned} \tag{2.31}$$

where F_k is the cumulative density function of ϑ_k.

Equation (2.31) fully characterizes the relationship between event counts and durations. In general, $F_k(T)$ is a complicated convolution of the underlying densities of τ_i, which makes it analytically intractable. However, a great simplification arises if τ_i are identically and independently distributed with a common distribution. The process is then in the form of a renewal process (Cox, 1962). In particular, assume that $\{\tau_1, \tau_2, \ldots\}$ are independently and identically exponentially distributed variables, all with density function

$$f(\tau) = \lambda e^{-\lambda \tau} \tag{2.32}$$

In order to establish the distribution function of $N(T)$ using (2.31) one first needs to derive the cumulative density function of $\vartheta_k = \sum_{i=1}^{k} \tau_i$. Given the assumption of independent waiting times, the distribution of this k-fold convolution can be derived using the calculus of Laplace transforms (See Feller, 1971). The Laplace transform $\mathcal{L}(s) = \mathrm{E}(e^{-sX})$ is defined for non-negative random variables. It shares many of the properties of the probability generating function defined for integer-valued random variables. In particular, $\mathcal{L}(s) = \mathcal{P}(e^{-s})$ and the Laplace transform of a sum of independent variables equals the product of the Laplace transforms.

The Laplace transform of the exponential distribution is given by

$$\begin{aligned} \mathcal{L}_\tau(s) &= \int_0^\infty e^{-s\tau} dF(\tau) \\ &= (1 + s/\lambda)^{-1} \end{aligned} \tag{2.33}$$

Under independence

$$\begin{aligned} \mathcal{L}_\vartheta(s) &= [\mathcal{L}_\tau(s)]^k \\ &= (1 + s/\lambda)^{-k} \end{aligned} \tag{2.34}$$

But (2.34) is the Laplace transform of a gamma distribution with parameters λ and k. Since k is an integer, the cumulative density $F_k(T)$ of the gamma distribution may be written as (Abramowitz and Stegun, 1968, p. 262; Feller, 1971, p. 11):

$$F_k(T) = 1 - e^{-\lambda T}\left(1 + \lambda T + \frac{(\lambda T)^2}{2!} + \ldots + \frac{(\lambda T)^{k-1}}{(k-1)!}\right) \tag{2.35}$$

Therefore,

$$\begin{aligned} P(N = k) &= F_k(1) - F_{k+1}(1) \\ &= \frac{e^{-\lambda}\lambda^k}{k!} \end{aligned} \tag{2.36}$$

We conclude that the Poisson distribution arises if the interarrival times are independently exponentially distributed; it requires both independence within the spell (i.e., no duration dependence) and independence between spells (i.e., no occurrence dependence).

2.2.7 Non-Poissonness

Clearly, the Poisson distribution requires strong independence assumptions with regard to the underlying stochastic process, and any violation of these assumptions in general invalidates the Poisson distribution. It will be shown how occurrence dependence or duration dependence can be modeled, and how both phenomena lead to count data distributions other than the Poisson.

Following Johnson and Kotz (1969, Chap. 9) and Heckman (1981), consider again the urn model that was introduced in Chap. 2.2.5. The urn has a red balls and b blue balls where a red ball stands for the occurrence of an event, and a blue ball for non-occurrence. The probability of an event is therefore given by the proportion $a/(a+b)$ of red balls in the urn. The experiment is repeated k consecutive times.

Different urn schemes for a given individual may be characterized by whether or not the composition of the urn changes in consecutive trials. The case of unchanged composition implies independent trials and this case has been treated in Chap. 2.2.5. It leads to a binomial distribution for the number of successes.

Now, assume instead that the composition of the urn is altered over consecutive trials. There exist three different possibilities. First, the composition changes as the consequence of previous success. This situation is referred to as "occurrence dependence". Second, the composition changes as the consequence of previous non-success. This situation is referred to as "duration dependence". Third, and finally, the composition may change for exogenous reasons independently of the previous process. This situation is referred to as "non-stationarity".

The first two situations, where previous outcomes have an influence on the current experiment, are also known as *contagion* in the statistics literature, while the notion of *state dependence* is more common in the econometrics literature (Heckman and Borjas, 1980, Heckman, 1981). *Positive* contagion indicates that the occurrence of an event makes further occurrences more likely. For *negative* contagion, the opposite holds. Both cases lead to a *contagious* distribution for the number of counts, the Poisson distribution being an example for a non-contagious distribution. Contagious distributions have originally been developed for the theory of accident proneness (Bates and Neyman, 1951).

Occurrence Dependence

Occurrence dependence can be formalized as follows (Johnson and Kotz, 1969, p. 229): Initially, there are a red balls and b blue balls in the urn. One ball is drawn at random. If it is a red ball representing a success, it is replaced together with s red balls. If it is a blue ball, the proportion $a/(a+b)$ is unchanged, i.e., the blue ball is replaced. If this procedure is repeated n times and X represents the total number of times a red ball is drawn, then X has a Pòlya-Eggenberger distribution (Johnson and Kotz, 1969, p. 231). If the number of red balls is increased after a success ($s > 0$), then an occurrence increases the probability of further occurrences and the urn model reflects positive contagion. Johnson and Kotz (1969, p. 231) show that the negative binomial distribution is obtained as a limiting form. (The negative binomial distribution and its properties are discussed in Chap. 2.3.1). For $s = 0$, the model reduces to the binomial model with independent trials. For $s = -1$, the urn scheme corresponds to a drawing without replacement, leading to a hypergeometric distribution. Thus, the hypergeometric distribution is a distribution for negative contagion.

Corresponding results can be obtained for stochastic processes in continuous time (see also Chap. 2.2.4). For instance, assume that

$$P\{N(0, t + \Delta) = k + 1 | N(0, t) = k\} = \lambda_k \Delta + o(\Delta)$$

This equation defines a pure birth process. If λ_k is an increasing function of k, we have positive occurrence dependence. A constant function gives the Poisson case without occurrence dependence. A decreasing function indicates negative occurrence dependence. It can be shown that the negative binomial model erises if λ_k increases linearly in k.

Duration Dependence

In the urn model for occurrence dependence, the composition of the urn was left unchanged when a blue ball, i.e., a failure, occurred. If failures matter, then the outcome of an experiment depends on the time (number of draws) that has elapsed since the last success. This dependence generates "duration dependence". Again, duration dependence can be analyzed either in discrete time as represented by the urn-model or in continuous time using the concept of (continuous) waiting times. The continuous time approach was already introduced in Chap. 2.2.6. Further details are provided in Chap. 2.6.

Non-Stationarity

Finally, the assumptions of the standard model may be violated because the composition of the urn changes over consecutive trials due to exogenous effects while being unaffected by previous trials. This is the case if the underlying process is *nonstationary*. Non-stationarity does not necessarily invalidate the Poisson distribution.

Heterogeneity

A genuine ambiguity of the relationship between the underlying stochastic process and the count data distribution arises if the population is heterogeneous rather than homogeneous, as was assumed so far. With heterogeneity, the probability of an occurrence becomes itself a random variable.

For instance, in reference to the urn model, individuals may possess distinct urns that differ in their composition of red and blue balls. Unobserved heterogeneity can be modeled through a population distribution of urn compositions. For sampling with replacement (i.e., no dependence), the composition of *individual* urns is kept constant over time and the trials are thus independent at the individual level. Although past events do not truly influence the composition of individual urns, they provide some information on the proportion of red and blue balls in an individual urn. By identifying individuals with a high proportion of red balls, past occurrences do influence (increase) the expected probability of further occurrences *for that individual.* The model is said to display 'spurious' or 'apparent' contagion.

Again, it can be shown that under certain parametric assumptions on the form of the (unobserved) heterogeneity, the negative binomial distribution arises as the limiting distribution. Recall that the negative binomial distribution may also arise as a limiting form of *true* positive contagion. This fact illustrates one of the main dilemmas of count data modeling: The distribution of the (static) random variable for counts cannot identify the underlying structural stochastic process if heterogeneity is present. This result is also expressed in an 'impossibility theorem' by Bates and Neyman (1951): In a cross section on counts it is impossible to distinguish between true and spurious contagion.

2.3 Further Distributions for Count Data

The main alternative to the Poisson distribution is the *negative binomial* distributions. Count data may be negative binomial distributed if they were generated from a contagious process (occurrence dependence, duration dependence) or if the rate, at wghich events occur, is heterogeneous. The *binomial* distribution also represents counts, namely the number of successes in independent Bernoulli trials with stationary probabilities, but it introduces an upper bound given by the number of trials n. This upper bound distinguishes it from the Poisson and negative binomial distributions. The *continuous parameter binomial* distribution is a modification of the binolial distribution with continuous parameter n. Finally, the *logarithmic* distribution is discussed because of its role as a mixing distribution for the Poisson distribution. Good further references for these distributions and their properties are Feller (1968) and Johnson and Kotz (1969).

2.3.1 Negative Binomial Distribution

A random variable X has a negative binomial distribution with parameters $\alpha \geq 0$ and $\theta \geq 0$, written $X \sim \text{Negbin}(\alpha, \theta)$, if the probability function is given by

$$P(X = k) = \frac{\Gamma(\alpha + k)}{\Gamma(\alpha)\Gamma(k+1)} \left(\frac{1}{1+\theta}\right)^{\alpha} \left(\frac{\theta}{1+\theta}\right)^{k} \quad k = 0, 1, 2, \dots \qquad (2.37)$$

$\Gamma(\cdot)$ denotes the gamma function such that $\Gamma(s) = z^{s-1}e^{-z}dz$ for $s > 0$. This two parameter distribution has probability generating function

$$\mathcal{P}(s) = [1 + \theta(1 - s)]^{-\alpha} \tag{2.38}$$

The mean and variance are given by

$$E(X) = \alpha\theta \tag{2.39}$$

and

$$\begin{aligned} \text{Var}(X) &= \alpha\theta(1 + \theta) \\ &= E(X)(1 + \theta) \end{aligned} \tag{2.40}$$

Since $\theta \geq 0$, the variance of the negative binomial distribution generally exceeds its mean (" overdispersion"). The overdispersion vanishes for $\theta \to 0$.

The negative binomial distribution comes in various parameterizations. From an econometric point of view, the following considerations apply. In order to be able to use the negative binomial distribution for regression analysis the first step is to convert the model into a mean parameterization, say

$$\lambda = \alpha\theta \tag{2.41}$$

where λ is the expected value. Inspection of (2.41) shows that there are two simple ways of doing this.

1. $\alpha = \lambda/\theta$. In this case, the variance function takes the form

   $$\text{Var}(X) = \lambda(1 + \theta)$$

   Hence, the variance is a linear function of the mean. This model is called "Negbin I" (Cameron and Trivedi, 1986).
2. $\theta = \lambda/\alpha$. In this case, the variance function takes the form

   $$\text{Var}(X) = \lambda + \alpha^{-1}\lambda^2$$

   A negative binomial distribution with quadratic variance function results. This model is called "Negbin II".

The probability functions associated with the two models are as follows:

$$\text{Negbin I:}\ \ P(X = k) = \frac{\Gamma(\lambda/\theta + k)}{\Gamma(\lambda/\theta)\Gamma(k + 1)} \left(\frac{1}{1 + \theta}\right)^{\lambda/\theta} \left(\frac{\theta}{1 + \theta}\right)^{k} \tag{2.42}$$

and

$$\text{Negbin II:}\ \ P(X = k) = \frac{\Gamma(\alpha + k)}{\Gamma(\alpha)\Gamma(k + 1)} \left(\frac{\alpha}{\alpha + \lambda}\right)^{\alpha} \left(\frac{\lambda}{\alpha + \lambda}\right)^{k} \tag{2.43}$$

Although these two types are the most widely used parameterizations in practice, others are possible. For instance, let

$$\alpha = \sigma^{-2}\lambda^{1-k} \text{ and } \theta = \sigma^2\lambda^k$$

As before, $E(X) = \lambda$. Substitution of α and θ into (2.40) gives

$$\text{Var}(X) = \lambda(1 + \sigma^2 \lambda^k)$$

Thus, for $k = 0$ this parameterization reduces to the negative binomial distribution with linear variance function while for $k = 1$, a quadratic variance function is obtained. Winkelmann and Zimmermann (1995) refer to this model as "Negbin$_k$".

Yet another parameterization is often found in the statistics literature (see e.g. DeGroot, 1986), where in the general expression (2.37), $1/(1+\theta)$ is replaced by p and $\theta/(1+\theta)$ is replaced by q. If α is an integer, say n, the distribution is called *Pascal* distribution, and it has the interpretation of a distribution of the number of failures that will occur before exactly n successes have occurred in an infinite sequence of Bernoulli trials with probability of success p. For $n = 1$, this distribution reduces to the *geometric* distribution.

$$P(X = k) = pq^k \, , \qquad k = 0, 1, 2, \ldots \tag{2.44}$$

To summarize, the main advantage of the negative binomial distribution over the Poisson distribution is that the additional parameter introduces substantial flexibility into the modeling of the variance function, and thus heteroskedasticity. In particular, it introduces overdispersion, a more general form of heteroskedasticity than the mean-variance equality implied by the Poisson distribution.

Computational Issues

The presence of the Gamma function in the negative binomial probability function can cause numerical difficulties in computing the probabilities on a computer. For instance, consider the Negbin I formulation where terms such as $\Gamma(\lambda/\theta + k)$ need to be evaluated numerically. According to the GAUSS reference manual (Aptech, 1994), the argument of the gamma function must be less than 169 to prevent numerical overflow. The overflow problem can be avoided when one uses the logarithm of the gamma function (as is usually the case in econometrics applications) where an approximation based on Stirling's formula can be used. But even then, the accuracy of the approximation decreases as the argument of the log-gamma function becomes large. Large arguments arise whenever θ is small and the negative binomial distribution approaches the Poisson distribution.

Fortunately, there is a relatively simple way to avoid this difficulty. In particular, the Gamma function follows the recursive relation $\Gamma(x) = (x - 1)\Gamma(x-1)$. Thus

$$\begin{aligned} \frac{\Gamma(\alpha+k)}{\Gamma(\alpha)} &= \frac{(\alpha+k-1)(\alpha+k-2)\cdots(\alpha+k-k)\Gamma(\alpha)}{\Gamma(\alpha)} \\ &= \prod_{j=1}^{k} (\alpha + j - 1) \end{aligned} \tag{2.45}$$

where it is understood that the product equals one for $k = 0$. By suitable change of index, the product can alternatively be expressed as

$$\frac{\Gamma(\alpha+k)}{\Gamma(\alpha)} = \prod_{j=0}^{k-1}(\alpha+j)$$

or as

$$\frac{\Gamma(\alpha+k)}{\Gamma(\alpha)} = \prod_{j=1}^{k}(\alpha+k-j)$$

Finally, taking logarithms of (2.45) we obtain

$$\ln\Gamma(\alpha+k) - \ln\Gamma(\alpha) = \sum_{j=1}^{k}\ln(\alpha+j-1) \tag{2.46}$$

Relationship to Other Distributions

Proposition 3. *For $X \sim \text{Negbin}(\alpha, \theta)$, let $\theta \to 0$ and $\alpha \to \infty$ such that $\theta\alpha = \lambda$, a constant. Then, the negative binomial distribution converges to the Poisson distribution with parameter λ.*

Proof: Consider the probability generating function of the negative binomial distribution, replace θ by λ/α, and take limits.

$$\begin{aligned}
\lim_{\substack{\alpha\to\infty\\ \theta\alpha\to\lambda}} \mathcal{P}(s) &= \lim_{\substack{\alpha\to\infty\\ \theta\alpha\to\lambda}} [1+\theta(1-s)]^{-\alpha} \\
&= \lim_{\alpha\to\infty}\left[1+\frac{\lambda(1-s)}{\alpha}\right]^{-\alpha} \\
&= e^{-\lambda(1-s)}
\end{aligned} \tag{2.47}$$

But this is exactly the probability generating function of a Poisson distribution with parameter λ. □

An alternative, and somewhat more cumbersome, derivation of this result can be based directly on the probability distribution function

$$\begin{aligned}
\lim_{\substack{\alpha\to\infty\\ \theta\alpha\to\lambda}} P(X=k) &= \lim_{\substack{\alpha\to\infty\\ \theta\alpha\to\lambda}} \frac{\Gamma(\alpha+k)}{\Gamma(\alpha)k!}\left(\frac{1}{1+\theta}\right)^{\alpha}\left(\frac{\theta}{1+\theta}\right)^{k} \\
&= \lim_{\alpha\to\infty} \frac{\Gamma(\alpha+k)}{\Gamma(\alpha)k!}\left(\frac{\alpha}{\alpha+\lambda}\right)^{\alpha}\left(\frac{\lambda}{\alpha+\lambda}\right)^{k} \\
&= \lim_{\alpha\to\infty}\left(\prod_{j=1}^{k}\frac{\alpha+j-1}{\alpha+\lambda}\right)\left(\frac{\alpha}{\alpha+\lambda}\right)^{\alpha}\frac{\lambda^k}{k!}
\end{aligned}$$

$$= \lim_{\alpha\to\infty} \left(\prod_{j=1}^{k} \frac{1+(j-1)/\alpha}{1+\lambda/\alpha} \right) \left(\frac{1}{1+\lambda/\alpha} \right)^{\alpha} \frac{\lambda^k}{k!}$$

$$= e^{-\lambda} \frac{\lambda^k}{k!}$$

where use was made of the product expression for the ratio of gamma functions and of the fact that $(\alpha+\lambda)^{-k} = \prod_{j=1}^{k}(\alpha+\lambda)^{-1}$.

Further characterization of the negative binomial distribution

The negative binomial distribution arises in a number of ways. It was mentioned in Chap. 2.2.7 that it is the limiting distribution of a sequence of non-independent Bernoulli trials. It also arises as a mixture distribution and as a compound distribution. For mixing, assume that $X \sim \text{Poisson}(\lambda)$ and that λ has a gamma distribution. The marginal distribution of X is then the negative binomial distribution. For compounding, assume that a Poisson distribution is compounded by a logarithmic distribution. The compound distribution is then the negative binomial distribution. Derivations of these two results are postponed until Chap. 2.5.1 and Chap. 2.5.2 where the general approaches of mixing and compounding are presented.

Proposition 4 (Sums of Negative Binomial Random Variables). *Assume that X and Y are independently negative binomial distributed with $X \sim$ Negbin $I(\lambda, \theta)$ and $Y \sim$ Negbin $I(\mu, \theta)$. It follows that the random variable $Z = X + Y$ is negative binomial distributed Negbin $I(\lambda+\mu, \theta)$.*

Proof: Recall that the generic probability generating function of the negative binomial distribution is given by $\mathcal{P}(s) = [1+\theta(1-s)]^{-\alpha}$. In Negbin I parameterization, we obtain

$$\mathcal{P}(s)^{(X)} = [1+\theta(1-s)]^{-\lambda/\theta}$$

and

$$\mathcal{P}(s)^{(Y)} = [1+\theta(1-s)]^{-\mu/\theta}$$

Thus

$$\begin{aligned}\mathcal{P}(s)^{(Z)} &= [1+\theta(1-s)]^{-\lambda/\theta}[1+\theta(1-s)]^{-\mu/\theta} \\ &= [1+\theta(1-s)]^{-(\lambda+\mu)/\theta} \qquad (2.48)\end{aligned}$$

Thus, negative binomial distributions of the type specified above are closed under convolution. □

This result depends critically on two assumptions: First, the Negbin I specification with linear variance function has to be adopted. Second, X and Y have to share a common variance parameter θ. In other words, the sum of two arbitrarily specified negative binomial distributions is in general **not** negative binomial distributed.

2.3.2 Binomial Distribution

A random variable X has a *binomial distribution* with parameters $n \in \mathbb{N}$, and $p \in (0,1)$, written $X \sim B(n,p)$), if

$$P(X=k) = \binom{n}{k} p^k (1-p)^{n-k} \qquad k = 0,1,\ldots,n. \tag{2.49}$$

The probability generating function is given by

$$\begin{aligned} \mathcal{P}(s) &= \sum_{k=0}^{n} s^k \binom{n}{k} p^k q^{n-k} \\ &= \sum_{k=0}^{n} \binom{n}{k} (ps)^k q^{n-k} \\ &= (q+ps)^n \;. \end{aligned} \tag{2.50}$$

and mean and variance are

$$\mathrm{E}(X) = np$$

and

$$\mathrm{Var}(X) = np(1-p),$$

respectively.

In estimation problems, the binomial parameter n is usually treated as given. Sometimes, however, one might wish to estimate n as a function of data as well. Under maximum likelihood, there are two possibilities. First, one can respect the integer nature of the parameter and maximize by way of a grid search. The resulting estimator won't have the standard properties of a maximum likelihood estimator. Alternatively, one can treat n as a continuous parameter. In this case, derivatives can be taken. Since

$$\begin{aligned} \binom{n}{k} &= \frac{n!}{k!(n-k)!} \\ &= \frac{\Gamma(n+1)}{\Gamma(k+1)\Gamma(n-k+1)} \end{aligned}$$

where $\Gamma(\cdot)$ denotes the gamma-function and $\Gamma(n+1) = n!$ if n is an integer, this involves computation of the digamma function. Alternatively, direct differentiation can be based on an approximation of the factorial representation using Stirling's formula

$$k! \approx (2\pi)^{1/2} k^{k+1/2} \exp(-k)\{1 + 1/12k\}$$

In either case, a logical difficulty arises with respect to the possible sample space of the underlying random variable X if n is a continuous non-negative parameter. Consider the following formal definition.

A random variable X has a *continuous parameter binomial* distribution with parameters $\alpha \in I\!R^+$, and $p \in (0,1)$, written $X \sim CPB(\alpha, p)$), if the nonnegative integer n in equation 2.49 is replaced by a continuous $\alpha \in I\!R^+$ where $k = 0, 1, \ldots, \tilde{n}$ and

$$\tilde{n} = \begin{cases} \text{int}[\alpha] + 1 & \text{if } \alpha \text{ non-integer} \\ \alpha & \text{if } \alpha \text{ integer} \end{cases}$$

(Johnson and Kotz, 1969, p.41, King, 1989b). When α is not an integer, the probabilities do not sum to one and the following normalization is used:

$$\tilde{p}_k = \frac{p_k}{\sum_{i=0}^{\tilde{n}} p_i}, \quad k = 0, 1, \ldots, \tilde{n}. \tag{2.51}$$

where

$$p_k = \frac{\Gamma(\alpha+1)}{\Gamma(k+1)\Gamma(\alpha-k+1)} p^k (1-p)^{\alpha-k}$$

However, this formulation has the defect that the expected value is not equal to αp, as the analogy to the binomial distribution would suggest. References that have ignored this point or were at least unclear about it include Guldberg (1931), Johnson and Kotz (1969), and King (1989b).

The discrepancy between the expected value and αp is relatively easy to derive for $0 < \alpha < 1$.

Proposition 5. *Let the random variable X have a continuous parameter binomial distribution $CPB(\alpha, p)$, where α, $p \in (0,1)$. Then*

$$\mathrm{E}(X) > \alpha p\,.$$

Proof: If $\alpha \in (0,1)$ then $k \in \{0,1\}$, and

$$p_0 = \frac{\alpha\Gamma(\alpha)}{\Gamma(1)\Gamma(\alpha+1)} p^0 (1-p)^{\alpha-0} = (1-p)^\alpha$$

$$p_1 = \frac{\alpha\Gamma(\alpha)}{\Gamma(2)\Gamma(\alpha)} p^1 (1-p)^{\alpha-1} = \alpha p (1-p)^{\alpha-1}$$

Moreover,

$$\begin{aligned}
\mathrm{E}(X) = \tilde{p}_1 &= \frac{\alpha p (1-p)^{\alpha-1}}{(1-p)^\alpha + \alpha p (1-p)^{\alpha-1}} \\
&= \alpha p \left[\frac{(1-p)^{\alpha-1}}{(1-p)^{\alpha-1}(1-p) + \alpha p (1-p)^{\alpha-1}} \right] \\
&= \alpha p \left[\frac{(1-p)^{\alpha-1}}{(1-p)^{\alpha-1}(1-p+\alpha p)} \right] \\
&= \alpha p \left[\frac{1}{1+(\alpha-1)p} \right] > \alpha p \qquad \square
\end{aligned}$$

A correct formula for the expected value of the continuous parameter binomial distribution for arbitrary α is derived in Winkelmann, Signorino, and King (1995). They start from the usual definition

$$\mathrm{E}(X) = \sum_{k=0}^{\tilde{n}} kCPB(k|\alpha, p). \tag{2.52}$$

Using the same method as above, $\mathrm{E}(X)$ can be calculated for successive intervals of α and generalized. For the first three unit regions, the expected values $\mathrm{E}(X|0 < \alpha \leq 1, p)$, $\mathrm{E}(X|1 < \alpha \leq 2, p)$, and $\mathrm{E}(X|2 < \alpha \leq 3, p)$, are given by:

$$\alpha p \left[\frac{1}{1 + (\alpha - 1)p}\right]$$

$$\alpha p \left[\frac{1 + (\alpha - 2)p}{1 + (\alpha - 2)p + \frac{(\alpha-2)(\alpha-1)}{2}p^2}\right]$$

and

$$\alpha p \left[\frac{1 + (\alpha - 3)p + \frac{(\alpha-3)(\alpha-2)}{2}p^2}{1 + (\alpha - 3)p + \frac{(\alpha-3)(\alpha-2)}{2}p^2 + \frac{(\alpha-3)(\alpha-2)(\alpha-1)}{6}p^3}\right]$$

respectively. The same pattern continues for subsequent intervals of α. Generalizing over all α and p yields

$$\mathrm{E}(X) = \alpha p \left[\frac{\Theta^{\tilde{n}-1}(\alpha - \tilde{n}, p)}{\Theta^{\tilde{n}}(\alpha - \tilde{n}, p)}\right] \tag{2.53}$$

where

$$\Theta^{\kappa}(\gamma, \rho) = \sum_{i=0}^{\kappa} \frac{(\gamma)_i}{i!}\rho^i \tag{2.54}$$

and $(\gamma)_i$ is Pochhammer's symbol, given by

$$(\gamma)_i = \gamma(\gamma + 1)(\gamma + 2) \cdots (\gamma + i - 1)$$
$$(\gamma)_0 = 1.$$

The difference between the expected value αp assumed in the literature and the correct expression in equation (2.53) is not large, but it is not zero, and it varies with the two parameters of the CPB. Winkelmann, Signorino, and King (1995) give further details.

2.3.3 Logarithmic Distribution

The random variable X has a logarithmic distribution if (Johnson and Kotz, 1969, p. 166)

$$P(X = k) = \alpha\theta^k/k \qquad k = 1, 2, \ldots, \; 0 < \theta < 1 \tag{2.55}$$

where $\alpha = -[\log(1-\theta)]^{-1}$. The probability generating function is given by

$$\begin{aligned}\mathcal{P}(s) &= \sum_{k=1}^{\infty} s^k \alpha \frac{\theta^k}{k} \\ &= \sum_{k=1}^{\infty} \alpha \frac{(\theta s)^k}{k} \\ &= -\alpha \ln(1-\theta s) \end{aligned} \tag{2.56}$$

where the last equality follows from a Taylor series expansion of $\ln(1-x)$ around 0:

$$\ln(1-x) = -\sum_{k=1}^{\infty} \frac{x^k}{k} \tag{2.57}$$

Alternatively, the probability generating function can be written using the explicit expression of the normalizing constant α as

$$\mathcal{P}(s) = [\log(1-\theta s)]/[\log(1-\theta)] \tag{2.58}$$

The mean and the variance are given by

$$\mathrm{E}(X) = \alpha\theta(1-\theta)^{-1} \tag{2.59}$$

and

$$\mathrm{Var}(X) = \alpha\theta(1-\alpha\theta)(1-\theta)^{-2} \; . \tag{2.60}$$

The distribution displays overdispersion for $0 < \alpha < 1$ (i.e., $\theta > 1 - e^{-1}$) and underdispersion for $\alpha > 1$ (i.e., $\theta < 1 - e^{-1}$).

In contrast to the previous distributions, the sample space of the logarithmic distribution is given by the set of *positive* integers. And in fact, it can be obtained as a limiting distribution of the truncated-at-zero negative binomial distribution (Kocherlakota and Kocherlakota, 1992, p.191). The likely reason for the logarithmic distribution being an ineffective competitor to the Poisson or negative binomial distributions is to be seen in its complicated mean function which factually, though not formally, prohibits the use of the distribution in a regression framework. For instance, Chatfield, Ehrenberg and Goodhardt (1966) use the logarithmic distribution to model the numbers of items of a product purchased by a buyer in a specified period of time, but they do not include covariates, i.e., they specify no regression. However, the logarithmic distribution plays a role as a compounding distribution (see Chap. 2.5.2).

2.3.4 Summary

The main properties of the described distributions for counts are summarized in Tab. 2.1.

Table 2.1. Distributions for Count Data

Distribution	Range	$\mathcal{P}(s)$	E(X)	Var(X)
Poisson	$0, 1, 2, \ldots$	$e^{-\lambda+\lambda s}$	λ	λ
Binomial	$0, 1, \ldots, n$	$(q+ps)^n$	np	$np(1-p)$
Negative Binomial	$0, 1, 2, \ldots$	$[1+\theta(1-s)]^{-\alpha}$	$\alpha\theta$	$\alpha\theta(1+\theta)$
Logarithmic	$1, 2, \ldots$	$-\alpha \ln(1-\theta s)$	$\dfrac{\alpha\theta}{1-\theta}$	$\dfrac{\alpha\theta(1-\alpha\theta)}{(1-\theta)^2}$

It is worth emphasizing that the first three distributions display a similar structure. In fact, they are related through various limiting forms that have been discussed in this chapter. The common structure of the distributions can be best captured by considering the following generic probability generating function (Johnson and Kotz, 1969, p. 138):

$$\mathcal{P}(s) = [(1+\omega) - \omega s]^{-m} \tag{2.61}$$

From (2.61) it follows directly that

$$\text{E}(X) = m\omega$$

and

$$\text{Var}(X) = m\omega(1+\omega)$$

The probability generating functions in Tab. 2.1 can be obtained as follows. For the negative binomial model, $\omega > 0$ and $m > 0$; for the binomial, $-1 < \omega < 0$ and $m < 0$. The Poisson distribution is obtained as the limiting intermediate case where $\omega \to 0$ and $m \to 0$ such that $\omega m = \lambda$.

Finally, the following figures compare the shape of the four probability functions for specific parameter values. In all figures, the expected value is set to 3.5 . Fig. 2.1 presents the Poisson distribution, the negative binomial distribution with $\text{Var}(X)/\text{E}(X) = 2$, the binomial distribution with $n = 10$ and the logarithmic distribution ($x \geq 1$). Fig. 2.2 shows the negative binomial distribution for varying degrees of dispersion ($\text{Var}(X)/\text{E}(X) = 1.5$: solid; and $\text{Var}(X)/\text{E}(X) = 3$: shaded).

The figures illustrate the different assumptions on the variance. Taking the Poisson distribution as reference distribution, the binomial distribution is more, and the negative binomial distribution is less concentrated around the mean. The concentration of the negative binomial distribution decreases with increasing variance mean ratio. Another useful comparison is the probability of zeros under the different models. While the probability of obtaining a zero is 1.3 percent under the binomial model, it is 14.6 percent under the negative binomial model with maximal dispersion, the Poisson model providing an intermediate case with 3 percent probability of obtaining a zero (The logarithmic distribution is defined only for positive integers). Finally, it is worth noticing that all distributions are skewed to theleft.

Fig. 2.1. Count Data Distributions (E(X) = 3.5)

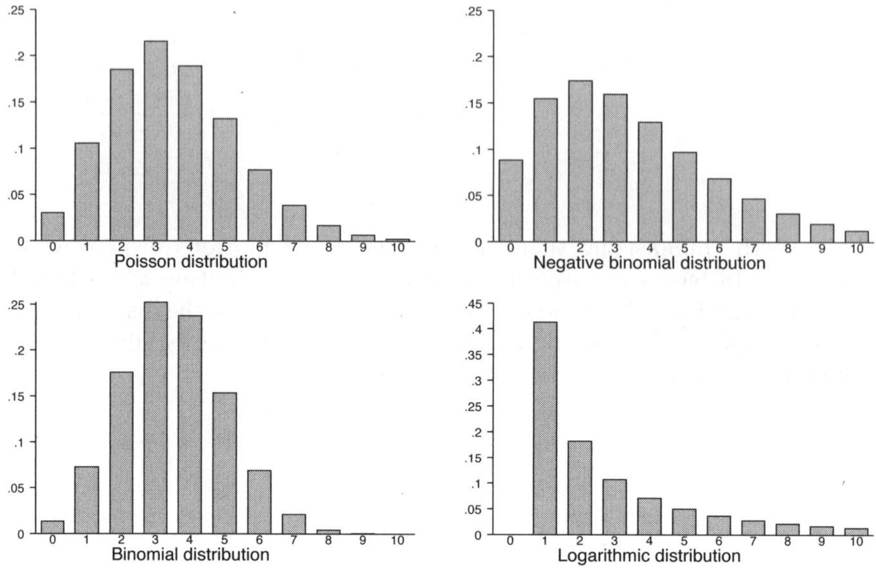

Fig. 2.2. Negative Binomial Distributions

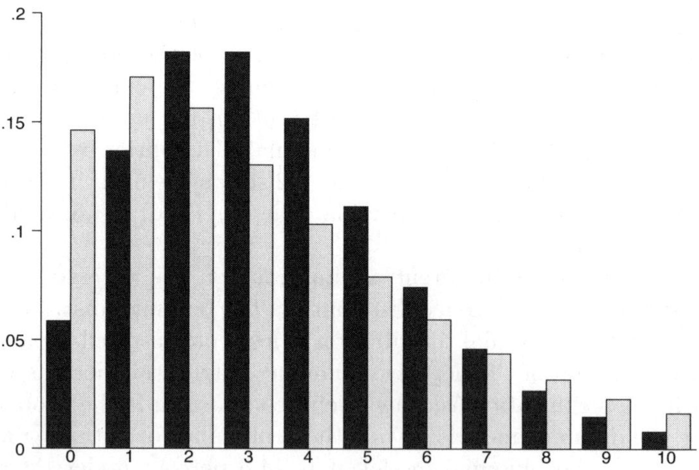

2.4 Modified Count Data Distributions

2.4.1 Truncation

If a count data distribution is observable not over the whole range of non-negative integers but rather only for a subset it is said to be truncated. For instance, if observations with zero outcomes are not observed, the distribution is "truncated-at-zero". In this case, "positive count data models" are required (Gurmu, 1991).

Truncated count data can be modeled as a two-part process. The first part consists of an untruncated latent distribution for X^*. The second part consists of a binary indicator variable c. The observed distribution for X is truncated if $c = 0$, and untruncated if $c = 1$. The generic model for truncation is then

$$X = \begin{cases} X^* \\ \text{unobserved} \end{cases} \text{if} \begin{cases} c = 1 \\ c = 0 \end{cases} \tag{2.62}$$

Further, assume that

$$c = \begin{cases} 1 \text{ if } X^* \in A \\ 0 \text{ if } X^* \notin A \end{cases} \tag{2.63}$$

that is, c is uniquely determined through the latent count variable X^*. The two most commonly encountered situations are:

1. A is the set of positive integers ("truncation at zero").
2. A is the set $\{0, \ldots, a\}$ where a is some positive integer ("truncation from above").

For instance, assume that c is defined as in (2.63) and X^* is Poisson distributed with parameter λ. For $A = \{1, 2, \ldots\}$

$$P(c = 1) = 1 - \exp(-\lambda)$$

and for $A = \{0, 1, \ldots, a\}$

$$P(c = 1) = F(a)$$

where F is the cumulative distribution function of X^*. In general,

$$P(X = k) = \frac{P(X^* = k|c = 1)}{P(c = 1)} \tag{2.64}$$

For the truncated-at-zero Poisson model, we have

$$P(X = k|X > 0) = \frac{\exp(-\lambda)\lambda^k}{k!(1 - \exp(-\lambda))}, \qquad k = 1, 2, 3 \ldots \tag{2.65}$$

with mean

$$\mathrm{E}(X|X > 0) = \frac{\lambda}{1 - \exp(-\lambda)} \tag{2.66}$$

and variance

$$\mathrm{Var}(X|X>0) = \mathrm{E}(X|X>0)\left(1 - \frac{\lambda}{\exp(\lambda)-1}\right). \tag{2.67}$$

Since λ (the mean of the untruncated distribution) is greater than zero, $0 < \exp(-\lambda) < 1$ and the truncated mean is shifted to the right. Moreover, the truncated-at-zero model displays underdispersion since $0 < 1-\lambda(\exp(\lambda)-1) < 1$.

2.4.2 Censoring and Grouping

A count data distribution is said to be censored if it is only partially observable: for a subset of outcomes the distribution is determined only up to an interval of outcomes. The leading example here is right-censoring, where all counts exceeding a certain threshold number k are reported in a category "k or more". For instance, such data are occasionally observed in household survey data (See Merkle and Zimmermann, 1992).

Denote the interval of partial observability by A. Then

$$P(X=k) = p_k \qquad \text{for } k \in I\!N \setminus A$$
$$P(X \in A) = \sum_{k\in A} p_k$$

Censoring can be seen as a special case of grouping. Assume that the set of non-negative integers is partitioned into J mutually exclusive and exhaustive subsets $A_1, \ldots, A_J$, and that each $A_j, j = 1, \ldots, J$ is the set of consecutive integers $\{a_j, a_j+1, \ldots, a_j+n_j\}$ such that $a_{j+1} = a_j + n_j + 1$ and $a_1 = 0$.

Hence, the set A_j to which a count belongs is known, but not the count itself. The resulting model is defined over the subsets with $P(A_j) = P(X \in A_j)$, where

$$P(X \in A_j) = \sum_{k\in A_j} P(X=k) \tag{2.68}$$

2.4.3 Altered Distributions

For discrete distributions, it is relatively straightforward to select one (or more) specific outcome and increase (or decrease) the probability of that outcome relative to the probability of the underlying model. The only two restrictions are the fundamental requirements for probabilities, namely that they are non-negative and sum up to one. Such a modeling strategy certainly can improve the ability of the probability model to describe actual discrete data. While this approach may appear ad-hoc at first glance, there are situations where adjustments to single probabilities can in fact be justified in terms of underlying structural processes. This idea will be followed up in a later chapter where regression models based on altered count data distributions are presented.

In practice, the most common alteration is to modify the probability of a zero relative to the underlying distribution. The resulting distributions are referred to as "zero-inflated" or "zero-deflated" count data distributions. For instance, the zero-inflated Poisson distribution can be written as

$$P(X=0) = \omega + (1-\omega)e^{-\lambda}$$

$$P(X=k) = (1-\omega)\frac{e^{-\lambda}\lambda^k}{k!} \qquad k = 1,2,\ldots$$

where $0 < \omega < 1$, or, more compactly, as

$$P(X=k) = \delta_{k=0}\omega + (1-\omega)\frac{e^{-\lambda}\lambda^k}{k!} \qquad k = 0,1,2,\ldots \tag{2.69}$$

where δ is a binary indicator taking the value 1 when $k=0$ and zero otherwise. "Zero-deflation" would be obtained for $0 > \omega > -(1-e^{-\lambda})^{-1}$.

Along the same lines one could adjust more than one probability. For instance, consider a zero-and-two inflated model that has been used in a study of completed fertility, i.e., the number of children born (Melkersson and Rooth, 2000): Now

$$P(X=0) = \omega_0 + (1-\omega_0-\omega_2)e^{-\lambda}$$

$$P(X=2) = \omega_2 + (1-\omega_0-\omega_2)\frac{e^{-\lambda}\lambda^2}{2!}$$

$$P(X=k) = (1-\omega_0-\omega_2)\frac{e^{-\lambda}\lambda^k}{k!} \qquad k = 1,3,4,\ldots$$

Obviously, the expected value and variance of such a distribution is quite different from the expected value and variance of the Poisson distribution.

2.5 Generalizations

Having to choose between one of the probability models introduced in the previous chapters, the researcher might encounter two problems:

- None of the distributions accounts for the full amount of the observed variation in the counts. In particular, the observed frequency distribution might exhibit extraordinary values for some categories, and/or the second moment assumption of the probability model might be too restrictive.
- Although one of the probability models does in fact correspond to the true data generating process, the process is unknown. In other words, the researcher has no *a-priori* information on the true model and he is left with the problem of choosing a particular distribution.

Both issues have been recognized and addressed by the literature leading to the development of so-called 'generalized' or 'modified' probability models. Two types of generalizations can be distinguished. The first type concentrates on additional, possibly more flexible, probability models. Consul (1989) lists generalizations based on the Poisson distribution. They include the "Neyman-type distributions, the Poisson binomial, Poisson negative binomial, Poisson geometric, Poisson Pascal, Poisson rectangular, Poisson log-normal, quasi Poisson, inflated Poisson, mixed Poisson, generalized Poisson, and double Poisson (...)" (Consul, 1989, p.3). Most of these models fall within the class of *compound* or *mixed* Poisson distribution families, which are presented in the next section. Alternatively, more general models have been derived from an application of birth processes.

The second type of generalization addresses the issue of selecting a specific model. Here, a *hyper-model* (or class of distribution families) encompasses several sub-models of interest. Examples are the *Katz* class of distributions and the class of *linear exponential families* which are introduced in Chap. 2.5.4 and Chap. 2.5.5, respectively. Both classes contain the most important distributions for count data – the Poisson, binomial and negative binomial distributions – in the form of either parametric (Katz) or functional restrictions (linear exponential family).

The distinction between the two types of generalizations blurrs sometimes, when generalized distributions nest more than one interesting sub-model. In these cases, they can be used either for the benefit of a more flexible model *per se*, or they can serve to discriminate between more restrictive sub-models.

2.5.1 Mixture Distributions

Mixture distributions play an important role in the modeling of counts (but their importance is by no means limited to count data). In general terms, mixtures are defined in the following way: consider various proper distribution functions F^j representing different random variables X^j, $j = 1, 2, \ldots$, and constants a_j with $a_j > 0\ \forall j$ and $\sum_j a_j = 1$. Then

$$F = \sum_{j=1}^{\infty} a_j F^j \tag{2.70}$$

is a proper distribution function and is called a mixture of the distributions $\{F^j\}$. The component distributions do not have to be defined over the same sample space $\mathbf{S}$. Let $\mathbf{S}_j$ denote the sample space of distribution j and let $\mathbf{S}$ denote the sample space of the mixture distribution F. Then $\mathbf{S}{=}\bigcup_j \mathbf{S}_j$.

There are various ways by which the general concept of mixing can be given more specific content. For instance, it can be used to give special weights to specific discrete values. For this purpose, one might mix for instance a Poisson distribution with a degenerate distribution putting unity probability mass at one point. The resulting mixture is an "inflated parameter probability" distribution.

Alternatively, F^j might be any parametric distribution function depending on a parameter θ. Moreover, assume that the parameter itself is a random variable with probability function $f(\theta)$. Thus, if the support of θ is discrete, we can write

$$F = \sum_{\theta \in \Theta} f(\theta) F(\theta) \tag{2.71}$$

whereas for continuous support, an integral replaces the summation

$$F = \int_{\theta \in \Theta} f(\theta) F(\theta) d\theta \tag{2.72}$$

Mixtures of this form are commonly expressed in terms of probability functions (rather than distribution functions). In the case of continuous mixing over a discrete probability function, we can write for instance

$$P(X = k) = \int_{\theta \in \Theta} P(X = k|\theta) f(\theta) d\theta \tag{2.73}$$

This last formulation makes it clear that mixing is really a randomization of a distribution parameter. In this scenario, two distinct distributions for X can be distinguished: the *conditional* distribution $P(X = k|\theta)$, and the *marginal* distribution $P(X = k)$. If the marginal distribution of θ is known, then the marginal distribution of X is obtained by integrating the joint distribution of X and θ over θ. An example for such an operation based on a Poisson distribution is given shortly. First, however, it is useful to study the mean and variance of the marginal distribution of X under mixing. Results can be established under very mild assumptions, whereas the derivation of the full marginal distribution requires knowledge of $f(\theta)$, a much stronger requirement.

Mean and variance of marginal distribution

By the law of the iterated expectation

$$\mathrm{E}(X) = \mathrm{E}_\theta[\mathrm{E}(X|\theta)] \tag{2.74}$$

and, using the variance decomposition theorem,

$$\mathrm{Var}(X) = \mathrm{E}_\theta[\mathrm{Var}(X|\theta)] + \mathrm{Var}_\theta[\mathrm{E}(X|\theta)] \tag{2.75}$$

A number of results follow. First, if the conditional distribution of $X|\theta$ is Poisson, then

$$\begin{aligned} \mathrm{E}_\theta[\mathrm{Var}(X|\theta)] &= \mathrm{E}_\theta[\mathrm{E}(X|\theta)] \\ &= \mathrm{E}(X) \end{aligned}$$

where the first equality uses the equi-dispersion property of the Poisson. Therefore,

$$\text{Var}(X) = \text{E}(X) + \text{Var}_\theta[\text{E}(X|\theta)] > \text{E}(X)\,,$$

and mixing introduces overdispersion at the marginal level.

Moreover, if we specify the mean and variance of the distribution of θ as $\text{E}(\theta) = \lambda$ and $\text{Var}(\theta) = \sigma_\theta^2$, then an application of (2.74) and (2.75) yields that $\text{E}(X) = \lambda$ and $\text{Var}(X) = \lambda + \sigma_\theta^2\lambda^2$. As the reader may recall, these expressions are equal to the mean and variance of the Negbin II model introduced in Chap. 2.3.1. This is not a coincidence, since the Negbin II model can be derived from mixing a Poisson distribution with a gamma distribution. However, it should be noted that the "semi-parametric" result derived from applying the law of iterative expectations is more general as it does not depend on the full density $f(\theta)$ but only on its first two moments.

Example for a fully parametric mixture model: Poisson-gamma

The leading example of a fully parametric mixture model for count data is the Poisson-Gamma mixture. Assume that $X \sim \text{Poisson}(\theta)$ where θ is gamma distributed with density function

$$f(\theta) = \frac{\beta^\alpha}{\Gamma(\alpha)}\theta^{\alpha-1}e^{-\theta\beta} \tag{2.76}$$

Mean and variance of the gamma distribution are given by $\text{E}(\theta) = \alpha/\beta$ and $\text{Var}(\theta) = \alpha/\beta^2$, respectively. Under the re-parameterization $\beta = \alpha/\lambda$, we obtain the desired specification where $\text{E}(\theta) = \lambda$ and $\text{Var}(\theta) = \alpha^{-1}\lambda^2$. Moreover, integration of $P(X = k, \theta) = P(X = k|\theta)f(\theta)$ over θ yields

$$\begin{aligned}
P(X = k) &= \int_0^\infty \frac{e^{-\theta}\theta^k}{k!}\frac{(\alpha/\lambda)^\alpha}{\Gamma(\alpha)}\theta^{\alpha-1}e^{-\frac{\theta\alpha}{\lambda}}d\theta \\
&= \frac{\alpha^\alpha}{\lambda^\alpha k!\Gamma(\alpha)}\int_0^\infty e^{-\theta(\frac{\lambda+\alpha}{\lambda})}\theta^{k+\alpha-1}d\theta \\
&= \frac{\alpha^\alpha\Gamma(k+\alpha)}{\lambda^\alpha k!\Gamma(\alpha)}\left(\frac{\lambda}{\lambda+\alpha}\right)^{k+\alpha} \\
&= \frac{\Gamma(\alpha+k)}{\Gamma(\alpha)\Gamma(k+1)}\left(\frac{\alpha}{\lambda+\alpha}\right)^\alpha\left(\frac{\lambda}{\lambda+\alpha}\right)^k
\end{aligned} \tag{2.77}$$

Thus, we obtain a negative binomial model of the Negbin II-variety (see equation (2.43)). In order to derive the Negbin I (or Negbin$_k$) model as a Poisson-gamma mixture, the underlying gamma distribution would need to be re-parameterized in a suitable manner.

This type of mixture distribution has an interesting econometric interpretation as the random variation around λ can be thought of as representing unobserved heterogeneity that is likely to arise in most practical situations. For instance, we could have written $\theta = \lambda u$ where λ is deterministic and u is a multiplicative error term. This formulation is equivalent if $u \sim \text{gamma}(\alpha, \alpha)$ and, therefore, $\mathrm{E}(u) = 1$ and $\mathrm{Var}(u) = \alpha^{-1}$.

Whether the gamma distribution is a good model for heterogeneity of the Poisson parameter is a different question. Other mixing distributions are presented later. The great advantage of the gamma distribution is that the integral over the joint distribution can be solved analytically, leading to a mixture distribution (i.e., the negative binomial distribution) in closed form.

2.5.2 Compound Distributions

Compound distributions are of the form

$$Z = \sum_{i=1}^{N} X_i \tag{2.78}$$

where N and $X_i, i = 1, \ldots, N$ are independent random variables. Compound distributions are sometimes also referred to as "stopped-sum distribution" (Santos Silva and Windmeijer, 2001). We say that the distribution of N is generalized by X.

The appeal of this framework is twofold. First, the derivation of the distribution of Z is relatively simple as long as certain independence assumptions are made. Second, and of equal importance, the structure of a compound distribution can be interpreted in terms of an underlying data generating process that has many applications of interest.

Example 1 Let N be the number of families moving from one country or region to another during a given time period, and let X be family size. Z gives then the number of individuals moving.

Example 2 Let N be the number of total events and X be a binary variable that takes the value "1" if the event is reported and "0" otherwise. Z gives then the number of reported events. Similarly, N could be the number of job offers and X a binary indicator that takes the value "1" if the offer is accepted and the value "0" otherwise. Z represents the number of accepted offers (=job changes).

Example 3 Let N be the number of spells of illness and X be the number of visits to a doctor in a given spell. Z gives then the total number of visits to a doctor.

Example 4 Let X be a degenerate random variable such that $P(X = 1) = 1$. Then $Z = N$.

The concept of compounding is not restricted to cases where N is a count random variable although this is of main interest here. One popular count data model, the hurdle Poisson model, can be interpreted as a Bernoulli distribution that is generalized by a truncated-at-zero Poisson distribution (Santos Silva and Windmeijer, 2001).

The main mathematical tool for dealing with compound distributions is again the probability generating function. The following theorem by Feller (1968) gives the essential result.

Theorem 1. *Let $\{X_k\}$ be a sequence of mutually independent random variables with common distribution $P(X_k = i) = f_i$ and probability generating function $\mathcal{P}^{(X)}(s) = \sum_{i=0}^{\infty} f_i s^i$. Consider the sum $Z_N = X_1 + X_2 + \ldots + X_N$, where the number of components N is a random variable independent of X with $P(N = n) = g_n$ and $\mathcal{P}^{(N)}(s) = \sum_{n=0}^{\infty} g_n s^n$. Then*

$$\mathcal{P}^{(Z_N)} = \mathcal{P}^{(N)}[\mathcal{P}^{(X)}(s)] \ . \tag{2.79}$$

Proof: The distribution of Z_N is obtained by the rule for conditional probabilities:

$$h_j = P(Z_N = j) = \sum_{n=0}^{\infty} P[(X_1 + \ldots + X_n) = j]P(N = n) \tag{2.80}$$

For given N, the rule for convolution applies:

$$\mathrm{E}(s^{Z_n}|N = n) = [\mathrm{E}(s^X)]^n \ , \tag{2.81}$$

and the probability generating function of Z_N equals the *marginal* expectation

$$\mathcal{P}^{(Z_N)} = \mathrm{E}(s^{Z_N}) = \sum_{n=0}^{\infty} g_n[\mathrm{E}(s^X)]^n = \mathcal{P}^{(N)}[\mathcal{P}^{(X)}(s)] \qquad \square \tag{2.82}$$

Of particular interest is the case where N has a Poisson distribution function. Z_N then has a compound Poisson distribution with probability generating function

$$\mathcal{P}(s) = e^{-\lambda+\lambda\mathcal{P}^{(X)}(s)} \tag{2.83}$$

Different results follow:

Corollary 1. .
Let X_i be identically and independently Bernoulli distributed with $B(1,p)$ and let N have a Poisson distribution function with $Po(\lambda)$. Then Z_N is Poisson distributed with parameter λp.

Proof: The probability generating function of the Bernoulli distribution has the form $\mathcal{P}(s) = q + ps$. The probability generating function of Z_N is then given by:

$$\begin{aligned}\mathcal{P}(s) &= e^{-\lambda+\lambda(q+ps)} \\ &= e^{-\lambda p+\lambda ps}\end{aligned} \tag{2.84}$$

Z_N is Poisson-distributed with expectation $\mathrm{E}(Z_N) = \lambda p$. The process of generalizing a Poisson distribution by a Bernoulli distribution is also called "binomial thinning".

Corollary 2. .
Let X have a logarithmic distribution *with parameter θ and let N have a Poisson distribution with parameter λ. Then Z_N is negative binomial distributed with parameters $-\lambda/\log(1-\theta)$ and $\theta/(1-\theta)$.*

Proof: The probability generating function of the logarithmic distribution is given by

$$\mathcal{P}(s) = -\alpha[\log(1-\theta s)] \tag{2.85}$$

with $\alpha = -[\log(1-\theta)]^{-1}$.

Thus, the probability generating function of Z_N is obtained as

$$\begin{aligned}
\mathcal{P}(s) &= \exp[-\lambda - \lambda\alpha\log(1-\theta s)] \\
&= \exp(-\lambda)(1-\theta s)^{-\alpha\lambda} \\
&= [\exp(1/\alpha)]^{-\alpha\lambda}(1-\theta s)^{-\alpha\lambda} \\
&= \left[\frac{1-\theta s}{1-\theta}\right]^{-\alpha\lambda} \\
&= \left[1 - \frac{\theta}{1-\theta}(1-s)\right]^{-\alpha\lambda}
\end{aligned}$$

This is the probability generating function of a negative binomial distribution with parameters $\alpha\lambda = -\lambda/\log(1-\theta)$ and $\theta/(1-\theta)$ (see (2.38)). □

Finally, we note that mixing and compounding are related concepts. For instance, consider a mixture distribution of the form (2.70), where

1. F^j are distribution functions of the $j-$fold convolutions of X, and
2. the a_j's are given by the probability function of N.

This is exactly the form of compounding described above. Alternatively, consider a parametric mixture distribution (2.72). Let X be a random variable with probability generating function $[\mathcal{P}^{(X)}(s)]^\theta$, where θ is a parameter. Suppose θ itself is a random variable with generating function $\mathcal{P}^{(\theta)}(s)$. Then, the probability generating function of the mixture distribution is given by:

$$\sum[\mathcal{P}^{(X)}(s)]^\theta p(\theta) = \mathcal{P}^{(\theta)}[\mathcal{P}^{(X)}(s)] \tag{2.86}$$

which is the generating function of a compound distribution.

2.5.3 Birth Process Generalizations

A pure birth process is defined by the transition probability (see Chap. 2.2.4)

$$P\{N(0,t+\Delta) = k+1|N(0,t) = k\} = \lambda\Delta + o(\Delta)$$

The transition probabilities can be used to construct the marginal distribution of the count data $N(T)$. This requires the solution of differential equations of the sort encountered in the context of the Poisson process (Chap. 2.2.3), which is always possible, if not analytically then numerically.

The main property of a pure birth process is that the probability of an event depends on the number of events that have occurred up to that moment, and not on *when* they occurred. The nature of the dependence can be kept very general. In fact, it can be shown that for any count data distribution there exists a sequence $\lambda_0, \lambda_1, \lambda_2, \ldots$ such that the count distribution is generated by the specified birth process (Faddy, 1997).

Thus, rather than specifying a parametric probability function directly, one can instead model the function $\lambda_k = f(k;\theta)$ parametrically and derive the corresponding probability function. A class of particular interest is generated by the function (Faddy, 1997)

$$\lambda_k = a(b+k)^c$$

where $a, b > 0$. This formulation nests the Poisson distribution (for $c = 0$) and the negative binomial distribution (for $c = 1$), and it allows for general types of overdispersion (for $c > 0$) and underdispersion (for $c < 0$) that are not linked to any particular existing parametric distribution.

The use of this model in regression analysis requires an expression for the mean. While the exact mean can in general not be computed analytically, Faddy (1997) derives the following approximation

$$\tilde{\mathrm{E}}(X) = b\left\{\left[1 + \frac{a(1-c)}{b^{1-c}}\right]^{\frac{1}{1-c}} - 1\right\} \tag{2.87}$$

The approximation is exact for $c = 0$ and for $c \to 1$. In order to set the (approximate) mean equal to a given value, say μ, one has to parametrize a accordingly, i.e, solve (2.87) for a :

$$a = \frac{(\mu+b)^{1-c} - b^{1-c}}{1-c}$$

This generalized count data distribution has two more parameters than the Poisson distribution. As in the standard Poisson model, μ can be expressed in terms of covariates, and the parameters of the model can be estimated by maximum likelihood.

2.5.4 Katz Family of Distributions

Distributions for non-negative integers can be uniquely represented by their *recursive probability ratios*

$$\frac{P(X=k)}{P(X=k-1)} = \frac{p_k}{p_{k-1}} = f(k,\theta) \quad k = 1, 2, \ldots \tag{2.88}$$

where θ is a vector of parameters. (2.88) is a first order difference equation of the form $p_k = f(k, \theta)p_{k-1}$.

Different recursive probability systems have been developed. The Katz family of distributions (Johnson and Kotz, 1969, p. 37) is the most prominent among them. The family provides a particularly useful tool for econometric modeling since it constitutes a generalization nesting several distributions for non-negative integers, while maintaining a parsimonious parameterization (two parameters). It is defined by the recursive probabilities

$$\frac{p_k}{p_{k-1}} = \frac{\omega + \gamma(k-1)}{k} \qquad k = 1, 2, \ldots \tag{2.89}$$

Since the right-hand-side has to be positive for all possible values of k, the following restrictions hold: a) $\omega > 0$, and b) $k \leq \omega/\gamma$ for $\gamma < 0$. The Poisson distribution is obtained for $\gamma = 0$, the negative binomial distribution for $0 < \gamma < 1$ and the binomial distribution for $\gamma < 0$ when $-\omega/\gamma$ is an integer. Tab. 2.2 compares the parameterizations:

Table 2.2. Sub-Models of the Katz System

Poisson	$\omega = \lambda\,,\ \gamma = 0$
Negative Binomial	$\omega = \alpha\left(\frac{1}{1+\theta}\right)\,,\ \gamma = \frac{1}{1+\theta}$
Geometric	$\omega = \gamma = \frac{1}{1+\theta}$
Binomial	$\omega = \frac{np}{1-p}\,,\ \gamma = -\frac{p}{1-p},\ y \leq n$

The mean of the Katz family of distributions can be calculated as follows: re-writing (2.89) as

$$kp_k = [\omega + \gamma(k-1)]p_{k-1},\ \ k = 1, 2, \ldots$$

and taking sums on both sides, one obtains (the derivation in Johnson and Kotz 1969, p. 37, contains an error: The summation in their formula (32) is with respect to j, not r):

$$\begin{aligned} \mathrm{E}(X) &= \sum_{k=1}^{\infty} kp_k \\ &= \omega \sum_{k=1}^{\infty} p_{k-1} + \gamma \sum_{k=1}^{\infty}(k-1)p_{k-1} \\ &= \omega + \gamma \mathrm{E}(X) \end{aligned}$$

and hence

$$E(X) = \omega/(1-\gamma) \tag{2.90}$$

The second noncentral moment is

$$\begin{aligned} E(X^2) &= \textstyle\sum_{k=1}^{\infty}[\omega + \omega(k-1) + \gamma(k-1)^2 + \gamma(k-1)]p_{k-1} \\ &= \omega + \gamma E(X^2) + (\omega+\gamma)E(X) \\ &= \omega(1+\omega)/(1-\gamma)^2 \end{aligned} \tag{2.91}$$

and the variance is given by

$$\begin{aligned} \text{Var}(X) &= E(X^2) - [E(X)]^2 \\ &= \omega/(1-\gamma)^2 \end{aligned} \tag{2.92}$$

The Katz system of distributions exhibits *equidispersion* for $\gamma = 0$, *overdispersion* for $0 < \gamma < 1$ and *underdispersion* for $\gamma < 0$.

Alternative recursive probability systems have been formulated. For instance, Yousry and Srivastava (1987) include a third parameter to obtain a *hyper-negative binomial distribution.* It is based on the recursive structure

$$\frac{p_{k+1}}{p_k} = \frac{\omega + \gamma k}{k + \psi} \qquad k = 0, 1, \ldots \tag{2.93}$$

For $\psi = 1$, the model corresponds to the Katz family. For $\gamma = 0$, the model reduces to the *hyper-Poisson distribution* mentioned in Johnson and Kotz (1969, p. 43).

Kulasekera and Tonkyn (1992) formulate a probability distribution for strictly positive integers with

$$\frac{p_{k+1}}{p_k} = \left(\frac{k+1}{k}\right)^{\alpha} q \qquad k = 1, 2, \ldots \tag{2.94}$$

where $q \in (0,1)$ and $\alpha \in \mathbb{R}$. It nests the shifted negative binomial, the logarithmic and the discrete Pareto distribution.

2.5.5 Linear Exponential Families

A second generalized class of distributions is the class of *linear exponential families*, though this class is not constrained to distributions for non-negative integers. It provides a unified framework for the analysis of many important distributions, including the normal distribution, the Poisson distribution, and the binomial distribution.

The interest in linear exponential family (LEF) stems mainly from certain results on robust (or *semi-parametric*) estimation that apply to such models. The density or probability functions of LEF distributions are of the form (see, for instance, McCullagh and Nelder (1989) and Gourieroux, Monfort and Trognon (1984a))

$$f(x;\psi) = c(x,\phi)\exp\{(x\psi - b(\psi))/\phi\}\ . \tag{2.95}$$

ψ is called the "natural" parameter and ϕ the "dispersion" parameter. The functional form (2.95) shows why the name "linear exponential" is appropriate: f is log-linear in its parameter ψ. As a consequence, the derivative of the logarithmic density with respect to ψ has a very simple form

$$\frac{\partial \ln f(x;\psi)}{\partial\psi} = \frac{x - b'(\psi)}{\phi} \tag{2.96}$$

This derivative plays an important role in estimation, as it is the "score" or "gradient" of the log-likelihood function.

It is a well known result in statistics that the expected gradient is zero. This result follows because differentiating the identity

$$\frac{\partial}{\partial\psi}\left(\int f(x;\psi)dx = 1\right)$$

yields (under suitable regularity conditions that ensure that integration and differentiation can be interchanged)

$$\begin{aligned}\int \frac{\partial f(x;\psi)}{\partial\psi}\,dx &= \int \frac{\partial \ln f(x;\psi)}{\partial\psi} f(x;\psi)\,dx \\ &= \mathrm{E}\left(\frac{\partial \ln f(x;\psi)}{\partial\psi}\right) \\ &= 0\end{aligned} \tag{2.97}$$

Applying this result to the right hand side expression of (2.96), we find that

$$\mathrm{E}(X) = b'(\psi) \tag{2.98}$$

The fact that the score of a linear exponential family is the difference between the random variable and its mean constitutes the reason why consistency of the maximum likelihood estimator requires only that the mean $b'(\psi)$ of the LEF is correctly specified (and that ψ is identified). Under independent sampling, the empirical gradient converges in probability to its expected value of zero. But this means that $\hat{\psi}$ converges in probability to the value where $\mathrm{E}(X) = b'(\psi)$. Note that this result is unaffected by the presence of a dispersion parameter ϕ. More details are given in the next chapter, when generalized linear models and robust Poisson regression are introduced.

We conclude by deriving the variance of a LEF distribution. To do this, one can use the results that the expected second derivative of a logarithmic density function is equal to the variance of its first derivative. For a LEF density,

$$\mathrm{E}\left(\frac{\partial^2 \ln f(x;\psi)}{\partial\psi^2}\right) = \frac{-b''(\psi)}{\phi} \tag{2.99}$$

Further, the variance of the first derivative is given by

$$\begin{aligned}\mathrm{E}\left(\frac{\partial \ln f(x;\psi)}{\partial \psi}\right)^2 &= \mathrm{E}\left(\frac{x - b'(\psi)}{\phi}\right)^2 \\ &= \frac{\mathrm{Var}(X)}{\phi^2}\end{aligned} \tag{2.100}$$

It follows that

$$\mathrm{Var}(X) = \phi b''(\psi) \tag{2.101}$$

The variance of X is the product of two components. One, $b''(\psi)$, is called the variance function. It can be written as a function of the mean $\mathrm{E}(X) = \mu$, since from (2.98) it holds that $\psi = (b')^{-1}(\mu)$. The second componet is a scaling parameter ϕ. For instance, the normal distribution assumes a scaling parameter $\phi = \sigma^2$ and a constant variance function $V(\mu) = 1$.

Tab. 2.3 gives the characteristics of some common univariate distributions contained in the linear exponential families. Further members are the gamma and the inverse Gaussian distributions.

Table 2.3. Linear Exponential Families

Distribution	$c(x,\phi)$	$b(\psi)$	ψ
Poisson	$1/x!$	$\exp(\psi)$	$\log(\lambda)$
Negative Binomial[a]	$\frac{\Gamma(\alpha+x)}{\Gamma(\alpha)\Gamma(x+1)}$	$-\alpha\log(1-e^{\psi})$	$\log\left(\frac{\theta}{1+\theta}\right)$
Geometric ($\alpha = 1$)	1	$-\log(1-e^{\psi})$	$\log\left(\frac{\theta}{1+\theta}\right)$
Binomial (n given)	$\binom{n}{x}$	$n\log(1+e^{\psi})$	$\log\left(\frac{p}{1-p}\right)$
Normal	$\frac{\exp(-\frac{x^2}{2\phi^2})}{\sqrt{2\pi\phi^2}}$	$\psi^2/2$	$\psi(\sigma=\phi)$

	$\mathrm{E}(X)$	$V(\mu)$
Poisson	λ	μ
Negative Binomial[a]	$\alpha\theta$	$\mu + \alpha^{-1}\mu^2$
Geometric ($\alpha = 1$)	θ	$\mu + \mu^2$
Binomial (n given)	np	$\mu(1-\mu)$
Normal	ψ	1

[a] for given α

2.5.6 Additive Log-Differenced Probability Models

Gourieroux and Monfort (1990) define the *additive-log-differenced probability class* (A.L.D.P.) which applies to situations, where the function f determining the ratio of recursive probabilities in (2.88) is separable:

$$f(x, \psi) = g(x)h(\psi) \tag{2.102}$$

It is a generalization since ψ can be estimated *without* specifying the functional form of $g(x)$.

Proposition 6. *If the probability distribution f(x) is a linear-exponential family defined over the non-negative integers, the ratio of recursive probabilities is separable.*

Proof: Linear exponential families are of the form

$$f(x; \psi) = c(x, \phi) \exp\{(x\psi - b(\psi))/\phi\}$$

The ratio of recursive probabilities is therefore given by:

$$\frac{f(x;\psi)}{f(x-1;\psi)} = c(x,\phi)/c(x-1,\phi)\exp(\psi) = g(x,\phi)h(\psi)\,.\square$$

Example 1.
The Poisson distribution is a linear exponential family with $c(x, \phi) = 1/x!$, $b(\psi) = \exp(\psi)$, and $\psi = \log(\lambda)$, where λ is equal to the expected value (mean parameterization). The ratio of recursive probabilities λ/x is separable into two functions $g(x) = 1/x$ and $h(\lambda) = \lambda$.

Example 2.
The geometric distribution is a linear exponential family with $c(x, \phi) = 1$, $b(\psi) = -\log(1 - \exp(\psi))$, and $\psi = \log(\lambda/(1+\lambda))$, again in mean parameterization with $E(X) = \lambda$. The ratio of recursive probabilities $\lambda/(1+\lambda)$ is separable into two functions $g(x) = 1$ and $h(\lambda) = \lambda/(1+\lambda)$.

The A.L.D.P. defines a class of probability distributions whose recursive probability ratio is separable. It is convenient to specify this distribution class in terms of log-ratios. From (2.88) and (2.102) it follows immediately that

$$\log f(x) - \log f(x-1) = \tilde{g}(x) + \tilde{h}(\psi) \tag{2.103}$$

where $\tilde{g} = \log(g)$ and $\tilde{h} = \log(h)$.

2.5.7 Summary

This section has introduced different types of generalizations. These generalizations had as a common point of departure the Poisson distribution with its restrictive assumptions and structure. More flexible probability models have been developed along two different routes.

The first approach formulated compound and mixture distributions the development of which often was motivated by a reconsideration of the data generating process. For instance, a compound Poisson distribution can describe the number of migrants between two geographical areas if the number of *families* moving is Poisson distributed and the number of *persons* in each family follows a binomial distribution. A Poisson mixture distribution may be appropriate if the Poisson parameter λ is measured with error.

The second approach directly formulated a more richly parameterized, and therefore more general, distribution model. An example is the Katz system of distributions. If the interest of the researcher rests less in the best possible fit to observed data but rather in the robustness of parameter estimates, distributions within the class of linear exponential families have desirable robustness properties.

2.6 Duration Analysis and Count Data

When looking at a sequence of events, most econometricians are more familiar with the concept of *waiting times* (or, in technical applications: *failure* times) and *duration models* than with the concept of event counts. See Allison (1984) for an excellent introduction to duration models. Lancaster (1990) provides a more advanced treatment. In this chapter it will be demonstrated that the two approaches are closely related. In particular, the distributions of cumulative waiting times uniquely determine the distribution of counts. This link has been exploited in Winkelmann (1995) in order to derive a new count data distribution, and improve the understanding of the properties of count data models in general (see also Lee, 1996). In particular, a new interpretation can be given to the phenomena of over- and underdispersion based on the duration properties of the underlying process.

The basic concepts of duration analysis were introduced in Chap. 2.2.6. They are repeated here for convenience. Let $\tau_i, i \in I\!N$ denote the interarrival, or waiting, times between the $(i-1)$-th and the i-th event. Then,

$$\vartheta_k = \sum_{k=1}^{k} \tau_t \, , \; k = 1, 2, \ldots \tag{2.104}$$

is the arrival time of the k-th event. By definition

$$N(T) < k \text{ if and only if } \vartheta_k > T \tag{2.105}$$

and

$$P(N(T) < k) = P(\vartheta_k > T) = 1 - F_k(T)\,, \tag{2.106}$$

where $F_k(T)$ is the cumulative density function of ϑ_k. Further,

$$\begin{aligned} P(N(T) = k) &= P(N(T) < k + 1) - P(N(T) < k) \\ &= F_k(T) - F_{k+1}(T)\,. \end{aligned} \tag{2.107}$$

Equation (2.107) provides the fundamental relation between the distribution of waiting times and the distribution of counts. The probability distribution of $N(T)$ can be obtained explicitly for all k from knowing the distribution of ϑ_k.

2.6.1 Distributions for Interarrival Times

The duration of stay in one state is modeled by a non-negative continuous random variable τ. The following definitions will be used in this section:

$f(t)$	density function of completed spells
$F(t) = P(\tau < t)$	distribution function of completed spells
$\bar{F}(t) = 1 - F(t)$	survivor function

An important concept for the analysis of durations is the *hazard rate* $\phi(t)$ which gives the instantaneous exit probability conditional on survival. Formally,

$$\phi(t) = \lim_{dt \to 0} \frac{P(t \le \tau < t + dt | \tau \ge t)}{dt} \tag{2.108}$$

Using Bayes rule

$$\begin{aligned} P(t \le \tau < t + dt | \tau \ge t) &= \frac{P(t \le \tau < t + dt, \tau \ge t)}{P(\tau \ge t)} \\ &= \frac{P(t \le \tau < t + dt)}{P(\tau \ge t)} \end{aligned}$$

Expressing the probabilities through cumulative density functions, dividing by dt and taking limits, we obtain

$$\phi(t) = \frac{f(t)}{1 - F(t)} = -\frac{d}{dt} \log \bar{F}(t) \tag{2.109}$$

The hazard function captures the underlying time dependence of the process. A decreasing hazard function implies that the spell is less likely to end the longer it lasts. This situation will be denoted as *negative duration dependence.* An increasing hazard function implies that the spell is the more likely to end the longer it lasts. This situation will be denoted as *positive duration dependence.* Clearly, positive duration dependence negatively affects the probability of long spells, while negative duration dependence does the opposite. No duration dependence corresponds to the case of a constant hazard. The hazard function of an interarrival time distribution does not need to be monotonic although many parametric distributions that are used in practice have a monotonic hazard function.

In the case of a constant hazard $\phi(t) = \lambda$, there is a unique underlying distribution which can be obtained directly as the solution to the differential equation

$$-\frac{d}{dt}\log \bar{F}(t) = \lambda \tag{2.110}$$

Hence, $\bar{F}(t) = Ae^{-\lambda t}$, and, using the initial condition $\bar{F}(0) = 1$,

$$\bar{F}(t) = e^{-\lambda t} \tag{2.111}$$

Thus, under the assumption of a constant hazard function we obtain that $F(t) = 1 - e^{-\lambda t}$ and $f(t) = \lambda e^{-\lambda t}$. These are the distribution and the density functions of the exponential distribution, respectivelt. The hazard (which equals the inverse of the expected value) is constant if and only if the distribution of completed spells is exponential.

A constant hazard function is closely related to the concept of absence of memory within the process:

Definition 3. *A process is called* memoryless *if the probability that a spell exceeds $s + t$, conditional on having lasted for t, is independent of t.*

Formally, this can be written as

$$P\{\tau > s + t | \tau > t\} = P\{\tau > s\} \quad \forall\, s, t \geq 0\ . \tag{2.112}$$

In other words, the past length of a spell is without influence on its future duration. Using Bayes' rule, this condition can be rewritten as

$$\frac{P\{\tau > s + t, \tau > t\}}{P\{\tau > t\}} = P\{\tau > s\} \tag{2.113}$$

or, since $\{\tau > t\}$ is contained in $\{\tau > s + t\}$,

$$P\{\tau > s + t\} = P\{\tau > t\}P\{\tau > s\}\ . \tag{2.114}$$

The survivor function of the exponential distribution is given by $P\{\tau > t\} = e^{-\lambda t}$. Since $e^{-\lambda(t+s)} = e^{-\lambda t}e^{-\lambda s}$, it follows that exponentially distributed waiting times is memoryless. If the spell lasted until time t, the distribution of the remaining elapsed time until completion is identical to the original lifetime distribution.

A distribution for discrete waiting times with similar properties is the geometric distribution. It is generated in the following way: Consider an infinite sequence of Bernoulli trials where the probability of a success in a particular trial is given by $p \in (0,1)$ and the probability of a failure by $q = 1 - p$. The index of this sequence may be considered as a time index, i.e., at every discrete time $t = 0, 1, \ldots$ an independent Bernoulli trial is performed. The (discrete) waiting time for a success is given by the total number of failures that occurred before the first success. This waiting time follows a geometric distribution with probability function (see e.g., DeGroot, 1986)

$$P(\tau = t) = p(1-p)^t \, , \; t = 0, 1, \ldots \tag{2.115}$$

Alternatively, setting $p = 1/(1+\theta)$, (2.115) may be rewritten as

$$P(\tau = t) = \left(\frac{1}{1+\theta}\right)\left(\frac{\theta}{1+\theta}\right)^t \tag{2.116}$$

The inverse distribution function $P(\tau > t)$ is given by:

$$\begin{aligned} 1 - \textstyle\sum_{i=0}^{t} p(1-p)^i &= (1-p)\,[1 - \textstyle\sum_{i=1}^{t} p(1-p)^{i-1}] \\ &= (1-p)^2\,[1 - \textstyle\sum_{i=2}^{t} p(1-p)^{i-2}] \\ &\;\;\vdots \\ &= (1-p)^t\,[1 - \textstyle\sum_{i=k}^{t} p(1-p)^{i-t}] \\ &= (1-p)^{t+1} \end{aligned} \tag{2.117}$$

For discrete waiting time distributions the hazard rate is defined by

$$\phi(t) = \frac{P(\tau = t)}{1 - P(\tau \leq t)} = \frac{P(\tau = t)}{P(\tau > t)} \tag{2.118}$$

Thus, the hazard function of the geometric distribution is given by $\phi(t) = p/(1-p)$. As for the exponential distribution, the hazard is constant and equal to the inverse expected value. Moreover, $P(\tau > t + j) = P(\tau \geq j)P(\tau > t)$ implying that the process starts permanently anew, conditional on survival.

Clearly, the assumption of a constant hazard is too restrictive in most applications. Distributions that allow for positive or negative duration dependence are, among others, the gamma and the Weibull distributions. Both distributions are characterized by a monotonic hazard function, either increasing or decreasing, the slope of which depends on the value taken by a specific parameter.

Until now the discussion has focussed on the properties for the distribution of the underlying interarrival times τ_i. The distributions of the waiting times ϑ_k are defined by a convolution operation. Two types of dependencies can be distinguished. The first type is duration dependence for the interarrival times, i.e., within the single spell. The second type is dependence between successive spells. This type is called occurrence dependence. In a model for repeated durations, such a dependence might be captured through time varying covariates or lagged variables. A process that excludes inter-spell dependence and assumes i.i.d. interarrival times is called a *renewal process.* Renewal processes may, however, display duration dependence and in the following section, results from renewal theory are used to provide important insights for the modeling of count data.

2.6.2 Renewal Processes

Useful references on renewal processes are Feller (1971), Cox (1962), and Lancaster (1990). Consider a stochastic process that is defined by a sequence of spells τ_i, where the end of one spell immediately leads to the start of a new spell. If $\{\tau_1, \tau_2, \ldots\}$ are independently and identically distributed variables, all with density function $f(\tau)$, the process is called a *renewal process.* Let $N(T)$ denote the number of renewals in $(0, T)$, i.e., the event counts before T. Its probability function in terms of the cumulative densities of the waiting times ϑ_k was given in (2.107). But $\vartheta_k = \sum_{i=1}^{k} \tau_i$. Given the assumption of independent renewals, the distribution of this k-fold convolution can be derived using the calculus of Laplace transforms (See Feller, 1971).

In Chap. 2.2.6 the case of independently and identically exponentially distributed renewals was considered. It was shown that the waiting time for the k-th occurrence had a gamma distribution with parameters λ and k. Furthermore, it was shown that in this case the count data distribution was of the Poisson form:

$$\begin{aligned} P(N = k) &= F_k - F_{k+1} \\ &= \frac{e^{-\lambda}\lambda^k}{k!} \end{aligned}$$

In the next section this result will be generalized. In particular, we will answer the question as to what happens if the distribution of the inter-arrival times is characterized by positive or negative duration dependence, while the assumption of independent renewals is maintained.

2.6.3 Duration Dependence and Dispersion

Even without fully specifying the distribution of τ, a limiting result can be obtained. Denote the mean and the variance of the waiting time distribution by $\mathrm{E}(\tau) = \mu$ and $\mathrm{Var}(\tau) = \sigma^2$, and the coefficient of variation by $v = \sigma/\mu$. Define the hazard function

$$\lambda(t) = \frac{f(t)}{1 - F(t)}$$

where $f(t)$ and $F(t)$ are the density function and the cumulative probability function of τ, respectively. The distribution displays negative duration dependence for $\mathrm{d}\lambda(t)/\mathrm{d}t < 0$ and positive duration dependence for $\mathrm{d}\lambda(t)/\mathrm{d}t > 0$. Assume that the hazard function is monotonic. Then

$$\frac{d\lambda(t)}{dt} \left.\begin{matrix} < \\ = \\ > \end{matrix}\right\} 0 \Longrightarrow v \left.\begin{matrix} > \\ = \\ < \end{matrix}\right\} 1$$

(See Barlow and Proschan, 1965, p. 33).

Theorem 2. .
Let $\{\tau_i\}$ be a sequence of independent, positive, identically distributed random variables and $N(t)$ the number of renewals between 0 and t. Assume that the densities of the waiting times τ_i have a monotonic hazard function. Then negative (positive) duration dependence of the waiting time densities causes over- (under) dispersion of the distribution of $N(t)$ as $t \to \infty$.

Proof: $N(t)$ is asymptotically normal distributed with

$$N(t) \overset{asy}{\sim} \text{normal}\left(\frac{t}{\mu}, \frac{\sigma^2 t}{\mu^3}\right) \tag{2.119}$$

as $t \to \infty$ (Cox, 1962, p.40). The ratio of variance to mean of the limiting distribution is given by

$$\frac{\text{variance}}{\text{mean}} \sim \frac{\sigma^2 t}{\mu^3}\frac{\mu}{t} = \frac{\sigma^2}{\mu^2} \tag{2.120}$$

Thus, the variance mean ratio is greater (less) than 1 if and only if the coefficient of variation of the waiting times $v = \sigma/\mu$ is greater (less) than 1. For positive duration dependence $v < 1$ and the count distribution is underdispersed. For negative duration dependence $v > 1$ and the count distribution is overdispersed. □

The exponential distribution has coefficient of variation $v = 1$, leading to equidispersion. This result is exact, whereas (2.120) is only a limiting result.

2.6.4 Gamma Count Distribution

It will be assumed that the waiting times τ_i are identically and independently gamma distributed. Dropping the index i the density can be written as

$$f(\tau;\alpha,\beta) = \frac{\beta^\alpha}{\Gamma(\alpha)}\tau^{\alpha-1}e^{-\beta\tau}, \qquad \alpha,\, \beta \in \mathbb{R}^+ \tag{2.121}$$

for $\tau > 0$. The waiting time has mean $\mathrm{E}(\tau) = \alpha/\beta$ and variance $\mathrm{Var}(\tau) = \alpha/\beta^2$. The hazard function $\lambda(\tau)$ obeys the equation

$$\frac{1}{\lambda(\tau)} = \int_0^\infty e^{-\beta u}\left(1 + \frac{u}{\tau}\right)^{\alpha-1} du \tag{2.122}$$

The gamma distribution admits no closed form expression for the tail probabilities and thus no simple formula for the hazard function. However, from

(2.122), it follows that $\lambda(\tau)$ is (monotonically) increasing for $\alpha > 1$, decreasing for $\alpha < 1$, and constant (and equal to β) for $\alpha = 1$.

Now, consider the arrival time of the k-th event

$$\vartheta_k = \tau_1 + \ldots + \tau_k, \qquad k = 1, 2, \ldots \tag{2.123}$$

where $\{\tau_i\}$ are i.i.d. gamma distributed. The reproductive property of the gamma distribution (Johnson and Kotz, 1970, p. 170) implies that ϑ_k is gamma distributed with density

$$f_k(\vartheta; \alpha, \beta) = \frac{\beta^{k\alpha}}{\Gamma(k\alpha)} \vartheta^{k\alpha-1} e^{-\beta\vartheta} \tag{2.124}$$

To derive the new count data distribution, we have to evaluate the cumulative distribution function

$$F_k(T) = \frac{1}{\Gamma(k\alpha)} \int_0^{\beta T} u^{k\alpha-1} e^{-u} du \qquad k = 1, 2, \ldots \tag{2.125}$$

where the integral is the incomplete gamma function. The right hand side will be denoted as $G(\alpha k, \beta T)$. Note that $F_0(T) = G(0, \beta T) = 1$. The number of event occurrences during the time interval $(0, T)$ has the two-parameter distribution function

$$P(N = k) = G(\alpha k, \beta T) - G(\alpha k + \alpha, \beta T) \tag{2.126}$$

for $k = 0, 1, 2, \ldots$, where α, $\beta > 0$.

If α is integer valued, (2.124) coincides with a distribution known in the statistical literature as the *Erlangian* distribution (Cox, 1962, p. 15). Integrating (2.125) by parts gives

$$G(\alpha k, \beta T) = 1 - e^{-\beta T} \left(1 + \beta T + \frac{(\beta T)^2}{2!} + \ldots + \frac{(\beta T)^{\alpha k - 1}}{(\alpha k - 1)!} \right) \tag{2.127}$$

Hence,

$$\begin{aligned} P(N = k) &= G(\alpha k, \beta T) - G(\alpha k + \alpha, \beta T) \\ &= e^{-\beta T} \sum_{i=0}^{\alpha-1} \frac{(\beta T)^{\alpha k + i}}{(\alpha k + i)!}, \qquad k = 0, 1, 2, \ldots \end{aligned} \tag{2.128}$$

For $\alpha = 1$, $f(\tau)$ is the exponential density and (2.128) simplifies to the Poisson distribution. It was noted above that the Poisson distribution is characterized by independently exponentially distributed waiting times.

For non-integer α, no closed-form expression is available for $G(\alpha k, \beta T)$ (and thus for $P(N = k)$). Numerical evaluations of the integral can be based on asymptotic expansions (See Abramowitz and Stegun, 1964, and Bowman and Shenton, 1988). Fig. 2.3 and 2.4 compare the probability functions of the gamma count distribution with a Poisson distribution of identical mean

($\mathrm{E}(N) = 2$) for two values of α. Depending on the value of α, the new model is more concentrated ($\alpha = 1.5$) or more dispersed ($\alpha = 0.5$) than the reference distribution.

Fig. 2.3. Probability Functions for Gamma Count and Poisson Distributions; $\alpha = 0.5$ (overdispersion): heavily shaded, Gamma count; lightly shaded, Poisson

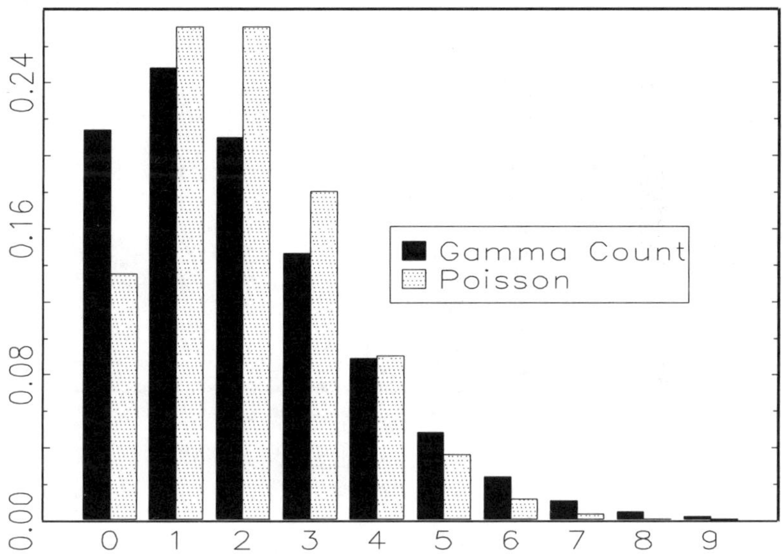

The expected value is given by

$$\mathrm{E}(N) = \sum_{i=1}^{\infty} G(\alpha i, \beta T) \tag{2.129}$$

For increasing T it holds that (See (2.119))

$$N(T) \overset{asy}{\sim} \text{normal}\left(\frac{\beta T}{\alpha}, \frac{\beta T}{\alpha^2}\right) \tag{2.130}$$

The limiting variance-mean ratio equals a constant $1/\alpha$. It follows that the gamma count distribution function (2.126) displays overdispersion for $0 < \alpha < 1$ and underdispersion for $\alpha > 1$. The same holds true for a unit time period where the moments are evaluated numerically. Fig. 2.5 and Fig. 2.6 show the variance mean ratio for various values of α and β.

Fig. 2.4. Probability Functions for Gamma Count and Poisson Distributions; $\alpha = 1.5$ (underdispersion): heavily shaded, Gamma count; lightly shaded, Poisson

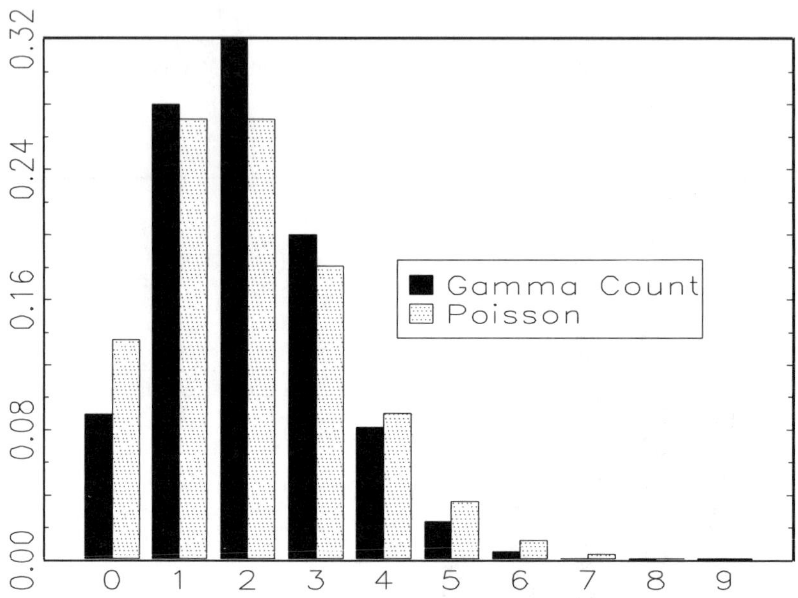

The underlying waiting times have a decreasing (increasing) hazard for $0 < \alpha < 1$ ($\alpha > 1$). Thus, negative duration dependence is associated with overdispersion, positive duration dependence with underdispersion. The intermediate case of no duration dependence, i.e., exponentially distributed waiting times, leads to the Poisson distribution with equal mean and variance.

2.6.5 Duration Mixture Models

The phenomena of a positive or negative relationship between duration and hazard in the aggregate does not need to reflect 'true' duration dependence but can also be due to a selection process: to take the example of negative duration dependence, individuals with duration of spells above average might have a (constant) hazard below average. Failure to account for this heterogeneity, for example by splitting up the population into sub-populations, results in spurious negative duration dependence. The problem of identifying true duration dependence was discussed in detail by Heckman and Singer (1984). It is closely related to the problem of distinguishing between occurrence dependence and unobserved heterogeneity in count data.

Fig. 2.5. Variance to Mean Ratio for Gamma Count Distribution; $0 < \alpha < 1$.

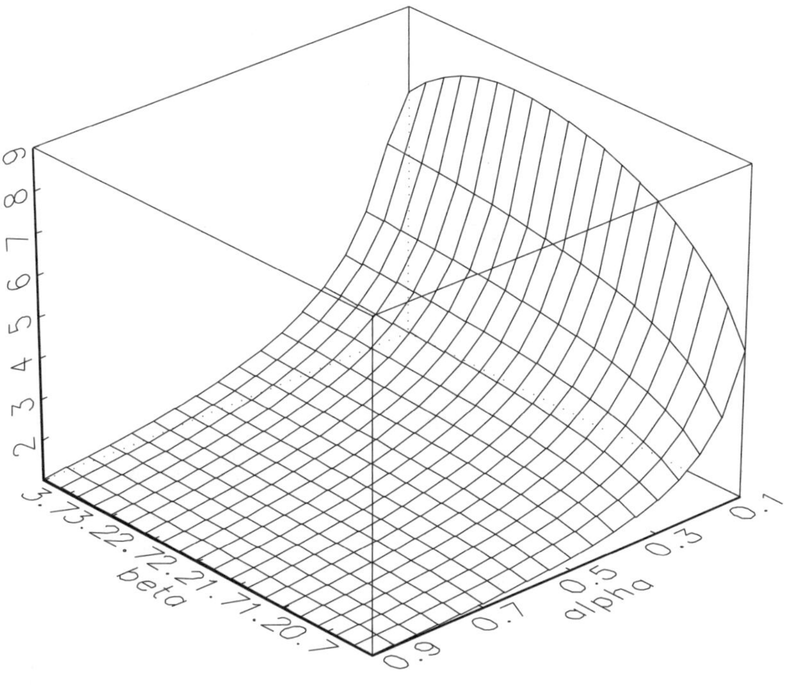

If one suspects the presence of unobserved heterogeneity, one possible solution is to assume that the heterogeneity follows a specific parametric distribution. This was already discussed in Chap. 2.5.1 where it was shown that if the Poisson parameter λ is a random variable with a gamma distribution, the number of events occurring in a given interval has a negative binomial distribution Negbin(α, θ). A corresponding result exists in the duration domain:

Proposition 7. *Assume that the Poisson parameter λ is a random variable with a gamma distribution. Then the waiting time for the first occurrence has an exponential-gamma mixture distribution and the hazard rate $\phi(t) = \alpha/(\beta + t)$ is a decreasing function of time.*

Proof: The probability function of the count is obtained in the usual way via integration, where we keep explicitly track of varying lengths of the time interval:

$$P(X = k; t) = \int_0^\infty \frac{(\lambda t)^k e^{-\lambda t}}{k!} \frac{\beta^\alpha}{\Gamma(\alpha)} \lambda^{\alpha-1} e^{-\lambda\beta} d\lambda$$

Fig. 2.6. Variance to Mean Ratio for Gamma Count Distribution; $\alpha > 1$.

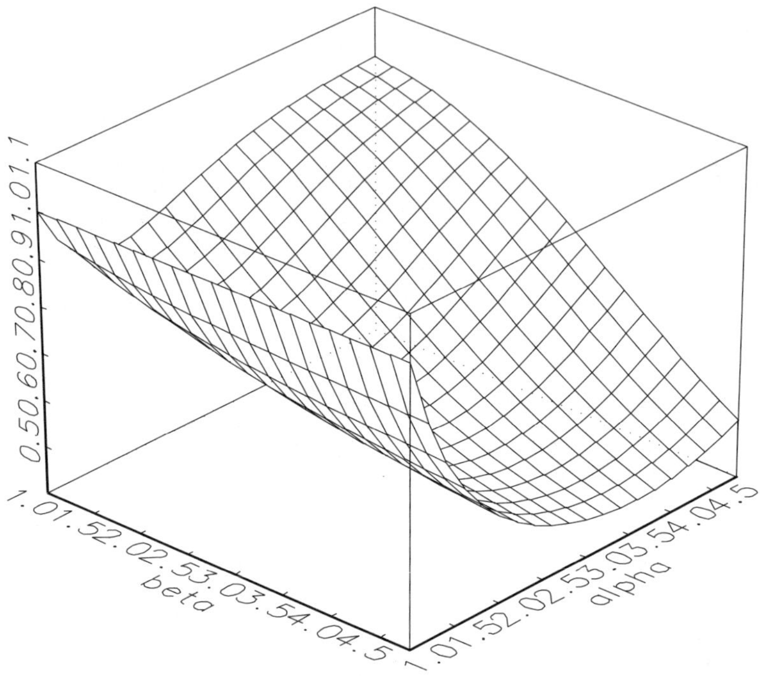

$$= t^k \frac{\Gamma(\alpha + k)}{\Gamma(\alpha)\Gamma(k+1)} \left(\frac{\beta}{\beta + t}\right)^{\alpha} \left(\frac{1}{\beta + t}\right)^{k}$$

But $P(X = 0; t)$ is the probability that no event has occurred before t, i.e., the survivor function at t:

$$\bar{F}(t) = \left(\frac{\beta}{\beta + t}\right)^{\alpha} \tag{2.131}$$

with density function

$$f(t) = -\frac{d}{dt}\bar{F}(t) = \alpha\beta^{\alpha}\left(\frac{1}{\beta + t}\right)^{\alpha+1} \tag{2.132}$$

and hazard rate

$$\phi(t) = -\frac{d}{dt}\log \bar{F}(t)$$

$$
\begin{aligned}
&= \frac{\alpha}{\beta + t} \\
&= \frac{\lambda}{1 + \lambda/\alpha t}
\end{aligned} \tag{2.133}
$$

where the last line follows from letting $\lambda = \alpha/\beta$. □

Thus, if $\alpha \to \infty$ and $1/\beta \to 0$ such that $\alpha/\beta = \lambda$ (i.e., the gamma mixture distribution has a mean of λ and a variance approaching zero) the hazard function collapses to the hazard function of the exponential distribution.

Incidentally, the density (2.132) can also be obtained directly by mixing the exponential density of the arrival time of the first event with a gamma distribution (see also Lancaster, 1990, Chap. 4):

$$
\begin{aligned}
f(t) &= \int_0^\infty \lambda e^{-\lambda t} \frac{\beta^\alpha}{\Gamma(\alpha)} \lambda^{\alpha-1} e^{-\lambda\beta} d\lambda \\
&= \frac{\beta^\alpha}{\Gamma(\alpha)} \int_0^\infty \lambda^\alpha \lambda e^{-\lambda(t+\beta)} d\lambda \\
&= \frac{\beta^\alpha}{\Gamma(\alpha)} \frac{\Gamma(\alpha+1)}{(t+\beta)^{\alpha+1}}
\end{aligned} \tag{2.134}
$$

With heterogeneity, the sample hazard is no longer constant but instead decreasing with time. As indicated above, this model describes a situation where each individual has a constant hazard that randomly varies between individuals according to a gamma law. The gamma disturbance captures unobserved heterogeneity. In terms of counts it leads to the negative binomial distribution with overdispersion. In terms of waiting times it leads to a decreasing overall hazard since the mean hazard among *survivors* at t is a decreasing function of t. A selection effect is taking place: individuals with larger hazard are likely to exit earlier, and the group of survivors is increasingly composed of individuals with relatively small ϕ's. The parameters α and β have opposite effects. An increase in α increases the hazard proportionally for all lengths of duration. The negative effect of β is reduced with increased duration.

3

Econometric Modeling – Basic Issues

3.1 Poisson Regression Model

The Poisson regression model is the benchmark model for count data in much the same way as the normal linear model is the benchmark for real-valued continuous data. Early references in econometrics include Gilbert (1982), Hausman, Hall and Griliches (1984), and Cameron and Trivedi (1986). Since then, both methodological contributions and applications of existing methods have proliferated (see also Cameron and Trivedi, 1998). A comprehensive account of the methodological state of the art constitutes the main focus this book. Applications are discussed in Chap. 7 where a selection of applied studies is presented by subject area, covering fields as diverse as accidents, health, demography, marketing and management, and labor mobility.

3.1.1 Specification of the Poisson Regression Model

Let Y denote the vector of regressands (endogenous, dependent, explained variables), and let X denote the matrix of regressors (exogenous, independent, or explanatory variables). Let n be the number of observations in the sample and k be the number of regressors (including, usually, a constant). The dimension of Y is $(n \times 1)$ and the dimension of X is $(n \times k)$. Furthermore, y_i denotes the i-th value of Y and x_i' denotes the i-th row of X. Hence x_i is a column vector of dimension $(k \times 1)$. In this notation, upper case letters are used for vectors or matrices that collect all observations, while lower case letters are used for single observations. In contrast to the previous section, we abandon the distinction between random variables (upper case letters) and their realizations (lower case letters).

The standard univariate Poisson regression model makes the following three assumptions:

Assumption 1

$$y_i|x_i \sim \text{Poisson}[\lambda(x_i)]$$

where $y_i|x_i$ stands for the conditional distribution of y_i given the realization x_i, and $\lambda(x_i) = \text{E}(y_i|x_i)$ (mean parameterization).

Assumption 2

$$\lambda(x_i;\beta) = \exp(x_i'\beta) \qquad i = 1,\ldots,n$$

where β is a $(k \times 1)$ vector of parameters.

Assumption 3

$(y_i|x_i), i = 1,\ldots,n$ are independently distributed.

Discussion

Assumptions 1 and **2** can be combined to obtain the following conditional probability distribution:

$$f(y_i) = \frac{\exp(-\exp(x_i'\beta))\exp(y_i x_i'\beta)}{y_i!} \qquad y_i = 0,1,2,\ldots \tag{3.1}$$

Note that $f(y_i)$ is shorthand notation for $P(Y_i = y_i|X_i = x_i)$. The Poisson distribution has only one parameter that simultaneously determines (conditional) mean and variance. Therefore

$$\text{E}(y_i|x_i) = \exp(x_i'\beta) \tag{3.2}$$

and

$$\text{Var}(y_i|x_i) = \exp(x_i'\beta) \tag{3.3}$$

(3.2) is referred to as *mean function*, *conditional expectation function*, or *regression function.* (3.3) is called the *skedastic function* or *variance function.* Hence, both logarithmic mean and logarithmic variance are linear functions of x and β.

In conjunction with **Assumptions 1** and **2**, **Assumption 3** allows for a straightforward application of maximum likelihood technology to estimate the parameters of the model. Maximum likelihood estimation is discussed below.

The probabilistic assumptions underlying the Poisson distribution have been discussed in the previous chapter. In a nutshell, events are assumed to occur truly randomly over time. In the context of the regression model, explanatory variables influence the dependent variable (the number of event counts in a time interval) through the intensity (or instantaneous occurrence rate) of the process. The heterogeneity of the latter is modeled as a deterministic function of the explanatory variables. This implies that, unlike for the normal linear regression model, the randomness of the Poisson model is intrinsic and not due to an additive stochastic error representing additional heterogeneity.

If the underlying stochastic process does not display the required randomness, or if there is not even a meaningful underlying stochastic process to think of, the Poisson regression model may still remain a valid approximation as well as a useful descriptive tool.

3.1.2 Ordinary Least Squares and Other Alternatives

The advantages and disadvantages of the Poisson regression model are best contemplated by addressing the practitioner's question "When and why should the Poisson regression model be used?". Clearly, a necessary requirement is that the dependent variable be a count. But this condition is not sufficient. Firstly, there are alternative count data models that take the nature of the dependent variable into account and that may be superior to the Poisson model. Possible specifications of alternative count data models and the selection of the right model are the main subject matter of this book.

Secondly, however, one could ignore the special nature of the dependent variable altogether and just apply standard regression models such as the normal linear model

$$y_i = x_i'\beta + e_i \qquad e_i \sim \mathrm{N}(0, \sigma^2) \tag{3.4}$$

Several objections against such an approach can be brought forward. (3.4) ignores the discrete nature of the dependent variable. Under the normal linear model, the probability of any particular outcome is zero whereas the probability of a discrete outcome is not zero in general. Model (3.4) allows for negative outcomes whereas counts are non-negative. And relatedly, the model violates **Assumption 2** that the conditional expectation function is log-linear. Thus, (3.4) will give an inconsistent estimator of β if the true data generating process follows the Poisson regression model. Finally, it ignores the heteroskedasticity inherent in count data (see equation (3.3)).

Log-Linear Model

These concerns can in part be addressed by conventional methods. Start with the conditional expectation function. We could specify

$$\ln y_i = x_i'\beta + u_i \qquad u_i \sim \mathrm{N}(0, \sigma^2) \tag{3.5}$$

where "ln" denotes the natural logarithm. In this model,

$$y_i = \exp(x_i'\beta + u_i) \tag{3.6}$$

has a log-normal distribution with conditional expectation

$$\mathrm{E}(y_i|x_i) = \exp(x_i'\beta + 1/2\sigma^2)$$

similar to the Poisson regression model (up to a scale factor $Q = \exp(1/2\sigma^2)$), and the values of y_i are restricted to the non-negative real line. As long as the model has an overall constant, we can redefine $\tilde{\beta}_0 = \beta_0 - 1/2\sigma^2$ and the two models have essentially the same conditional expectation function.

The log-normal distribution implies a different variance function, though. In particular, it holds that

$$\text{Var}(y_i|x_i) = \phi[\text{E}(y_i|x_i)]^2$$

where $\phi = e^{\sigma^2} - 1$. Generally speaking, estimated standard models of the log-normal and Poisson models won't be comparable, and heteroskedasticity consistent standard errors should be computed.

A more fundamental problem with the log-normal approach is that "zero" counts are inadmissible, as the logarithm is defined only for positive outcomes. On the other hand, count data have a positive probability of the zero outcome. Indeed, zeros frequently command most of the intrinsic research interest, and dropping these observations seems not advisable. Ad-hoc solutions have been proposed, such as adding a constant c to each count. Common choices for c are 0.1 and 0.5 (see King, 1988). In this case, the model is written as

$$\ln(y_i + c) = x_i'\beta + u_i$$

or, equivalently.

$$\ln y_i = \ln(\exp(x_i'\beta + u_i) - c)$$

Estimation of such a model in general will introduce bias, as illustrated in Fig. 3.1. In this figure, models with and without adjustment ($c = 0.5$) are drawn for $u_i = 0$ in the $\ln y_i/x_i$ space (assuming that there is only one regressor and $\beta_0 = 0$ and $\beta_1 = 1$). As is apparent from Fig. 3.1, the slope parameter of the adjusted models exceeds the true slope of unity the more, the closer the value of x comes to its logical lower bound $\ln(c)$.

King (1988) reports results from a Monte Carlo analysis where the adjusted log-linear model is applied to artificial data from a Poisson regression model. He finds substantial bias for the parameter estimates when the log-linear model is used instead of the Poisson regression model. The bias does not disappear with increasing sample size. The log-linear model tends to over-estimate the slope parameters when positive, and to underestimate the slope parameters when negative, i.e., the parameters are biased away from zero. By introducing bias and ignoring the discrete nature of the data, this model is quite unsatisfactory and its use cannot be recommended.

Non-Linear Least Squares

Part of the problem arises because we have considered a model with multiplicative error $\varepsilon_i = \exp(u_i)$ (See equation (3.6)). Consider the alternative model

Fig. 3.1. Bias in the Log-Linear Model When a Constant is Added in Order to Deal With Zero Counts

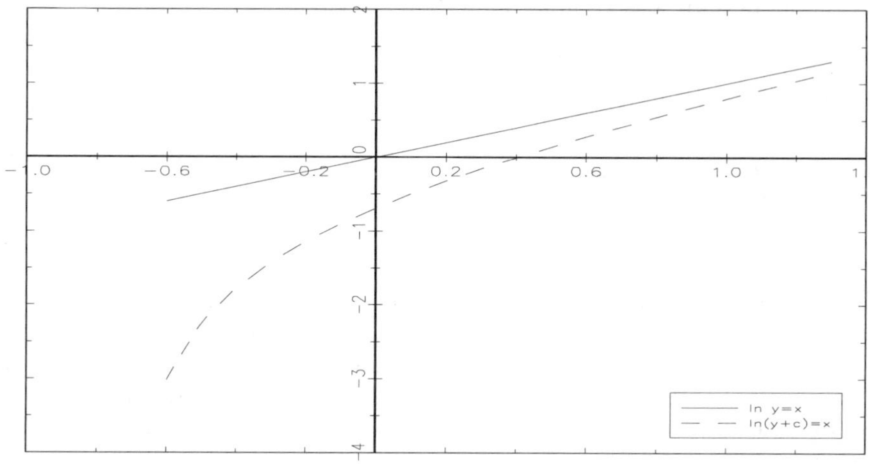

$$y_i = \exp(x_i'\beta) + v_i \qquad v_i \overset{\text{i.i.d}}{\sim} \mathrm{N}(0, \sigma^2) \tag{3.7}$$

This model has the same conditional expectation function as the Poisson regression model. As the logarithm of y_i needs not to be taken, the problem with zero counts disappears. Indeed, from an estimation point of view, model (3.7), if estimated by maximum likelihood (which is the same as non-linear least squares in this case), gives a consistent estimator of β if the true world is Poisson (estimation is discussed in greater detail later in the book). It is not efficient, as it ignores the heteroskedasticity inherent in the Poisson regression model. However, the model could be modified such that $\sigma_i^2 = \exp(x_i'\beta)$, in which case iteratively weighted non-linear least squares would give the same results as Poisson maximum likelihood. The main problem with this model is that, while offering no advantage in terms of ease of estimation or interpretation of parameters, it fails to take into account the non-negative and integer-valued nature of the dependent variable. The model cannot be used to predict the probability of single outcomes and allows for the possibility that predictions take on negative values with positive probability.

Ordered Probit and Logit

Two non-count data models that overcome these shortcomings are the ordered logit and ordered probit models. Both are models for experiments in which outcomes are measured on an ordinal scale. An example is a survey question that sollicits the agreement or disagreement with a certain proposition (such as: X is a good teacher) using the responses *strongly disagree / disagree /*

neutral / agree / strongly agree. The five possible outcomes can be coded, for instance, as 0,1,2,3, and 4, respectively, although the coding is arbitrary as long as it preserves the ordering.

The models are based on an underlying latent model

$$y_i^* = x_i'\beta + \varepsilon_i$$

with the observation mechanism

$$\begin{aligned} y_i &= 0 \text{ if } y_i^* < \alpha_0 \\ y_i &= 1 \text{ if } \alpha_0 \leq y_i^* < \alpha_1 \\ y_i &= 2 \text{ if } \alpha_1 \leq y_i^* < \alpha_2 \\ &\vdots \end{aligned}$$

where α_j are "threshold values". Depending on the assumptions for ε_i, the ordered probit ($\varepsilon_i \sim N(0, \sigma^2)$) or ordered logit ($\varepsilon_i \sim$ logistic) arises. Given α, β and x_i, the probability of each of the 5 possible outcomes is determined and α and β can be estimated by maximum likelihood. Clearly, ordinal models can also be used for counts as long as the number of different counts observed in the sample is not too large (the number of threshold parameters that require estimation increases with the observed sample space by one-to-one; for more details on these models, see Greene, 2000).

Ordered models in general provide a better fit to the data than pure count data models. The threshold parameters give the flexibility to align predicted and actual frequencies. However, their use for modeling count data has a number of serious deficiencies.

- They are theoretically implausible as a model for counts. They are not based on the concept of an underlying count process.
- Counts are cardinal rather than ordinal. Hence, under the ordinal approach, the sequence "2, 5, 50" is assumed to carry the same information as the sequence "0, 1, 2" which is not the case for count data. Ordinal models disregard this information and cannot be efficient.
- One reason of having parametric models in the first place is the ability of predicting the probability of arbitrary counts. While genuine count data models can do that, ordered models can only predict outcomes that are actually observed in the sample.

In addition, these models in general imply a mean function that is different from the mean function of the standard count model. In the general case, the mean function of ordered models (not the latent model) is highly complex. Consider the simplest case of a binary 0/1 variable. For example, in the binary logit model

$$P(y_i = 0) = \frac{1}{1 + \exp(x_i'\beta)}$$

and

$$E(y_i|x_i) = \frac{\exp(x_i'\beta)}{1+\exp(x_i'\beta)}$$

For the Poisson model, instead,

$$P(y_i = 0) = \exp(-\exp(x_i'\beta))$$

and

$$E(y_i|x_i) = \exp(x_i'\beta)$$

The two are fundamentally different mean functions. Of course, this does not imply that the Poisson model is necessarily the superior model, as its mean function may be misspecified as well. However, it suggests that the use of ordered models for count data, and the interpretation of the results, has to proceed with necessary caution. In practice, few applications of ordered models to count data are known.

To summarize, the Poisson regression model has many virtues when one wishes to model a count dependent variable. The Poisson model accounts for the discrete and non-negative nature of the data. It attributes positive probability to the outcome "zero". And it allows inferences to be drawn on the probability of a particular outcome. The Poisson regression model naturally accounts for the heteroskedastic and skewed distribution associated with a non-negative random variable. The more the mean of the dependent variable approaches zero and thus the lower bound of its sample space, the smaller the variance. Finally, the Poisson model has a simple structure and the parameters of the model can be estimated with relative ease.

3.1.3 Generalized Linear Model

Generalized linear models provide an alternative context for the Poisson regression model. Standard references are Nelder and Wedderburn (1972) and McCullagh and Nelder (1989). Firth (1991) provides a concise exposition. In historical perspective, the use of generalized linear models has mainly been associated with the discipline of biometrics rather than that of econometrics.

Generalized linear models (GLMs) are based on probability distributions that are members of the linear exponential families. LEFs were introduced in Chap. 2.5.5. Their density function can be written as

$$f(y;\psi,\phi) = c(y,\phi)\exp\{(y\psi - b(\psi))/\phi\}$$

with mean

$$E(y) = \mu = b'(\psi)$$

and variance

$$\mathrm{Var}(y) = b''(\psi)\phi\,.$$

(For a derivation of these results, see Chap. 2.5.5). In order to model the effect of regressors $x_i = (x_{i1} \cdots x_{ik})'$, let $b'(\psi_i) = g(x_i, \beta)$. This is a natural starting point as $b'(\psi_i)$ is the (conditional) expectation function of the model. How should the function $g(x_i, \beta)$ look like? Firstly, generalized linear models always assume that

$$g(x_i, \beta) = g(x_i'\beta)$$

i.e., the mean of the distribution is a function of a linear combination of the regressors. The linear combination $x_i'\beta$ is sometimes called a "linear predictor".

Models differ by the choice of the underlying density and the choice of the g-function. A case of particular relevance, sometimes referred to as "canonical link", occurs when $g(\cdot) = b'(\cdot)$. In this case, we obtain

$$\mathrm{E}(y_i|x_i) = g(x_i'\beta)$$

and $\psi_i = x_i'\beta$. Recall from Tab. 2.3 that for the Poisson distribution, $b(\psi) = \exp(\psi)$. Hence, $b'(\psi) = \exp(\psi)$ and with $\psi_i = x_i'\beta$ the log-linear Poisson regression model of Chap. 3.1.1 is a generalized linear model in canonical parameterization.

3.1.4 Interpretation of the Parameters

The exponential form of the conditional expectation function implies that the necessary increase in $x_i'\beta = \beta_1 x_{i1} + \beta_2 x_{i2} + \ldots + \beta_k x_{ik}$ to obtain a one unit increase in $\mathrm{E}(y_i|x_i)$ is smaller, the further one moves away from zero. To put it differently, the level change in $x_i'\beta$ required for a given percentage change in $\mathrm{E}(y_i|x_i)$ is kept constant. As a consequence the partial derivative of $\mathrm{E}(y_i|x_i)$ with respect to any element of x_i depends on the value of $x_i'\beta$:

$$\frac{\partial \mathrm{E}(y_i|x_i)}{\partial x_{ij}} = \exp(x_i'\beta)\beta_j = \mathrm{E}(y_i|x_i)\beta_j \qquad j = 1, \ldots, k \tag{3.8}$$

These marginal effects obviously differ between individuals. Sometimes, it may be informative to calculate the marginal effects for a representative individual, such as the sample average of the explanatory variables. Alternatively, one can calculate the expected (or average) marginal effect

$$\mathrm{E}_x\left[\frac{\partial \mathrm{E}(y_i|x_i)}{\partial x_{ij}}\right] = \beta_j \exp(x_i'\beta)$$

which can be estimated consistently by $\beta_j(1/n)\sum_{i=1}^n \exp(x_i'\beta)$.

It is more common, and simpler, though, to consider the *relative change* in $\mathrm{E}(y_i|x_i)$ associated with a small change in x_{ij} since this is constant and equal to β_j:

$$\frac{\partial \mathrm{E}(y_i|x_i)/\mathrm{E}(y_i|x_i)}{\partial x_{ij}} = \beta_j \tag{3.9}$$

If x_i is in logarithmic form, β_j has the interpretation of an elasticity, giving the percentage change in $\mathrm{E}(y_i|x_i)$ per percentage change in x_{ij}.

Sometimes, we are interested in assessing the effect of a (discrete) unit change in x_{ij} on the expected value of y_i. That is, we want to compare the expected value of y_i for x_{ij} and $x_{ij}+1$, respectively. In this case, the calculus method gives only an approximation of the relative change. The exact formula is given by (assuming for notational convenience a model with constant and one regressor only - adding other regressors does not change the result):

$$\begin{aligned}\frac{\mathrm{E}(y_i|x_{i2}+1)-\mathrm{E}(y_i|x_{i2})}{\mathrm{E}(y_i|x_{i2})} &= \frac{\exp[\beta_1+\beta_2(x_{i2}+1)]-\exp(\beta_1+\beta_2 x_{i2})}{\exp(\beta_1+\beta_2 x_{i2})} \\ &= \exp(\beta_2)-1\end{aligned}$$

A special case is a dummy variable taking values 0 or 1. Hence, the relative impact of a dummy variable on the expected count is $\exp(\beta_2)-1$. A linear Taylor series approximation of $\exp(\beta_2)-1$ around $\beta_2^0=0$ yields

$$\begin{aligned}\exp(\beta_2)-1 &\approx \exp(\beta_2^0)-1+\exp(\beta_2^0)(\beta_2-\beta_2^0) \\ &= \beta_2\end{aligned}$$

Thus, β_2 gives the approximate relative impact of a dummy variable for small β_2.

These results are similar to those encountered in the standard log-linear model. However, there is a conceptual difference that removes a certain ambiguity in the interpretation of the Poisson parameters, an ambiguity that was first noted by Goldberger (1968) for the log-linear model (see also Winkelmann, 2001). There, $\mathrm{E}(\ln y_i|x_i)=x_i'\beta$, from which it does not follow that $\mathrm{E}(y_i|x_i)=\exp(x_i'\beta)$. It is only under some additional assumptions that an expression such as $\exp(\beta_j)-1$ correctly identifies the relative change in $\mathrm{E}(y_i|x_i)$ due to a unit change in x_{ij}. The situation in the Poisson regression model is much more straightforward. However, estimation is still an issue. As pointed out by Goldberger (1968) for the log-linear model, estimating $\exp(\beta_j)-1$ by $\exp(b_j)-1$, where b_j is the maximum likelihood estimator, though consistent, introduces small sample bias. An improved estimator could follow the methods suggested by Goldberger (1968) and Kennedy (1981).

Interactive Effects

Interactive terms are used to model complementarities between variables. For instance, let

$$\mathrm{E}(y|x_1,x_2)=\exp(\beta_1 x_1+\beta_2 x_2+\beta_{12}x_1 x_2) \tag{3.10}$$

In this model, $x_1 x_2$ is an interactive term, and β_{12} measures its effect on the linear predictor. For instance, for positive β_{12}, the impact of a given increase in x_1 on $\mathrm{E}(y|x_1,x_2)$ is the larger the greater value of x_2. An underappreciated aspect of the exponential regression model, first pointed out by Mullahy

(1999), is that even if $\beta_{12} = 0$ does the magnitude of the effect of x_1 on $\mathrm{E}(y|x_1, x_2)$ depend on x_2. A zero coefficient of the interactive *term* does not imply the absence of an interactive *effect*, as defined by the cross derivative $\partial^2 \mathrm{E}(y|x_1, x_2)/\partial x_1 \partial x_2$. In the exponential regression model, this cross derivative can be written as

$$\frac{\partial^2 \mathrm{E}(y|x_1, x_2)}{\partial x_1 \partial x_2} = \mathrm{E}(y|x_1, x_2) \times [\beta_{12}(1 + \beta_1 x_1 + \beta_2 x_2 + \beta_{12} x_1 x_2) + \beta_1 \beta_2] \tag{3.11}$$

A non-zero value indicates the presence of an interactive effect. It is seen from (3.11) that the interactive effect depends on particular values of x_1 and x_2 as well as all three parameters β_1, β_2 and β_{12}. This is unlike in the linear model $\mathrm{E}(y|x_1, x_2) = \beta_1 x_1 + \beta_2 x_2 + \beta_{12} x_1 x_2$ where

$$\frac{\partial^2 \mathrm{E}(y|x_1, x_2)}{\partial x_1 \partial x_2} = \beta_{12}$$

and a proper test for an interaction is $H_0 : \beta_{12} = 0$. In the exponential model, by contrast, a test for zero interaction involves the non-linear null hypothesis (for given values of x_1 and x_2):

$$H_0 : \beta_{12}(1 + \beta_1 x_1 + \beta_2 x_2 + \beta_{12} x_1 x_2) + \beta_1 \beta_2 = 0$$

Clearly then, even if $\beta_{12} = 0$, the interactive effect is non-zero for any values of x_1 and x_2 as long as the last term $\beta_1 \beta_2$ is non- zero. Moreover, if β_{12} is restricted to zero a-priori, the sign of the interactive effect is equal to the sign of $\beta_1 \beta_2$, and a test for no interaction should consider the non-linear hypothesis $H_0 : \beta_1 \beta_2 = 0$. The correct interpretation of interactive effects has not been recognized in the previous applied literature. Mullahy (1999) gives further results for dummy variables and differences-in-differences estimators in the exponential model.

Marginal Probability Effects

So far, the discussion has focussed exclusively on marginal mean effects, i.e., the question how the mean, or conditional expectation function, varies as any of the explanatory variables changes, ceteris paribus. This focus is natural. It very closely resembles the approach commonly taken in linear regression models for continuous data. However, the focus on conditional expectations is arguably also overly restrictive, as it misses some of the richness inherent in modelling discrete data, and count data in particular. The discreteness implies that statements about single probabilities are meaningful. Such statements may be of substantive interest.

The question then becomes: how does the distribution (or probability function) respond to a small ceteris paribus change in an explanatory variable. The answer to this question is given by the "marginal probability effects". In the exponential Poisson regression model

$$\frac{df(k;\lambda)}{dx} = \frac{df(k;\lambda)}{d\lambda}\frac{d\lambda}{dx} = f(k;\lambda)(k-\lambda)\beta \tag{3.12}$$

where $f(k;\lambda)$ is the Poisson probability function and $\lambda = \exp(x'\beta)$. We see that the marginal probability effects are quite restrictive. This follows directly from the simple structure of the Poisson model. Observe that

$$\text{sgn}(df(k;\lambda)/dx_j) = -\text{sgn}(\beta_j) \text{ iff } k < \lambda$$

$$\text{sgn}(df(k;\lambda)/dx_j) = \text{sgn}(\beta_j) \text{ iff } k > \lambda$$

Increasing the value of the dependent variable, k, over its support, starting at zero, it must be the case that the marginal probability effects are either initially positive, turning negative after a certain value of the dependent variable; or they are initially negative, turning positive after a certain value. One may refer to this result as a "single-crossing" property of the Poisson model.

Depending on the question one wants to address, models allowing for more flexible marginal probability effects may be desirable. Such models will be discussed later one, one very prominent example being the class of hurdle models (Chap. 4.6). Generally speaking, single-index models (such as the Poisson or negative binomial regression models) will always be restrictive, as the pattern of marginal probability effects is fully determined by the functional form of the underlying probability function - it cannot be modelled flexibly based on covariates and corresponding parameter values.

More flexible models have multiple indices through which the covariates influence the probability function. In such models, changes of an explanatory variable may have different effects in different parts of the outcome distribution. Also, by implication, marginal mean effects are then no longer directly linked to marginal effects for higher order moments. For example, the effect of a variable on the variance is independent of the effect of that variable on the mean (and it becomes meaningful to study questions such as "has an explanatory variable an increasing or decreasing effect on inequality in the outcome variable").

3.1.5 Period at Risk

Count data measure the number of times a certain event occurs during a given time interval. The length of this interval is sometimes called "risk-period", or "exposure". In the standard Poisson model, it is assumed that the risk period is the same for all observations. Under this assumption, it can be normalized to unity without loss of generality, and $\exp(x_i'\beta)$ is the expected value of y_i per time interval (such as year, month, or week).

However, in other cases, the risk-period varies across observations. For instance, McCullagh and Nelder (1989) analyze the number of reported damage incidents by ship type. Aggregate months of service vary from 45 months for one ship type to 44,882 months for another. Clearly, one would expect the number of incidents to increase with aggregate months of service. In a similar vein, Diggle, Liang and Zeger (1994) use data from a randomized experiment in order to compare the number of epileptic seizures during a 8-week pre-treatment observation period with the number of epileptic seizures during 2-week post-treatment observation period. Finally, Barmby, Nolan and Winkelmann (2001) analyze the number of days absent from work for a sample of workers some of whom are contracted for 4 days of work per week while others are contracted for 5 days of work per week.

Differences in exposure need not be limited to calendar time. For instance, Rose (1990) analyses the determinants of air-traffic incidents. In her case, the different size of operation between the various carriers is expressed by the number of scheduled departures per year (in thousands). Bauer et al. (1998) are interested in the number of workplace accidents, using firm level data for Germany. Again, one would expect that the number of accidents increases with the size of the risk-group, here the number of workers.

The benchmark case for dealing with exposure is to assume proportionality. In the above examples, McCullagh and Nelder (1989) model the expected number of ship damage incidents *per* aggregate month of operation, while Rose (1990) models the expected number of air-traffic incidents *per* 1,000 departures. If we denote the individual level of exposure by t_i, we can write

$$\frac{\mathrm{E}(y_i|x_i)}{t_i} = \exp(x_i'\beta) \tag{3.13}$$

or, equivalently,

$$\begin{aligned}\mathrm{E}(y_i|x_i) &= t_i \exp(x_i'\beta) \\ &= \exp(x_i'\beta + \ln t_i)\end{aligned} \tag{3.14}$$

Thus, under proportionality, a doubling in exposure time doubles the expected count. In the second line of (3.14), $\ln t_i$ is sometimes referred to as "logarithmic offset".

Alternatively, one might want to give the proportionality assumption free for test. A simple possibility is to include $\ln t_i$ as a regressor without restricting its coefficient to unity:

$$\mathrm{E}(y_i|x_i, t_i) = \exp(x_i'\beta + \gamma \ln t_i) \tag{3.15}$$

The restriction $H_0 : \gamma = 1$ can then be tested with standard methods. Alternatively, one can reparameterize using $\theta = \gamma - 1$. The mean function then reads

$$\mathrm{E}(y_i|x_i, t_i) = \exp(x_i'\beta + \theta \ln t_i + \ln t_i) \tag{3.16}$$

Logarithmic time of exposure is included twice, both as offset and as regressor, and the test for proportionality now simplifies to testing $H_0 : \theta = 0$.

Yet another variant of this test exists if time of exposure can take on only two values, t_1 and t_2. This case is presented in Barmby, Nolan and Winkelmann (2001). Define a dummy variable $D = 1$ if $t = t_1$ and $D = 0$ if $t = t_2$. Then, in the regression model

$$\mathrm{E}(y_i|x_i, D_i) = \exp(x_i'\beta + \delta D_i) \tag{3.17}$$

the test for proportionality to exposure reduces to $H_0 : \delta = \ln(t_1/t_2)$. To establish the equivalence, note that

$$\ln t_i = \ln t_2 + (\ln t_1 - \ln t_2)D_i$$

Thus, under strict proportionality,

$$\mathrm{E}(y_i|x_i, D_i) = \exp(\ln t_2 + x_i'\beta + (\ln t_1 - \ln t_2)D_i)\,,$$

i.e., $\delta = \ln t_1 - \ln t_2 = \ln(t_1/t_2)$ ($\ln t_2$ is absorbed into the overall constant).

Endogenous Exposure Time

An interesting alternative class of models arises if the time of exposure is endogenous. A potential source of endogeneity could be that exposure time depends on the occurrence or non- occurrence of events. One plausible scenario is that of a "blockage": there is no exposure, i.e., a zero risk of occurrence or at least of measuring an occurrence, for some given time interval following an occurrence. An example for such a situation occurs in the modelling of fertility: after a birth, no further birth can occur for a period of about 11 months, adding the time of pregnancy and uterine involution.

Feller (1971, p. 372) discusses a blockage time model where events are generated by a Poisson process, and after each event occurrence, a fixed bloackage time ξ sets in. Feller shows that 1) the number of recorded events is not Poisson distributed; and 2) that the number of recorded events is underdispersed. Such a model would be well suited to think about the pregnancy example above. Indeed, it is typically found that fertility data are underdispersed (Winkelmann and Zimmermann, 1994, see also Chap. 7.5 and the empirical fertility distribution in Tab. 1.1)

Alternatively, we can model the blockage duration itself as a random variable. An example from labor economics would be the number of unemployment episodes. A second unemployment episode requires termination of the first one, and hence, the time of blockage can be thought of as random.

A manageable result can be obtained under the following assumptions: Let the events arrive according to a Poisson process, and assume that the blockage time is independently distributed with exponential density function. Since the Poisson process is memoryless, the waiting time from the moment where blockage ends until the next occurrence is also exponentially distributed. Therefore, the distribution of the interarrival time between two events is a convolution of two exponential distributions. If we assume, admittedly unrealistically, that both distributions share a common parameter λ, the interarrival time has a gamma distribution with parameter $\alpha = 2$ (see Chap. 2.6.4). Moreover, the distribution of the number of events is of Erlangian type, with

$$f(y) = e^{-\lambda}\left(\frac{\lambda^{2y}}{(2y)!} + \frac{\lambda^{2y+1}}{(2y+1)!}\right) \qquad y = 0, 1, \ldots \tag{3.18}$$

(See also 2.128). Fig. 3.2 plots the mean and variance of this distribution for $0.1 < \lambda < 5$. As for fixed blockage, this distribution is under-dispersed. Clearly, it would be desirable in future work to lift the restriction that the parameters of the two exponential distributions be the same.

Fig. 3.2. Mean and Variance of Exponential Blockage Model for $0.1 < \lambda < 5$

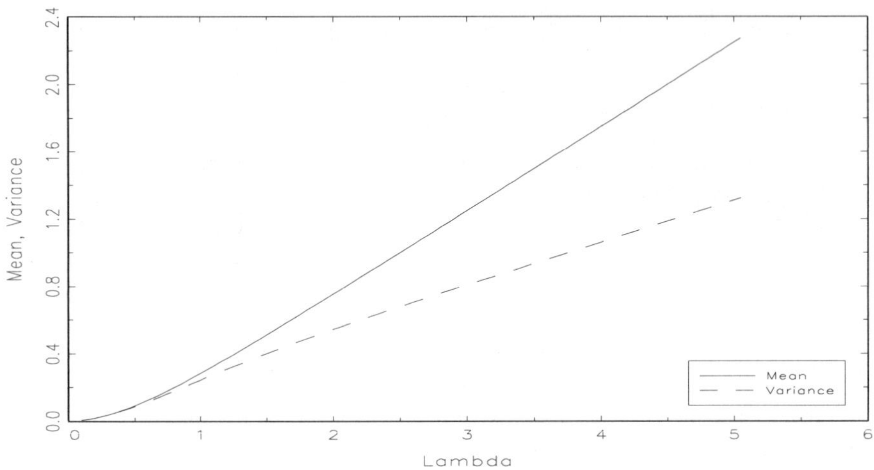

3.2 Estimation

This section is concerned with the problem of estimating β, the (kx1) vector of regression coefficients in the Poisson regression model. Most of this chapter deals with the maximum likelihood method as it is the most common method to estimate count data models. The maximum likelihood principle states that the parameter should be chosen as to maximize the probability that the specified model has generated the observed sample. Alternative principles that are discussed as well are pseudo maximum likelihood and generalized method of moments.

3.2.1 Maximum Likelihood Estimation

Given an independent sample of n pairs of observations (y_i, x_i), the joint probability distribution of the sample is given by the product of the individual conditional probability distributions:

$$f(y_1, \ldots, y_n | x_1, \ldots, x_n; \beta) = \prod_{i=1}^{n} f(y_i | x_i; \beta) \tag{3.19}$$

Understood as a function of the parameters, (3.19) is called *likelihood function*, and we write

$$\begin{aligned} L &= L(\beta; y_1, \ldots, y_n, x_1, \ldots, x_n) \\ &= L(\beta; Y, X) \end{aligned} \tag{3.20}$$

The maximum likelihood estimator is defined as

$$\hat{\beta} = \underset{\beta}{\operatorname{argmax}}\ L(\beta; Y, X)$$

As the logarithm is a monotonic transformation, maximization of the likelihood function is equivalent to maximization of the logarithmic, or log-likelihood function $\ell = \ln L$. In general, this transformation simplifies matters considerably, as it replaces products by sums. Moreover, it allows the use of the central limit theorem when studying the properties of the maximum likelihood estimator. The log-likelihood function for the Poisson regression model takes the form

$$\begin{aligned} \ell(\beta; Y, X) &= \ln \prod_{i=1}^{n} f(y_i|x_i; \beta) \\ &= \sum_{i=1}^{n} \ln f(y_i|x_i; \beta) \\ &= \sum_{i=1}^{n} \ell(\beta; y_i, x_i) \\ &= \sum_{i=1}^{n} - \exp(x_i'\beta) + y_i x_i'\beta - \ln(y_i!) \end{aligned} \tag{3.21}$$

The maximizing value for β, denoted as $\hat{\beta}$, is derived by computing the first derivatives of the log-likelihood function (the gradient vector; for rules of matrix differentiation, see Greene, 2000)

$$g(\beta; y, x) = \frac{\partial \ell(\beta; y, x)}{\partial \beta} = \sum_{i=1}^{n} [y_i - \exp(x_i'\beta)] x_i$$

and then solving the first order conditions for a maximum

$$g(\beta; y, x) = \sum_{i=1}^{n} [y_i - \exp(x_i'\beta)] x_i = 0 \tag{3.22}$$

Note that as long as x_i includes a constant, the first order conditions impliy that $\sum_{i=1}^{n} \hat{u}_i = 0$, where $\hat{u}_i$ is an implicit residual defined as

$$\begin{aligned} \hat{u}_i &= y_i - \hat{\mathrm{E}}(y_i|x_i) \\ &= y_i - \exp(x_i'\hat{\beta}) \end{aligned}$$

For the non-constant regressors, (3.22) can be interpreted as a set of orthogonality conditions:

$$\sum_{i=1}^{n} \hat{u}_i x_{ij} = 0, \; j = 2, \ldots, k$$

Equation (3.22) gives the necessary conditions for a maximum. If, in addition, the matrix of second derivatives, the Hessian matrix, is negative definite for all values of β, the solution to (3.22) is called the *maximum likelihood estimator.* The Hessian matrix of the Poisson log-likelihood function is given by

$$\begin{aligned} H(\beta; y, x) &= \frac{\partial^2 \ell(\beta; y, x)}{\partial\beta\partial\beta'} \\ &= -\sum_{i=1}^{n} \exp(x_i'\beta) x_i x_i' \end{aligned} \tag{3.23}$$

H is negative definite, and the log-likelihood function of the Poisson regression model is globally concave.

The Newton-Raphson Algorithm

Since (3.22) is non-linear in β, the system of k equations has to be solved using an iterative algorithm. A common choice that works well for concave objective functions is the Newton-Raphson method. It can be motivated as follows. Given any initial parameter estimate, say $\hat{\beta}^0$, we can obtain a second-order approximation of $\ell(\beta)$ around $\hat{\beta}^0$:

$$\ell^*(\beta) = \ell(\hat{\beta}^0) + g(\hat{\beta}^0)'(\beta - \hat{\beta}^0) + \frac{1}{2}(\beta - \hat{\beta}^0)' H(\hat{\beta}^0)(\beta - \hat{\beta}^0) \approx \ell(\beta)$$

Now, we can maximize $\ell^*(\beta)$ (rather than $\ell(\beta)$) with respect to β, yielding a new parameter value which we call $\hat{\beta}^1$. The first order condition of this simpler problem is

$$g(\hat{\beta}^0) + H(\hat{\beta}^0)(\hat{\beta}^1 - \hat{\beta}^0) = 0$$

or

$$\hat{\beta}^1 = \hat{\beta}^0 - [H(\hat{\beta}^0)]^{-1} g(\hat{\beta}^0)$$

Thus, for arbitrary starting value $\hat{\beta}^0$, the Newton-Raphson updating rule is given by

$$\hat{\beta}^{t+1} = \hat{\beta}^t - [H(\hat{\beta}^t)]^{-1} g(\hat{\beta}^t) \qquad t = 0, 1, \ldots \tag{3.24}$$

where $g(\cdot)$ denotes the gradient and $H(\cdot)$ the Hessian of the Poisson log-likelihood function.

The iterative procedure ends when a predefined convergence criterion is satisfied. Possible criteria are the change in the value of the estimate $\hat{\beta}^{t+1} - \hat{\beta}^t$, the change in the log-likelihood $\ell(\hat{\beta}^{t+1}) - \ell(\hat{\beta}^t)$, or the value of the gradient at the estimate $g(\hat{\beta}^t)$. Convergence occurs when any of these values, or a combination of them, are close to zero (say, smaller than 10^{-5} in absolute value).

Numerical Derivatives

The algorithm presented in (3.24) could be based on analytical first and second derivatives of the log-likelihood function. A common alternative is the use of numerical derivatives. Numerical derivatives are the preferred option whenever analytical derivatives are difficult to establish. But even in cases such as the Poisson regression model, where the derivation of gradient and Hessian is relatively simple, numerical derivatives may lower the risk of programming errors. A downside is that numerical derivatives are considerably more time consuming. However, progress in computing speed has reduced the importance of this argument, unless data sets and the number of parameters are very large (for instance, calculation time for a numerical Hessian is a quadratic function of the size of the matrix).

The standard formulas for numerical derivatives are $[f(b_i + h_i) - f(b_i)]/h_i$ if forward calculation is chosen, or $[f(b_i + h_i/2) - f(b_i - h_i/2)]/h_i$ if centered calculation is used. Methods differ in the way they determine h_i. For instance, StataCorp. (1997) uses an algorithm where h_i is selected such that $\varepsilon_1(|f(b_i)| + \varepsilon_1) \leq |f(b_i + h) - f(b_i)| \leq \varepsilon_2(|f(b_i)| + \varepsilon_2)$ for $\varepsilon_1 < \varepsilon_2$.

Properties of Maximum Likelihood Estimators

The maximum likelihood estimator $\hat{\beta} = \text{argmax } L(\beta)$ is in general a non-linear function of the dependent variable. Therefore, analytical small sample results are unavailable. Provided a number of regularity conditions are satisfied, it can be shown that the maximum likelihood estimator is:

- consistent
- asymptotically normal
- asymptotically efficient

The most important assumption is that the model is correctly specified. The true density is denoted as $f_0(y_i, x_i)$. Thus, there exist a β_0 such that

$$\prod_{i=1}^{n} f(y_i|x_i;\beta_0) = \prod_{i=1}^{n} f_0(y_i|x_i) \tag{3.25}$$

For further regularity conditions, see, for instance, Cramer (1986).

The approximate distribution of $\hat{\beta}$ is given by

$$\hat{\beta} \overset{\text{app}}{\sim} N(\beta_0, [-\text{E}H(\beta_0)]^{-1}) \tag{3.26}$$

where $-\mathrm{E}H(\beta_0) \stackrel{\text{def}}{=} I$ is the Fisher information matrix of the sample. The main steps of a proof are as follows. As a starting point, consider a first-order Taylor series approximation of $g_n(\hat{\beta})$ around the true parameter vector β_0 (The subscript n is a reminder that the gradient depends on the sample size):

$$g_n(\hat{\beta}_n) \approx g_n(\beta_0) + H_n(\beta_0)(\hat{\beta}_n - \beta_0)$$

Since $g_n(\hat{\beta}_n)$ is zero by definition of a maximum likelihood estimator, we have

$$\hat{\beta}_n - \beta_0 \approx -H_n(\beta_0)^{-1} g_n(\beta_0)$$

or

$$\sqrt{n}(\hat{\beta}_n - \beta_0) \approx \left(-\frac{1}{n} H_n(\beta_0)\right)^{-1} \frac{1}{\sqrt{n}} g_n(\beta_0)$$

But for independent observations

$$g_n(\beta_0) = \left[\frac{\partial \ell}{\partial \beta}\right]_{\beta_0} = \sum_{i=1}^{n} \left[\frac{\partial \ell_i}{\partial \beta}\right]_{\beta_0}$$

$$H_n(\beta_0) = \left[\frac{\partial^2 \ell}{\partial \beta \partial \beta'}\right]_{\beta_0} = \sum_{i=1}^{n} \left[\frac{\partial^2 \ell_i}{\partial \beta \partial \beta'}\right]_{\beta_0}$$

Thus, gradient and Hessian are sums of independent components and law of large numbers and central limit theorems can be invoked. For increasing n, both $g_n(\beta_0)$ and $H_n(\beta_0)$ converge to their first moments $\mathrm{E}(g_n) = 0$ and $\mathrm{E}(H_n) = -I_n$. The first equality holds since

$$\begin{aligned}
\mathrm{E}(g_i) &= \mathrm{E}_y \left[\frac{\partial \ell_i}{\partial \beta}\right]_{\beta_0} \\
&= \int \frac{f'(y_i)}{f(y_i)} f(y_i) \mathrm{d}y_i \\
&= \int f'(y_i) \mathrm{d}y_i = 0
\end{aligned}$$

The last step follows from differentiating the identity $\int f(y)\mathrm{d}y_i = 1$. Furthermore, it can be shown that $n^{-1/2} g_n(\beta_0)$ converges to a normal distribution with mean zero and variance $n^{-1} I_n$. The derivation of the variance is based on the following considerations. First, the Hessian matrix of a single observation can be written as

$$\begin{aligned}
H_i(\beta_0) &= \frac{\partial g_i(\beta_0)}{\partial \beta} \\
&= \frac{f''(y_i)}{f(y_i)} - \frac{f'(y_i)}{f(y_i)} \frac{f'(y_i)}{f(y_i)}
\end{aligned}$$

Second, taking expectations, we see that the first term on the right disappears because

$$\mathrm{E}\left(\frac{f''(y_i)}{f(y_i)}\right) = \int f''(y_i)\mathrm{d}y_i = 0$$

and we have

$$\begin{aligned}\mathrm{E}H_i(\beta_0) &= -\mathrm{E}\left(\frac{f'(y_i)}{f(y_i)}\frac{f'(y_i)}{f(y_i)}\right)\\ &= -\mathrm{Var}(g_i(\beta_0))\\ &= -I_i\end{aligned}$$

The expected Hessian is equal to minus the variance of the gradient, where both are evaluated at the true parameter β_0. This result is known as the "information matrix equality". For a sample of n independent observations, we get

$$\begin{aligned}\mathrm{Var}(g_n(\beta_0)) &= \sum_{i=1}^{n}\mathrm{Var}(g_i(\beta_0))\\ &= \sum_{i=1}^{n} I_i = I_n\end{aligned}$$

Putting pieces together, we find that

$$\frac{1}{\sqrt{n}}g_n(\beta_0) \xrightarrow{d} N(0, n^{-1}I_n) \tag{3.27}$$

(where $\xrightarrow{d}$ stands for "converges in distribution") and

$$\sqrt{n}(\hat{\beta}_n - \beta_0) \approx \left(-\frac{1}{n}H_n(\beta_0)\right)^{-1}\frac{1}{\sqrt{n}}g_n(\beta_0) \xrightarrow{d} N(0, nI_n^{-1})$$

Finally,

$$\hat{\beta} \overset{\text{app}}{\sim} N(\beta_0, I_n^{-1})$$

(where $\overset{\text{app}}{\sim}$ stands for "approximately distributed as"). In practice, β_0 and thus $I(\beta_0)$ are not known. For the purpose of inference, the true variance covariance matrix of $\hat{\beta}$ can be replaced by a consistent estimator. A first possibility is to evaluate minus the expected Hessian at the parameter estimate (instead of the true value β_0):

$$\widehat{\mathrm{Var}}_1 = [-\mathrm{E}H_n(\hat{\beta})]^{-1}$$

It is frequently the case that the Hessian matrix is a highly nonlinear function of y, making it practically impossible to take expectations. A second alternative estimator for the variance covariance matrix takes the inverse of minus the *actual* Hessian matrix evaluated at the maximum likelihood estimate

$$\widehat{\mathrm{Var}}_2 = [-H_n(\hat{\beta})]^{-1}$$

Sometimes, even the computation of the Hessian is complicated and maximization routines rely on gradient oriented methods only. Then, the relation $\mathrm{Var}(g_n) = I_n$ may be exploited to estimate the variance of $\hat{\beta}$ using the summed outer product of the gradient

$$\widehat{\mathrm{Var}}_3 = \left[\sum_{i=1}^{n} g_i(\hat{\beta})g_i(\hat{\beta})'\right]^{-1}$$

Example: Constant-Only Poisson Model

If the only regressor is a constant, we find that

$$g_n = \sum_{i=1}^{n} -1 + \frac{y_i}{\lambda}$$

$$H_n = \sum_{i=1}^{n} -\frac{y_i}{\lambda^2}$$

From the first order condition the ML estimator is given by $\hat{\lambda} = \bar{y}$. In this case, the ML estimator coincides with least squares and method of moments estimators. The variance of the estimator can be computed by either of the three methods.

Expected Hessian.

$$\widehat{\mathrm{Var}}_1(\hat{\lambda}) = \left[-\mathrm{E}\left[\sum_{i=1}^{n} -\frac{y_i}{\lambda^2}\right]_{\hat{\lambda}}\right]^{-1} = \frac{\hat{\lambda}}{n}$$

Actual Hessian.

$$\widehat{\mathrm{Var}}_2(\hat{\lambda}) = -\left[\sum_{i=1}^{n} -\frac{y_i}{\lambda^2}\right]_{\hat{\lambda}}^{-1} = \left[\frac{\sum_{i=1}^{n} y_i}{\lambda^2}\right]_{\hat{\lambda}}^{-1} = \frac{\hat{\lambda}}{n}$$

Variance of Gradient.

$$\widehat{\mathrm{Var}}_3(\hat{\lambda}) = \left[\mathrm{Var}\sum_{i=1}^{n}\left(\frac{y_i}{\lambda} - 1\right)\right]^{-1} = \left[\frac{n\mathrm{Var}(y_i)}{\lambda^2}\right]_{\hat{\lambda}}^{-1} = \frac{\hat{\lambda}}{n}$$

Thus for all three methods, $\hat{\lambda} \overset{\text{app}}{\sim} N(\lambda, \hat{\lambda}/n)$. This is not surprising, since we know that for *i.i.d* sampling from any distribution $\bar{y} \overset{\text{app}}{\sim} N(\mathrm{E}(y), \mathrm{Var}(y)/n)$. The identity of the three variance estimators no longer holds in the Poisson regression model, as we will see next.

Asymptotic Distribution of the Poisson Regression Coefficients

Applying the previous results to the correctly specified Poisson regression model, it follows that

$$\sqrt{n}(\hat{\beta}_{\text{psn}} - \beta_0) \xrightarrow{d} N\left(0, n\left[\sum_{i=1}^{n} \exp(x_i'\beta_0)x_i x_i'\right]^{-1}\right) \tag{3.28}$$

Alternatively, we can say that

$$\hat{\beta} \overset{\text{app}}{\sim} N\left(\beta_0, \left[\sum_{i=1}^{n} \exp(x_i'\beta_0)x_i x_i'\right]^{-1}\right) \tag{3.29}$$

Since the Hessian matrix does not depend on y, the Fisher information and Hessian estimators of the variance coincide, resulting in the variance estimator

$$\widehat{\text{Var}}_1(\hat{\beta}) = \left[\sum_{i=1}^{n} \exp(x_i'\hat{\beta})x_i x_i'\right]^{-1} \tag{3.30}$$

where the unknown true parameter β_0 has been replaced by the consistent estimator $\hat{\beta}$. Similarly, one could estimate the variance of $\hat{\beta}$ by summing over the outer product of the gradient

$$\widehat{\text{Var}}_2(\hat{\beta}) = \left[\sum_{i=1}^{n} (y_i - \exp(x_i'\hat{\beta}))^2 x_i x_i'\right]^{-1} \tag{3.31}$$

Thus, in finite samples, $\widehat{\text{Var}}_2$ and $\widehat{\text{Var}}_2$ generally differ. Asymptotically, these two formula are the same provided the model is correctly specified because then $\text{Var}(y_i|x_i) = \text{E}(y_i - \exp(x_i'\beta_0))^2 = \exp(x_i'\beta_0)$.

3.2.2 Pseudo-Maximum Likelihood

White (1982) considers a situation where the model is not correctly specified: There exists no β such that

$$\sum_{i=1}^{n} \log[f(y_i|x_i, \beta)] = \sum_{i=1}^{n} \log[f_0(y_i|x_i)] \tag{3.32}$$

where f is the specified conditional density and $f_0(y|x)$ is the true conditional density. Obtaining parameter estimates by maximizing a misspecified log-likelihood

$$\ell(\beta; y, x) = \sum_{i=1}^{n} \log[f(y_i|x_i, \beta)]$$

is a method that is conventionally referred to as *quasi maximum likelihood estimation* (QML).

The consequences of misspecification can be best stated considering the asymptotic distribution of a quasi maximum likelihood estimator (QMLE). Under the assumption of independent sampling, the QMLE has in fact a well defined limiting distribution given by (See White, 1982)

$$\sqrt{n}(\tilde{\beta} - \beta^*) \xrightarrow{d} N(0, n\mathcal{J}^{-1}\mathcal{I}\mathcal{J}^{-1}) \tag{3.33}$$

The quasi maximum likelihood estimator $\tilde{\beta}$ is a consistent estimator for a pseudo-true value β^*, where β^* minimizes the Kullback-distance between the specified model and the true model:

$$\mathcal{K}\{f_0(y|x), f(y|x;\beta)\} = \frac{1}{n}\mathrm{E}_0\left[\log \frac{f_0(y|x)}{f(y|x;\beta)}\right] \tag{3.34}$$

The asymptotic covariance is given by $\mathcal{J}^{-1}\mathcal{I}\mathcal{J}^{-1}$ where the matrices $\mathcal{J}$ and $\mathcal{I}$ denote minus the expected Hessian and the variance of the gradient, respectively:

$$\mathcal{J} = -\mathrm{E}_0\left(\frac{\partial^2 \ell(\beta; y, x)}{\partial\beta\partial\beta'}\right)$$

$$\mathcal{I} = \mathrm{E}_0\left(\frac{\partial \ell(\beta; y, x)}{\partial\beta}\frac{\partial \ell(\beta; y, x)}{\partial\beta'}\right)$$

If the model is correctly specified, $\beta^* = \beta_0$ and $\mathcal{J} = \mathcal{I}$. Thus, the limiting distribution of the QML collapses to the limiting distribution of the maximum likelihood estimator with covariance matrix $\mathcal{J}^{-1} = \mathcal{I}^{-1}$.

Under misspecification, little can be said about the relationship between the QMLE $\tilde{\beta}$ and the maximum likelihood estimator $\hat{\beta}$. The two main criteria for 'good' estimators are consistency and efficiency. The QMLE is in general inconsistent (for β_0) and inefficient. However, Gourieroux, Monfort and Trognon (1984a) give a condition under which $\tilde{\beta} \xrightarrow{p} \beta_0$. It refers to the special case, where the mean is correctly specified, i.e., there exists a $\beta = \beta_0$ such that

$$\mu(\beta_0) = \mu_0(x),$$

it holds that β is *first order identifiable*, i.e., $\mu(x,\beta) = \mu(x,\beta_0)\ \forall x$ implies that $\beta = \beta_0$, and the quasi-likelihood function f is a member of the linear exponential family. Gourieroux, Monfort and Trognon (1984a) refer to this situation as *pseudo maximum likelihood estimation* (PMLE). They show that a quasi maximum likelihood estimator is consistent for β_0 if and only if the distribution family of the estimated model is a linear exponential family (and thus QML=PML).

In this case, the only problem is with the estimated standard errors. The two forms of the information matrix, $\mathcal{J}$ and $\mathcal{I}$, are equivalent only if the probability model $f(y;\beta)$ is true. When this assumption is incorrect, $\mathcal{J}$ and $\mathcal{I}$ are not equal and inference derived using $\hat{\mathcal{J}}$ to estimate the information matrix of the estimated parameter vector $\hat{\beta}$ will lead to incorrect inferences. The bias in inference is given by $\Delta = \mathcal{J}^{-1} - \mathcal{J}^{-1}\mathcal{I}\mathcal{J}^{-1}$.

Example: Standard Error of the Mean in the Constant-Only Poisson Model

In the previous chapter, expressions for the expectes Hession and the variance of the gradient were derived for the constant-only Poisson model. In current notation,

$$\mathcal{J} = \frac{n}{\lambda}$$

and

$$\mathcal{I} = \frac{n\text{Var}(y_i)}{\lambda^2}$$

Now observe that if the true model is the Poisson probability function, $\text{Var}(y_i) = \lambda$, and $\mathcal{J}$ is identical to $\mathcal{I}$. Suppose instead that the true model is not the Poisson density. Then it need not be true that $\text{Var}(y_i) = \lambda$, and the correct expression for the asymptotic variance of the quasi-maximum likelihood estimator $\hat{\lambda}$is given by

$$\mathcal{J}^{-1}\mathcal{I}\mathcal{J}^{-1} = \frac{\text{Var}(y_i)}{n}$$

Hence, if $\text{Var}(y_i) > \lambda$, ignoring misspecification would under-estimate the correct standard errors. If $\text{Var}(y_i) < \lambda$ the correct variance of the sample mean would be less than is indicated by $\mathcal{J}^{-1}$.

Discussion

The Poisson distribution is a linear exponential family. Thus, erroneously assuming the Poisson distribution if the true model is, for instance, the negative binomial distribution, yields a PMLE and consistent estimates of the mean parameters. However, the standard errors are wrong, leading to spurious inference. However, asymptotically valid standard errors can be computed, and this method is further investigated in Chap. 4.10.3 when we talk about *robust Poisson regression.* Note that these results do require a correctly specified mean function. If, for instance, relevant variables are excluded, the resulting estimator is, in the terminology used here, a QMLE and not a PMLE.

3.2.3 Generalized Method of Moments

Estimation by the Generalized Method of Moments (GMM) has become increasingly popular in recent years. The origins of this method can be found in Hansen (1982). GMM is a very general estimation method and, in some sense, maximum likelihood and quasi-maximum likelihood can be seen as special cases of GMM. The main practical importance of GMM stems from the fact that it works in situations where conventional methods fail, in particular in the area of panel data models with weakly exogenous regressors (Chap.

5.2.5). Moreover, GMM provides a natural framework in which to conduct specification tests by way of testing for overidentifying restrictions (See, for instance, Santos Silva and Windmeijer, 1999).

Let θ be a $(p \times 1)$ vector of parameters that is to be estimated, and assume that there are l moment restrictions

$$\mathrm{E}[m_i(y_i, x_i; \theta)] = 0 \tag{3.35}$$

An example for a set of moment restrictions from instrumental variable estimation in a linear regression model is $\mathrm{E}(z_i u_i) = 0$ where $u_i = y_i - x_i'\beta$ and z_i is a vector of instruments. The GMM estimator $\hat{\theta}$ minimizes the quadratic form

$$m(y, x; \theta)' A_n m(y, x; \theta) \tag{3.36}$$

where

$$m(y, x; \theta) = \sum_{i=1}^{n} m_i(y_i, x_i; \theta) \tag{3.37}$$

In the panel case, the simple sum over i in (3.37) has to be replaced by a double sum over both i and t. m is an $(l \times 1)$ vector of empirical moment restrictions and A_n is an $(l \times l)$ positive definite symmetric weighting matrix such that $\lim_{n \to \infty} A_n = A$. For $l = p$, the model is just identified and the empirical moment conditions can be solved directly so that the objective function at $\hat{\theta}$ will have a value of zero. However, the real advantage of GMM arises in situations of overidentification where $l > p$, i.e., the number of moment conditions exceeds the number of parameters. In this case, the various moment conditions will usually be conflicting and the minimization of (3.36), for a particular choice of weighting matrix A_n to be discussed below, combines the information provided by the different moment conditions in an optimal way.

In order to differentiate the objective function with respect to θ, note that

$$\frac{\partial m' A_n m}{\partial \theta} = 2 \left[\frac{\partial m'}{\partial \theta} \right] A_n m$$

Thus, the p first order conditions of the GMM estimator can be written as

$$D' A_n m = 0 \tag{3.38}$$

where

$$D = \sum_{i=1}^{n} \frac{\partial m_i(y_i, x_i; \theta)}{\partial \theta'}$$

is a $(l \times p)$ matrix. Numerical methods are usually required in order to solve the first-order conditions. If there is a unique vector θ satisfying

$$\mathrm{E}[m_i(y_i, x_i; \theta)] = 0$$

the model is identified, and $\hat{\theta}$ is the GMM estimator of θ.

Under mild regularity conditions, $\hat{\theta}$ is consistent and normally distributed with asymptotic covariance matrix (see, for instance, Davidson and McKinnon, 1993, Chap. 17)

$$\mathrm{Cov}(\hat{\theta}) = (D'AD)^{-1}D'AWAD(D'AD)^{-1}$$

where W is the covariance matrix of the specified moment restrictions

$$W = \mathrm{E}[m(y_i, x_i; \theta)m(y_i, x_i; \theta)']$$

If $A = W^{-1}$ the estimator is asymptotically efficient in the class of the given moment restrictions, and the asymptotic covariance matrix simplifies to

$$\mathrm{Cov}(\hat{\theta}) = D'W^{-1}D$$

W in general depends on θ. In order to obtain a consistent estimator for $\hat{W}$, one can for instance obtain a consistent estimate $\hat{\theta}$ using any positive definite weighting matrix such as the identity matrix. When observations are independent over i the covariance matrix $\hat{W}$ can be calculated as

$$\hat{W} = \frac{1}{n}\sum_{i=1}^{n} m_i(\hat{\theta})m_i(\hat{\theta})'$$

In a next step, the objective function $m'\hat{W}^{-1}m$ is minimized in order to obtain $\hat{\theta}$. The covariance matrix can be consistently estimated by $\widehat{\mathrm{Cov}}(\hat{\theta}) = D'\hat{W}^{-1}D$. Rather than doing just two steps, the process can be iterated to convergence (which can make a big difference in finite samples).

It was mentioned before that one of the advantages of GMM is that it offers a simple specification test for the validity of the overidentifying restrictions. If the model that led to the moment equations is incorrect, at least some of the sample moment conditions will by systematically violated. This provides the basis for a test. The statistic is simply the minimum of the criterion function evaluated at $\theta = \hat{\theta}$ and divided by the sample size n (division by n is not required if m is defined as the sample mean of the empirical moment conditions rather than the sample sum as in (3.37)). Hansen (1982) showed that under the null hypothesis $H_0 : \mathrm{E}[m_i(y_i, x_i; \theta)] = 0$, this test statistic has a chi-squared distribution with degrees of freedom equal to $l - p$. This test is clearly non-directional as a rejection of the null hypothesis could be due to any of the moment conditions, ranging from failure of higher order moment conditions (if employed) to misspecified mean functions.

The treatment in this chapter is quite general, and the "art" of GMM estimation lies in deriving appropriate moment conditions. An illustration is given in a later chapter on panel data estimation (Chap. 5.2.5).

3.2.4 Bias Reduction Techniques

> "The few finite sample corrections that have been proposed remain extensively ignored by empiricists (...)" (Maasoumi, 1992, p. 2)

The maximum likelihood technique provides, if correctly specified, consistent parameter estimates. This means that for large n the probability of the estimates being 'close' to the true value gets larger and larger. However, maximum likelihood estimation in many cases is subject to small sample bias. This section is concerned with one part of that bias, the so-called first order bias that goes to zero at speed $O(n^{-1})$. We will discuss methods that remove such bias.

In general, small sample bias arises if the score function $g(\beta)$ is non-linear. It was shown before that for any score

$$\mathrm{E}(g(\beta))|_{\beta=\beta_0} = 0$$

where β_0 is the true parameter. The score g is a random variable, since it depends on y. Moreover, it must be the case that $\partial g(\beta)/\partial\beta < 0$ as long as the log likelihood function is globally concave. If $g(\beta)$ is non-linear in β then, for finite n,

$$\mathrm{E}(\hat{\beta}) \neq \beta \tag{3.39}$$

where $\hat{\beta}$ denotes the value that solves the equation $g(\beta) = 0$, i.e., the maximum likelihood estimator.

To make this result intuitively plausible, consider a convex score function $(\partial^2 g(\beta)/(\partial\beta)^2 > 0)$ and assume that the distribution of $g(\beta_0)$ is symmetric (i.e., mean and mode are zero). In this situation, $g(\beta_0) > 0$ implies that $\hat{\beta} > \beta$, and $g(\beta_0) < 0$ implies that $\hat{\beta} < \beta$. Moreover, for a given $\Delta = g(\beta_0)$,

$$|\hat{\beta}(\Delta) - \beta_0| > |\hat{\beta}(-\Delta) - \beta_0|$$

The maximum likelihood estimator overestimates the true value for a given positive value of the gradient (at the true value) by more than it underestimates it in case of a negative gradient that is equal in absolute value. As a consequence an upward bias results. If the score function is concave in β, the opposite result follows.

The Poisson maximum likelihood estimator is based on a non-linear score function:

$$g(\beta) = \sum_{i=1}^{n}(y_i - \exp(x_i'\beta))x_i$$

Therefore, it is biased in finite samples. Whether the score function is convex or concave depends on the values of x. This section introduces a method that removes the first order bias $b(\beta)$ from the Poisson estimates. The first order bias may implicitly be defined as

$$\mathrm{E}(\hat{\beta} - \beta) = b(\beta) + o(n^{-2})$$

where $o(n^{-2})$ denotes terms that are at most of order in probability n^{-2}. The first order bias $b(\beta)$ has to be calculated depending on the model under investigation. Assume that this has been done. Then, there are two approaches for removing the first order bias from the maximum likelihood estimator. First, $\hat{\beta}$ may be calculated as usual, and the bias is subtracted after estimation (see, for instance, McCullagh and Nelder, 1989, Chap. 15.2). Alternatively, the bias may be removed by artificially introducing a bias in the score equation (Firth, 1992). Consider the following modified score equation g^* :

$$g^*(\beta) = g(\beta) - \mathcal{I}(\beta)b(\beta) \tag{3.40}$$

where $\mathcal{I}$ is the Fisher information. Taking roots of g^*, instead of g, yields an estimator β^*. In general, this estimator is not equal to the maximum likelihood estimator $\hat{\beta}$. First note that from (3.40)

$$g^*(\hat{\beta}) = -\mathcal{I}(\hat{\beta})b(\hat{\beta}) \tag{3.41}$$

Expanding g^* around $\hat{\beta}$ and evaluating at $\beta = \beta^*$ yields:

$$g^*(\beta^*) \approx g^*(\hat{\beta}) + H^*(\hat{\beta})(\beta^* - \hat{\beta})$$

But the left hand side is zero by definition. Further, $H^*(\beta) = -\mathcal{I}(\beta) + O(n^{-1})$. Finally, using (3.41)

$$-\mathcal{I}(\hat{\beta})b(\hat{\beta}) - \mathcal{I}(\hat{\beta})(\beta^* - \hat{\beta}) \approx 0$$

or

$$\beta^* \approx \hat{\beta} - b(\hat{\beta})$$

Thus, introducing a weighted first order bias term in the score function gives bias corrected roots, i.e., bias corrected estimates.

Firth (1992) shows that, for the log-linear Poisson model, the bias-corrected score can be written as

$$g^*(\beta; y, x) = \sum_{i=1}^{n}(y_i + h_i/2 - \exp(x_i'\beta))x_i' \tag{3.42}$$

where h_i is the ith diagonal element of the matrix $H = WX(X'WX)^{-1}X'$. W is defined as in the previous section as a matrix having the individual variances on its main diagonal and zeros elsewhere.

The following Monte Carlo experiments illustrate the existence, and the consequences of the removal of the first order bias. Four different sample sizes were considered ($n = 10, 20, 50$, and 100). For each sample size, 1000 vectors of y were drawn from a Poisson distribution with fixed mean-vector λ. λ was constructed as $\lambda_i = -1 + x_i$, where x_i are independent drawings from a normal distribution with mean 0 and variance 4. The true slope coefficient, on which the discussion will focus, is thus $\beta_0 = 1$. For each of the 4×1000 datasets, a Poisson regression with and without bias correction was performed. Some characteristics of the empirical distributions of $\hat{\beta}$ and β^* are given in Tab. 3.1

Table 3.1. Bias Reduced Poisson Estimates

		ML-Poisson Estimates $\hat{\beta}$	Bias Reduced Estimates β^*
n=10	Mean	1.0290	1.0029
	Standard Error	0.2333	0.2222
	Minimum	0.4027	0.3815
	Maximum	2.4715	2.1479
n=20	Mean	1.0357	1.0062
	Standard Error	0.1712	0.1578
	Minimum	0.6158	0.6069
	Maximum	1.7512	1.6238
n=50	Mean	1.0022	0.9996
	Standard Error	0.0601	0.0598
	Minimum	0.8366	0.8349
	Maximum	1.2251	1.2212
n=100	Mean	0.9984	0.9960
	Standard Error	0.0487	0.0483
	Minimum	0.8062	0.8050
	Maximum	1.1515	1.1475

Except for n=100, the mean of β^* is closer to the true value than the mean of the MLE $\hat{\beta}$. Always, the maximal deviations from the true value in both directions are smaller for β^*. The bias corrected estimator has also a smaller standard error. Thus, the Monte Carlo evidence suggests that using the bias reduction provides some gain in bias reduction as well as in precision, in particular when the sample size is moderate. The dependence of the effect of the correction on sample size does not come as a surprise, since we know that the Poisson maximum likelihood estimator is consistent in this set-up.

3.3 Sources of Misspecification

In Chap. 3.1.1 the specification of the Poisson regression model was introduced in three steps: the distributional assumption, the regression (mean function), and the assumption of independent sampling. A *misspecification* is a violation of any of the three assumptions. They can be dealt with in this order, although specific types of misspecification may affect not only one assumption, but two or three at a time.

The close relation between regression, variance function, and distribution is a particular feature of the Poisson regression model. Under the Poisson assumption, the equality of conditional mean and variance implies the loss of one degree of freedom as compared to, for instance, the normal linear model. Thus, a violation of the variance function always implies a violation of the distributional assumption. The violation that has obtained most attention is *overdispersion*, a situation where the variance exceeds the mean, conditional on covariates (see, among others, Cameron and Trivedi, 1990, Dean and Lawless, 1989, and Ganio and Shafer, 1992).

3.3.1 Mean Function

Recall that the mean function of the Poisson regression is specified as

$$\mathrm{E}(y_i|x_i) = \lambda_i = \exp(x_i'\beta)$$

where x is a vector of individual covariates. Let λ_0 denote the true mean function and E_0 the expectation with respect to the true density. Let

$$\mathrm{E}_0(y_i|x_i) = \lambda_i^0 = f(x_i, \beta_0) \tag{3.43}$$

The mean function is misspecified if, for arbitrary x_i, there exists no β such that $\lambda_i = f(x_i, \beta_0)$. Possible sourves of misspecification include:

- Explanatory variables are omitted, and these omitted variables are not independent of x.
- The predictor is not linear in β.
- Explanatory variables enter the predictor via some transformation $f(x)$, rather than linearly.
- The function is misspecified. For instance, the true mean function may be linear rather than log-linear.

In practice, there can be little hope that the mean function is correctly specified in all respects. However, it will be useful to maintain the hypothesis of a correctly specified model and test for one particular misspecification at a time.

3.3.2 Unobserved Heterogeneity

The situation of unobserved heterogeneity arises if the explanatory variables do not account for the full amount of individual heterogeneity. This situation can be interpreted as a defect of the mean function, in particular a problem of omitted variables: assume that the true model is

$$\tilde{\lambda}_i = \exp(x_i'\beta + z_i'\gamma) \tag{3.44}$$

However, z_i are unobserved by the econometrician. Can there be any hope that the Poisson maximum likelihood estimator based on the (incorrect) conditional expectation function $\lambda_i = \exp(x_i'\beta)$ maintains some or all of its usual desirable properties?

It is useful to recall the results for the normal linear model. There, omitted variable bias arises whenever x_i and z_i are correlated. However, without correlation, omitted variables do not cause a problem. In fact, omitted variables are, apart from measurement error in y, a standard argument for introducing a stochastic relation with additive error term to begin with.

The Poisson regression model is of a different stochastic nature. The model has no additive error term. Y is a random variable because the count process is intrinsically stochastic, given an intensity λ that is measured without error. Hence, a model with unobserved heterogeneity cannot be Poisson distributed.

Based on (3.44), we can write

$$\tilde{\lambda}_i = \lambda_i u_i \,, \quad u_i > 0 \,, \tag{3.45}$$

where $\lambda_i = \exp(x_i'\beta)$ and $u_i = \exp(z_i'\gamma)$. It is assumed that u_i and x_i are independent. Since u_i is unobserved by assumption, it has to be treated as a random variable. Hence, the Poisson parameter $\tilde{\lambda}_i$ is now a random variable itself. The two sources of variation in the Poisson parameter $\tilde{\lambda}_i$ interact in a multiplicative way: The first source is systematic and due to the observed variables x_i, while the second source is unobserved and caused by an individual random effect u_i independent of x_i. Let $\varepsilon_i = \ln u_i$. Then $\tilde{\lambda}_i = \exp(x_i'\beta + \varepsilon_i)$ and we see that the error is additive on a logarithmic scale.

Without loss of generality $\mathrm{E}(u_i)$ can be normalized to 1 as long as the model has an overall constant. Furthermore, we assume that the error has constant variance σ_u^2. It follows that $\tilde{\lambda}_i$ has mean

$$\mathrm{E}(\tilde{\lambda}_i|\lambda_i) = \lambda_i$$

and variance

$$\mathrm{Var}(\tilde{\lambda}_i|\lambda_i) = \lambda_i^2\sigma_u^2 \,.$$

Since u_i and thus $\tilde{\lambda}_i$ are unobserved, conditioning is not possible, and the marginal distribution $f(y_i)$ is obtained by integrating the joint distribution over u:

$$f(y_i) = \int f(y_i|u_i)g(u_i)\,du_i \tag{3.46}$$

Knowledge of the full distribution of u_i is required in order to perform the integration. However, the implications of unobserved heterogeneity for the first two *marginal* moments of the distribution of y_i can be studied without knowledge of the full distribution of u_i as long as the first two moments of u_i are known. Consider the following theorem:

Theorem: Let y and u denote random variables such that y has finite mean $\mathrm{E}(y)$ and variance $\mathrm{Var}(y)$. Then

$$\mathrm{E}(y) = \mathrm{E}_u[\mathrm{E}(y|u)] \tag{3.47}$$

and

$$\text{Var}(y) = \text{E}_u[\text{Var}(y|u)] + \text{Var}_u[\text{E}(y|u)] \geq \text{Var}[\text{E}(y|u)] \tag{3.48}$$

A proof is given in Hogg and Craig (1978, p. 349).

Unobserved Heterogeneity in the Poisson Regression Model

Let $y|x, u$ have a Poisson distribution with parameter $\tilde{\lambda} = \exp(x'\beta)u$, where u is distributed independently of x with mean 1 and variance σ_u^2. The distribution of interest is $y|x$ (since we observe y and x but not u.) The previous theorem implies that $\text{Var}(y|x) > \text{E}(y|x)$, i.e., unobserved heterogeneity of this form causes overdispersion of the conditional model for $y|x$ relative to the Poisson model

Proof: Applying (3.47) we obtain $\text{E}(y|x) = \exp(x'\beta)\text{E}(u) = \lambda$. Applying (3.48) we obtain

$$\begin{aligned}\text{Var}(y|x) &= \text{E}_u(\tilde{\lambda}|x) + \text{Var}_u(\tilde{\lambda}|x)\\ &= \lambda + \sigma_u^2\lambda^2\end{aligned}$$

For $\sigma_u^2 > 0$, $\text{Var}(y|x) > \text{E}(y|x)$. □

Example

The result can be illustrated using a simple example. Assume that the population can be split up into two groups of equal size, each of which is Poisson distributed with $\lambda_1 = 0.5$ and $\lambda_2 = 1.5$, respectively. The expected value in the population is then given by

$$\begin{aligned}\text{E}(y) &= 0.5 * 0.5 + 0.5 * 1.5\\ &= 1\end{aligned}$$

For the variance in the population we have

$$\begin{aligned}\text{Var}(y) &= 0.5\text{Var}_1 + 0.5\text{Var}_2 + 0.5(\text{E}_1 - \text{E}(y))^2 + 0.5(\text{E}_2 - \text{E}(y))^2\\ &= 0.5 * 0.5 + 0.5 * 1.5 + 0.5 * (-0.5)^2 + 0.5 * 0.5^2\\ &= 1.25\end{aligned}$$

Thus, the variance of population is greater than its mean, and the population cannot be Poisson distributed.

The possibility of unobserved heterogeneity and overdispersion is not limited to the Poisson distribution or Poisson regression model. Consider instead a binomial distribution with unobserved heterogeneity: let $y|n, p$ be binomial distributed with mean np and variance $np(1-p)$. Furthermore, assume that p is itself a random variable with mean π and variance ν. Then, an application of rules (3.47) and (3.48) gives

$$\mathrm{E}(y) = n\mathrm{E}p = n\pi$$

and

$$\begin{aligned}\mathrm{Var}(y) &= \mathrm{E}[np(1-p)] + \mathrm{Var}(np) \\ &= n(n-1)\nu + n\pi(1-\pi)\end{aligned}$$

Thus, in general, $\mathrm{Var}(y) > n\pi(1-\pi)$ and we have overdispersion relative to the binomial model. The exception is $n = 1$ (the Bernoulli case), where the first term of the variance function disappears and the mean fully determines the variance.

In a way, one can also draw an analogy to the usual linear model. If $y = x_1\beta_1 + x_2'\beta_2 + u$ where x_1 and x_2 are independent, then $\mathrm{Var}(y|x_1) > \mathrm{Var}(y|x_1, x_2)$. Ignoring x_2 does not cause a bias in estimating β_1, but the error term variance increases since the new error term is $x_2'\beta_2 + u$.

Spell-Specific Heterogeneity

Gourieroux and Visser (1997) have introduced the concept of *spell-specific heterogeneity*, modeling the counts as outcome of an underlying sequence of exponentially distributed spells (waiting times until the next occurrence). They use the fact that the probability that at most $k-1$ events occurred in a given interval $(0, T)$ equals the probability that the arrival time of the k-th event, given by the sum of the k waiting times τ_k between consecutive events, exceeds T. Moreover, assume that the waiting times τ_k $k = 1, 2, \ldots$ follow independent exponential distribution functions with parameters

$$\tilde{\lambda} = \tilde{\lambda}(x, u, \eta_k) \tag{3.49}$$

In addition to the two individual specific factors, the observed (x_i) and the unobserved (u_i), an additional spell-specific (unobserved) heterogeneity factor η_k is introduced. Gourieroux and Visser (1997) show that the underlying count data distribution, derived from a convolution operation and a local approximation of the characteristic function, can display both under- and overdispersion.

3.3.3 Measurement Error

Let $y|x$ be Poisson distributed with mean $\exp(x'\beta)$. One possible way of introducing measurement error in explanatory variables is to assume that rather than observing x, we observe

$$z = x + \varepsilon$$

where ε are assumed to be independent of x with mean 0 and covariance matrix Ω. Guo and Li (2000, 2001) study the consequences of such a set-up, and possible remedies. First, they note that measurement error leads to overdispersion for the observed model $f(y|z)$, much as unobserved heterogeneity does. We can write

$$f(y|z) = \int f(y|x)g(x|z)\mathrm{d}x$$

Unless $g(x|z) = 0$ almost everywhere (or $\mathrm{E}(y|z) = 1$), it can be shown that

$$\mathrm{E}(y|z) < \mathrm{Var}(y|z)$$

even though $\mathrm{E}(y|x) = \mathrm{Var}(y|x)$. Clearly, thus, the Poisson model is inappropriate. However, in contrast to standard unobserved heterogeneity, the problem goes beyond considerations of efficiency and consistent estimation of the covariance matrix for valid inference. Rather, the Poisson estimator $\hat{\beta}$ is inconsistent in general. To see the thrust of the argument, re-write the log-likelihood function of the falsely assumed Poisson model $f(y|z)$

$$\frac{1}{n}\sum_{i=1}^{n}[-\exp(z_i'\beta) + y_i z_i\beta - \ln(y_i!)]$$

using $z_i = x_i + \varepsilon_i$, as

$$\frac{1}{n}\sum_{i=1}^{n}[-\exp(x_i'\beta) + y_i x_i\beta - \ln(y_i!)]$$

$$+\frac{1}{n}\sum_{i=1}^{n}[-\exp(x_i'\beta)[\exp(\varepsilon_i'\beta) - 1] + y_i\varepsilon_i\beta]$$

Consistent parameter estimation could be based on the first (unobserved) term of the log-likelihood function. Since the likelihood function of the Poisson model with measurement error adds a second term, its maximization in general will not yield a consistent estimator. Note that the second term converges to

$$\mathrm{E}_x[-\exp(x'\beta)](\mathrm{E}_\varepsilon[\exp(\varepsilon'\beta)] - 1) = \mathrm{E}_z[-\exp(z'\beta)] + \mathrm{E}_x[\exp(x'\beta)]$$

This suggests that to obtain a consistent estimator, one can possibly maximize

$$\frac{1}{n}\sum_{i=1}^{n}[-\exp(z_i'\beta) + y_i z_i\beta - \ln(y_i!)] - \frac{1}{n}\sum_{i=1}^{n}[-\exp(z_i'\beta)] - \mathrm{E}_x[\exp(x'\beta)]$$

$$= \frac{1}{n}\sum_{i=1}^{n}[y_i z_i\beta - \ln(y_i!)] - \mathrm{E}_x[\exp(x'\beta)]$$

given that $\mathrm{E}_x[\exp(x'\beta)]$ is known or can be estimated first. Guo and Li (2000) refer to the estimator that maximizes this modified log-likelihood function by solving

$$\frac{1}{n}\sum_{i=1}^{n} y_i z_i - \mathrm{E}_x[x\exp(x'\beta)] \equiv 0$$

as *corrected score estimator*. The crucial question is now how to obtain $\mathrm{E}_x[\exp(x'\beta)]$. For example, when ε is multivariate normal with mean 0 and covariance Ω, then

$$\mathrm{E}_x[\exp(x'\beta)] = \mathrm{E}_z\left[\exp\left(z'\beta - \frac{\beta'\Omega\beta}{2}\right)\right]$$

which can be consistently estimated by

$$1/n\sum_{i=1}\exp\left(z_i'\beta - \frac{\beta'\Omega\beta}{2}\right),$$

so that the corrected likelihood function can be written as

$$\frac{1}{n}\sum_{i=1}^{n}\left[y_i z_i\beta - \ln(y_i!) - \exp\left(z_i'\beta - \frac{\beta'\Omega\beta}{2}\right)\right]$$

See Guo and Li (2000) and Nakamura (1990) for further details on this approach. Guo and Li (2001) study measurement erro in the negative binomial model. Also note that no clear cut results on the direction of the bias are available (not necessarily attenuation). The methods also can be extended to deal with proxy variables rather then measurement error. Finally, Cameron and Trivedi (1998) discuss the case of multiplicative measurement error in explanatory variables.

3.3.4 Dependent Process

The Poisson distribution is appropriate if events occur completely randomly over time. The Poisson model is misspecified if the process that generates events is not "memoryless", that is, when the probability of an occurrence between t and $t+h$ depends on the past process. Few attempts have been made in the literature to directly model the dependence structure. One such model is a "contagious" process that leads to the negative binomial distribution. Another is the Winkelmann (1995) model for duration dependence in count processes. Starting point is the observation that the Poisson distribution is characterized by independently exponentially distributed interarrival times of events. The key feature of the exponential distribution is its constant hazard function, i.e., lack of memory. Winkelmann (1995) shows how a count data distribution can be derived based on a more general distribution of interarrival times, the gamma distribution. In particular, the gamma distribution allows for both negative duration dependence, leading to overdispersion, and positive duration dependence, leading to underdispersion.

3.3.5 Selectivity

While there is a vast literature on sample selection in the linear model, the analysis of count data models with selectivity is less well developed. Sample selection occurs if the data are generated such that the researcher does not observe the underlying count variable y^* but rather a "selected" count y. The types of selection considered include truncation and censoring, but also underreporting (see Chap. 3.3.7). If sample selection is ignored the estimator of the regression parameters is generally inconsistent.

Two patterns of selectivity can be distinguished. Firstly, observations can be censored or truncated depending on the outcome of y^*. For instance, many survey questionnaires are "top-coded", introducing a category of the type "x or more" events. In this case, the data are censored from above. Secondly, observations can be censored or truncated depending on the outcome of another variable c, that may or may not be independent of y^*. The literature refers to this case as "incidental" truncation or censoring (Greene, 2000). We denote this situation as "endogenous" selectivity. An example is the study of credit card defaults. Incidental truncation occurs since some individuals have no access to credit cards. Some information on the joint distribution of y^* and c is required in this situation. Such models are discussed, among others, in Greene (1994, 1997), Terza (1998), and Winkelmann (1998).

3.3.6 Simultaneity and Endogeneity

The basic Poisson regression model was introduced as a single equation regression model for cross section data. In a next step, unobserved heterogeneity was allowed for. In contrast to the linear model where unobserved heterogeneity is automatically taken into account, we showed that the Poisson model was inappropriate in this situation and had to be generalized. The generalization was based on the key assumption that unobserved heterogeneity and regressors x were statistically independent.

This assumption is likely to be violated in many applications, in particular, when regressors are simultaneously determined and hence endogenous. The prime example for endogenous regressors is an endogenous treatment effect, where individuals self-select into treatment, and those who take the treatment are systematically different from the control group. If this selection is correlated with the outcome, either directly or through unobservables, the assumption of statistical independence between regressors and the error term will break down and standard estimation methods like maximum likelihood will not generally be consistent. The generic solution to this problem is a nonlinear instrumental variable approach as outlined in Mullahy (1997a) and in Windmeijer and Santos Silva (1997). Alternatively, one may attempt to directly model the endogenous regressor and employ two-stage estimation techniques. This is described in Terza (1998).

A problem of a different sort is the modeling of multivariate counts. For instance, Chap. 5.1.1 introduces a bivariate Poisson model. While this model does not give rise to a simultaneous system (since none of the dependent variables appears as a regressor), it constitutes what has come to be known in the literature on linear models as the seemingly unrelated regression model.

In particular, the bivariate Poisson model allows for a non-trivial correlation structure between the two or more endogenous count variables. If correlation is present, joint estimation will lead to a more efficient estimator than separate estimation.

3.3.7 Underreporting

Although the issue of underreporting can be interpreted in terms of selectivity, it leads to models of a different type. Count data are underreported if only a fraction of the total events is reported. If y^* denotes the total number of events, and y the number of reported events, then clearly $y \leq y^*$. In a different interpretation, y^* denotes the number of potential events and y the number of actual events. For instance, in labor economics, y^* could be the number of job offers during a given period of time, in which case y as the number of accepted offers and $y^* - y$ is the number of rejected offers. Both interpretations have the same formal structure. In particular,

$$y = \sum_{i=1}^{y^*} B_i \tag{3.50}$$

where B_i is an indicator variable taking the value 1 if an event is reported (or a job is accepted) and zero otherwise. In statistical terms, the distribution of y is referred to as a convolution (or stopped-sum distribution; see Chap. 2.5.2). The distribution of y depends on the joint distribution for y^* and the B_i's. In general, closed form results are only available under strong independence assumptions.

Three generic types of underreporting have been discussed in the literature, each giving rise to a different count data model:

1. Random underreporting. Here, the B_i's are independently and identically Bernoulli distributed with parameter p (Winkelmann, 1996).
2. Logistic underreporting. The probability of reporting $P(B_i = 1)$ is a logistic function of covariates (Winkelmann and Zimmermann, 1993, Mukhopaday and Trivedi, 1995).
3. Count amount model. Events are associated with a nonnegative size variable (for instance a purchase amount), and recorded only if this variable exceeds a specific minimum threshold (Van Praag and Vermeulen, 1993).

3.3.8 Excess Zeros

The idea of adjusting the probability of zero outcomes for count distributions goes at least back to Johnson and Kotz (1969). See also Mullahy (1986). An early application in a regression framework is Lambert (1992). She introduces a zero-inflated Poisson model where with probability ω the only possible observation is 0, and with probability $1 - \omega$ a Poisson(λ) random variable is observed. Both ω and λ may depend on covariates. The overall probability of a zero outcome is then

$$
\begin{aligned}
f(0) &= \omega + (1-\omega)e^{-\lambda} \\
&= e^{-\lambda} + \omega(1-e^{-\lambda})
\end{aligned}
\tag{3.51}
$$

This probability is strictly greater than the Poisson probability of 0 as long as $\omega > 0$. Excess zeros, like unobserved heterogeneity and dependence, lead to overdispersion. Hence, excess zeros provide yet another explanation for this frequently observed phenomenon. Moreover, the model has an interesting interpretation. Lambert (1992) studies the number of defects in manufacturing. Here, the count generating process is decomposed into a "perfect state" where defects are extremely rare and an "imperfect state" where defects are possible but not inevitable.

An economic interpretation is given in Crépon and Duguet (1997b) in a study of R&D productivity. Assume that the count variable is the number of patent applications lodged by a firm during a given period of time. Here, firms face the strategic choice whether or not to apply for patents in general. Only when this choice is made in the affirmative is it that the number of actual discoveries becomes relevant. Again, no applications may result for firms that decided to patent but had no discoveries during a particular period.

3.3.9 Variance Function

It was noted before that a variance violation implies a distributional violation. The opposite does not follow, since the distributional violation might originate in higher order moments. Here, the possibility of such higher order violations is left unexplored and the focus is on the variance function. A rationale for this approach is that most properties of the model and of the estimator are established through asymptotic results which require assumptions on the first two moments only.

The variance function of the benchmark Poisson model is $\text{Var}(y_i|x_i) = \text{E}(y_i|x_i) = \exp(x_i'\beta)$. Count data of this sort are said to be equi-dispersed. The two alternatives are *overdispersion* or *underdispersion*. In the former situation, the conditional variance exceeds the conditional mean; in the latter, the conditional mean exceeds the conditional variance. The following causes for non-Poissoness of the variance function can be distinguished:

- Unobserved individual heterogeneity causes overdispersion. This case has been discussed in Chap. 3.3.2.
- Spell specific heterogeneity as defined by Gourieroux and Visser (1997) may result in either over- or underdispersion.
- True positive contagion causes overdispersion; true negative contagion causes underdispersion.

- Non-stationarity has an ambiguous effect. If non-stationarity can be modeled as a convolution of independent Poisson distributions, the convolution is again Poisson distributed.

Over- and underdispersion exist if the function mapping the conditional mean into the conditional variance is not the identity function. In general, this may be an arbitrary function, possibly involving further individual attributes z_i:

$$\mathrm{Var}(y_i|x_i, z_i) = f[\mathrm{E}(y_i|x_i), z_i] \tag{3.52}$$

or, assuming that the mean function is correctly specified

$$\mathrm{Var}(y_i|x_i, z_i) = f[\exp(x_i'\beta), z_i] \tag{3.53}$$

The quadratic variance function

$$\mathrm{Var}(y_i|x_i) = \exp(x_i'\beta) + \sigma^2[\exp(x_i'\beta)]^2 \qquad \sigma^2 \in \mathbb{R}^+ \tag{3.54}$$

has received most attention in the literature. It arises naturally if the model has unobserved heterogeneity with constant variance.

Generalizations have proceeded in two directions. First, the range of σ^2 may include negative values, allowing for underdispersion (King, 1989b). Some constraints on the parameter space are required since the left side, a variance, has to be kept positive. Second, σ^2 may be parameterized in terms of explanatory variables. These variables usually coincide with those appearing in the mean function, although this is not a formal requirement. In count data models where mean and variance are intrinsically linked it would be difficult, however, to justify that some variables z_i affect the variance but not the mean. Thus, one common parameterization is

$$\tilde{\sigma}_i^2 = \sigma^2 \exp(x_i'\gamma) \tag{3.55}$$

The potential problem with this approach is that the model tends to be overparameterized. Winkelmann and Zimmermann (1991a) avoid this problem by imposing proportionality: $\gamma = (k-1)\beta$, where $(k-1)$ is a proportionality factor. Then

$$\begin{aligned}\tilde{\sigma}_i^2 &= \sigma^2 \exp[(k-1)x_i'\beta] \\ &= \sigma^2[\exp(x_i'\beta)]^{k-1}\end{aligned} \tag{3.56}$$

Therefore,

$$\mathrm{Var}(y_i|x_i) = \mathrm{E}(y_i|x_i) + \sigma^2[\mathrm{E}(y_i|x_i)]^{k+1} \qquad \sigma^2 \in \mathbb{R}^+,\ k \in \mathbb{R}\,. \tag{3.57}$$

σ^2 is the *dispersion*-parameter and k the non-linearity parameter. Compared to the Poisson variance assumption, this specification has two additional parameters. It allows for linear, convex as well as concave relationships between the variance and the mean. For $k = 0$ and positive σ^2, this is the variance function of the Negbin I model. For $k = 1$ and positive σ^2, this is the variance function of the Negbin II model. Fig. 3.3 shows some possible shapes of the variance function.

Fig. 3.3. Variance-Mean Relationships for Different k's and σ^2's

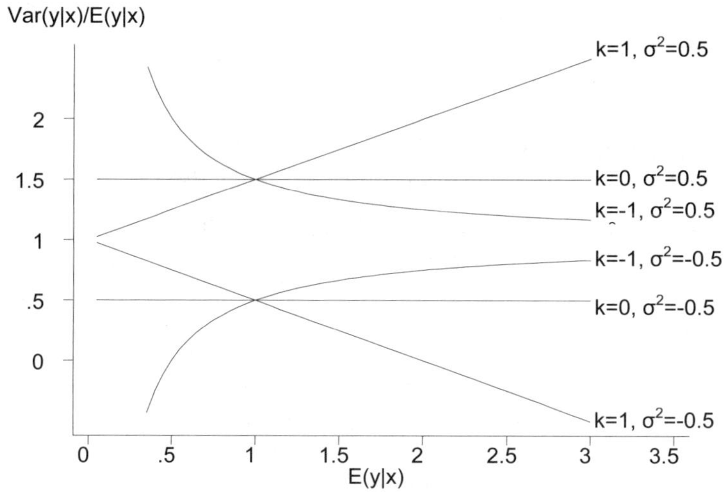

3.4 Testing for Misspecification

Since many of the misspecifications listed in the previous section lead to a violation of the assumption of equal conditional variance and mean, the variance function provides a natural starting point for misspecification tests. Frequently one has a specific alternative in mind that allows for a more general variance function and, at the same time, nests the Poisson variance function through a parametric restriction. In this situation, the three classical tests, the likelihood ratio, the Wald, and the Lagrange multiplier test can be used.

A related class of 'misspecification' tests addresses the inclusion of irrelevant variables in the mean function. The difference is that these restrictions are tested within a given parametric distribution family that is specified up to a parameter vector β which may contain zeros. As a consequence, these tests can rely on standard maximum likelihood technology and do not need a special treatment.

Tests that discriminate between non-nested models have a shorter history in econometrics. And yet, non-nested models frequently arise in applied work with count data. Examples include the hurdle Poisson and the zero-inflated Poisson models, or Poisson models with linear and log-linear conditional expectation functions. A simple asymptotic test for non-nested models has been introduced by Vuong (1989). See also Santos Silva (2001). A simulation based approach offers an alternative to discriminate between non-nested models. Finally, the Hausman (1978) test can be applied in some specific situations.

3.4.1 Classical Specification Tests

Testing a Poisson model against a more general parametric model is straightforward if the former is contained in the latter through a parametric restriction. The two models are said to be "nested". Examples for restrictions are:

1. Poisson versus negative binomial: $H_0 : \sigma_u^2 = 0$.
2. Linear restrictions on regression coefficients: $H_0 : R\beta = q$.
3. Non-linear restrictions on regression coefficients: $H_0 : R(\beta) = q$.

Assume that estimation is by the method of maximum likelihood. Tests for the validity of H_0 can be based on any one of the three following principles:

1. Likelihood-ratio test
2. Wald test
3. Lagrange multiplier test

These are asymptotic tests. Their small sample properties (size and power) are generally unknown. Asymptotically, all three tests are equivalent. The tests are directional, implying that if the null hypothesis is rejected, there is a well defined alternative.

Likelihood Ratio Test

Let $\hat{\ell}_r$ denote the value of the log-likelihood function evaluated at the restricted maximum likelihood estimates (for instance, the Poisson model), and $\hat{\ell}_u$ the value of the log-likelihood function evaluated at the unrestricted maximum likelihood estimates (for instance, the negative binomial model), and let k denote the number of restrictions (k=1 in a test of the Poisson model against the negative binomial model). Then, under H_0 (if the restriction is correct):

$$\text{LR} = -2(\hat{\ell}_r - \hat{\ell}_u) \sim \chi^2_{(k)} \tag{3.58}$$

where $\chi^2_{(k)}$ is a chi-squared distribution with k degrees of freedom. The test is based on the difference of two log-likelihood values, or, equivalently, the log of a likelihood-ratio, hence its name. If the restriction is lifted, the value of the log-likelihood function (evaluated at the maximum likelihood estimates) must increase. If the increase is "large", where large is defined as any test statistic that exceeds the critical value $\chi^2_{\alpha,k}$, the null hypothesis is rejected.

This test is sometimes criticized because it requires separate estimation of two models, the restricted model and the unrestricted model. However, given modern computing power, this criticism has lost some of its original weight.

One problem with the likelihood ratio test of the Poisson against the negative binomial model is that under the null hypothesis the true parameter is on the boundary of the parameter space. If a parameter, such as a variance, is bounded from below at H_0, the estimate must be greater than or equal to H_0

and vice versa. The asymptotic normality of the MLE can no longer hold under H_0. Chernoff (1954) and Lawless (1987b) address this problem. Chernoff (1954) shows that under H_0, the likelihood ratio statistic has a distribution with probability mass of 0.5 at 0 and a $0.5\chi^2_{(1)}$ distribution for positive values. This adjustment is not required when testing restrictions on regression coefficients.

Wald Test

Point of departure for the Wald test is the asymptotic distribution of the maximum likelihood estimator in the unrestricted model. In contrast to the likelihood-ratio test, estimation of a single model is sufficient:

$$\hat{\theta} \overset{\text{app}}{\sim} N(\theta_0, \widehat{\text{Var}}(\hat{\theta}))$$

where $\hat{\theta}$ consists of the estimated regression coefficients plus any additional parameters, such as $\hat{\sigma}^2_u$. For any linear combination of the parameter vector $R\hat{\theta} - q$ it follows that

$$R\hat{\theta} - q \overset{\text{app}}{\sim} N(R\theta_0 - q, R\widehat{\text{Var}}(\hat{\theta})R')$$

Under $H_0 : R\theta_0 - q = 0$ and therefore

$$\text{W} = (R\hat{\theta} - q)'[R\widehat{\text{Var}}(\hat{\theta})R']^{-1}(R\hat{\theta} - q) \tag{3.59}$$

has a chi-squared distribution with degrees of freedom equal to the number of restrictions if the null hypothesis is correct. If the number of restrictions is one, the statistic simplifies to the squared "t"-statistic. Dividing W in (3.59) by the degrees of freedom produces an "F"-statistic. The quotation marks indicate that, under maximum likelihood estimation, asymptotic results apply, whereas the t-distribution and the F-distribution are finite sample distributions.

To give an example, assume that estimation of a negative binomial model produces an estimate $\hat{\sigma}^2$ with estimated asymptotic variance $\widehat{\text{Var}}(\hat{\sigma}^2)$. The Poisson model requires $\sigma^2 = 0$. Hence, the Wald test of H_0 : Poisson(λ) against H_1: negative binomial with mean λ and variance $\lambda + \sigma^2\lambda^2$ is based on the "t"-statistic $(\hat{\sigma}^2 - 0)/\sqrt{\widehat{\text{Var}}(\hat{\sigma}^2)}$. Again, the parameter is at the boundary of the parameter space, and hence inference should use a one-sided alternative, based on half a standard normal distribution $N(0, 1)$.

Finally, if the restriction is of a non-linear nature, we can substitute

$$\text{Var}[R(\hat{\theta}) - q] = \left[\frac{\partial R(\hat{\theta})}{\partial \hat{\theta}'}\right] \widehat{\text{Var}}(\hat{\theta}) \left[\frac{\partial R(\hat{\theta})}{\partial \hat{\theta}}\right]$$

for the variance (an application of the "delta" rule), and $R(\hat{\theta}) - q$ for $R\hat{\theta} - q$ in equation (3.59). Under H_0, the resulting statistic is again approximately chi-squared distributed with k degrees of freedom.

Lagrange Multiplier Test

Instead of computing both models und performing a likelihood ratio test, or computing the alternative model only and performing a Wald test, the Lagrange multiplier (LM) test avoids the computation of the alternative model altogether. This test is also known as "score" test. In this context, the score vector is the vector of first derivatives (or gradient) of the log-likelihood function with respect to the parameters.

Let $\ln L_u$ be the log-likelihood function of the unrestricted model. Then $\hat{\theta}_u$ solves the first-order conditions

$$\left.\frac{\partial \ln L}{\partial \theta}\right|_{\theta_u} = \left.\frac{\partial \ell}{\partial \theta}\right|_{\theta_u} = 0 \tag{3.60}$$

Alternatively, we could evaluate the score vector of the unrestricted model at the maximum likelihood estimator $\hat{\theta}_r$ obtained under the restricted model:

$$\left.\frac{\partial \ell}{\partial \theta}\right|_{\theta_r} \neq 0 \tag{3.61}$$

Unless the restriction is true, this expression differs from zero. Hence, the restriction is rejected if the score (3.61) is "far" from zero. If the null hypothesis is true (i.e., $\hat{\theta}_r = \theta_0$), it was shown in Chap. 3.2.1 , equation (3.27), that the score vector has an asymptotic normal distribution with mean zero and variance covariance matrix equal to the Fisher information matrix. Hence, a test can be based on the quadratic form

$$\mathrm{LM} = \left(\frac{\partial \ell}{\partial \theta}\right)'_{\hat{\theta}_r} [I(\hat{\theta}_r)]^{-1} \left(\frac{\partial \ell}{\partial \theta}\right)_{\hat{\theta}_r} \tag{3.62}$$

which has a chi-squared distributed with degrees of freedom equal to the number of restrictions under the null hypothesis. (The variance of the score can be estimated along the common three ways, see Chap. 3.2.1; See Greene (2000) for an explanation why this test statistic is related to a Lagrange multiplier.).

As an example, one could use the score/LM test for testing a linear restriction on the regression coefficient. Let $\hat{\beta}_r$ be the maximum likelihood estimator obtained from estimating the Poisson regression with restriction imposed. Let $\hat{\Lambda}_r = \exp(X\hat{\beta}_r)$ be the $(n \times 1)$ vector of predicted conditional means in the restricted model. The score/LM statistic is given by

$$\mathrm{LM} = (Y - \hat{\Lambda}_r)'X(X'\mathrm{diag}\hat{\Lambda}_r X)^{-1}X'(Y - \hat{\Lambda}_r) \tag{3.63}$$

Under the null hypothesis, this statistic is chi-squared distributed with r degrees of freedom.

Usually, however, score/LM tests are used in a different context: rather than testing restrictions on regression coefficients, score/LM procedures have been developed to test certain aspects of the stochastic specification.

A leading example is Lee (1986) who derives a Lagrange multiplier for the Poisson regression model against the more general Katz models. The Katz family contains the negative binomial model as a special case, and we shall illustrate the derivation of the LM statistic for the Poisson against the Negbin II model. Recall that the probability distribution function is given by

$$f(y) = \frac{\Gamma(\alpha+y)}{\Gamma(\alpha)\Gamma(y+1)}\left(\frac{\alpha}{\alpha+\lambda}\right)^{\alpha}\left(\frac{\lambda}{\alpha+\lambda}\right)^{y}$$

The relevant restriction is $H_0 : \alpha^{-1} = 0$. The derivatives of the log density with respect to α and λ are given in Chap. 4.2.2. Unfortunately, brute force evaluation of $\partial\ell/\partial\alpha|_{\alpha^{-1}=0}$ does not work in this case, and a more elaborate derivation is needed (See also Cameron and Trivedi, 1998, Chap. 5).

Using the product representation of the gamma ratio, the probability distribution function for a single observation can be written as

$$\begin{aligned} f(y) &= \prod_{j=1}^{y}\frac{j+\alpha-1}{j(1+\alpha/\lambda)}(1+\lambda/\alpha)^{-\alpha} \\ &= f(0)\prod_{l=1}^{y}\frac{\alpha+(y-l)}{(1+\alpha/\lambda)(y-l+1)} \\ &= f(0)\prod_{l=1}^{y}\frac{\lambda+a\lambda(y-l)}{(1+a\lambda)(y-l+1)} \end{aligned}$$

where $f(0) = (1+\lambda/\alpha)^{-\alpha}$, $l = y-j+1$ and $a = \alpha^{-1}$. The derivative of the log-density with respect to a is given by

$$\frac{\partial\ell}{\partial a} = \left\{\sum_{l=1}^{y}\frac{\lambda(y-l)}{\lambda+a\lambda(y-l)} - \frac{\lambda}{1+a\lambda}\right\} + \frac{\partial\ln f(0)}{\partial a} \tag{3.64}$$

Under $H_0 : a = 0$,

$$\begin{aligned} \left.\frac{\partial\ell}{\partial a}\right|_{a=0} &= \left\{\sum_{l=1}^{y}(y-l) - \lambda\right\} + \left.\frac{\partial\ln f(0)}{\partial a}\right|_{a=0} \\ &= \sum_{j=1}^{y}(j-l) - \lambda y + \left.\frac{\partial\ln f(0)}{\partial a}\right|_{a=0} \end{aligned}$$

The sum is simply $y(y-1)/2$. Since for any proper distribution the expected score is zero, we have that $\mathrm{E}(\partial\ln f(y)/\partial a) = \mathrm{E}(\partial h(y)/\partial a + \partial\ln f(0)/\partial a) = 0$, and thus $\mathrm{E}(\partial\ln f(0)/\partial a) = \partial\ln f(0)/\partial a = -\mathrm{E}(h(y))$. Putting things together,

$$\begin{aligned} \left.\frac{\partial\ell}{\partial a}\right|_{a=0} &= y(y-1)/2 - \lambda y - \mathrm{E}[y(y-1)/2 - \lambda y] \\ &= y(y-1)/2 - \lambda y - (\lambda^2/2 - \lambda^2) \\ &= \frac{1}{2}[(y-\lambda)^2 - y] \end{aligned}$$

For a random sample of size n, and with $\hat{\lambda}_i = \exp(x_i'\hat{\beta})$ where $\hat{\beta}$ is the estimated vector of regression coefficients under the null hypothesis (i.e., the Poisson maximum likelihood estimator), the sample score can be written as

$$\left.\frac{\partial \ell_n}{\partial a}\right|_{a=0} = \frac{1}{2}\sum_{i=1}^{n}(y_i - \hat{\lambda}_i)^2 - y_i \tag{3.65}$$

In addition, a consistent estimator for the variance can be obtained using the information matrix equality as

$$\frac{1}{n}\mathrm{E}\left(\frac{\partial \ell_n}{\partial a}\right)^2_{\hat{\beta},a=0} = \sum_{i=1}^{n}\frac{1}{2\hat{\lambda}_i^2}$$

and the square root of the scalar test statistic is given by

$$\mathrm{LM} = \left[\sum_{i=1}^{n}\frac{1}{2\hat{\lambda}_i^2}\right]^{-1/2}\frac{1}{2}\sum_{i=1}^{n}(y_i - \hat{\lambda}_i)^2 - y_i$$

Under H_0 the score has an asymptotic standard normal distribution (since it is the square root of a chi-squared distribution with one degree of freedom) and a test for overdispersion is a one sided test with critical value z_α. Since the same statistic is obtained for the more general Katz family, it can also be used to test for underdispersion. Rejection of equidispersion against underdispersion requires a test statistic smaller than $-z_\alpha$.

Score tests have been developed in the literature on count data models for all sorts of null and alternative hypotheses. For instance, Gurmu (1991) derives a score test for overdispersion in the positive Poisson regression models, Gurmu and Trivedi (1992) consider overdispersion in truncated Poisson regression models, while van den Broek (1995) develops a score test for extra zeros in the Poisson model. As in Chap. 2.4.3 the zero altered Poisson model can be written as

$$P(y_i = 0) = \omega + (1-\omega)e^{-\lambda_i}$$

$$P(y_i = k) = (1-\omega)\frac{e^{-\lambda_i}\lambda_i^k}{k!} \qquad k = 1, 2, \ldots$$

This model collapses to the standard Poisson model for $\omega = 0$. Let $\hat{\lambda}_i = \exp(x_i'\hat{\beta})$ where $\hat{\beta}$ are the usual Poisson estimates. van den Broek shows that under $H_0 : \omega = 0$,

$$\mathrm{LM} = \frac{\left(\sum_{i=1}^{n}(\delta(y_i = 0) - e^{-\hat{\lambda}_i})/e^{-\hat{\lambda}_i}\right)^2}{\left(\sum_{i=1}^{n}(1 - e^{-\hat{\lambda}_i})/e^{-\hat{\lambda}_i}\right) - n\bar{y}} \tag{3.66}$$

where δ takes the value 1 if the condition is true and 0 otherwise, has a chi-squared distribution with 1 degree of freedom.

Information Matrix Test

The information matrix test (White, 1982) is based on the sample analogue of the identity

$$\mathrm{E}\left(\frac{\partial \ell}{\partial \theta}\frac{\partial \ell}{\partial \theta'}\right) = -\mathrm{E}\left(\frac{\partial^2 \ell}{\partial \theta \partial \theta'}\right)$$

which can be rewritten as

$$\mathrm{E}\left(\frac{\partial \ell}{\partial \theta}\frac{\partial \ell}{\partial \theta'} + \frac{\partial^2 \ell}{\partial \theta \partial \theta'}\right) = 0$$

In the context of count data models with unobserved heterogeneity, it has been noted by Lee (1986) that a result originally due to Chesher (1984) on the potential equality between information matrix tests and score/Lagrange multiplier tests applies to the Poisson regression model as well. From Chap. 3.2.1, the sample analogues for the Hessian and the variance of the outer product of the score vector of the Poisson regression model are given by the two expressions $\sum_{i=1}^n \hat{\lambda}_i x_i x_i'$ and $\sum_{i=1}^n (y_i - \hat{\lambda}_i)^2 x_i x_i'$, respectively. The difference between the two matrices depends on elements $(y_i - \hat{\lambda}_i)^2 - \hat{\lambda}_i$. In particular, the sample matrix difference is, up to a factor of 1/2, equal to the sample score evaluated at the restricted value (3.65) if the information test precedure is applied to the intercept parameter β_0. This follows, since $\sum_{i=1}^n (y_i - \hat{\lambda}_i)^2 - \hat{\lambda}_i = \sum_{i=1}^n (y_i - \hat{\lambda}_i)^2 - y_i$. Hence the interpretation of the information matrix test as a test for neglected heterogeneity (See also Moffatt, 1997a).

3.4.2 Regression Based Tests

In the same way as a residual analysis in the linear model with normally distributed errors can reveal heteroskedasticity, the Poisson residuals may indicate a violation of equidispersion. The analysis may proceed either graphically, or through auxiliary regressions. Define $\hat{\mathrm{E}}(y_i) = \hat{\lambda}_i$ and $\widehat{\mathrm{Var}}(y_i) = (y_i - \hat{\lambda}_i)^2 = \hat{u}_i^2$. Plotting $\hat{\mathrm{E}}(y_i)$ against $\widehat{\mathrm{Var}}(y_i)$ should produce points scattered around the 45° line. Alternatively the regression (See Cameron and Trivedi 1986, 1990)

$$\widehat{\mathrm{Var}}(y_i) = \theta \hat{\mathrm{E}}(y_i) + \nu_i \tag{3.67}$$

should yield an estimate $\hat{\theta}$ close to 1. The regression

$$\frac{\widehat{\mathrm{Var}}(y_i)}{\hat{\mathrm{E}}(y_i)} = \theta_1 + \theta_2 \hat{\mathrm{E}}(y_i) + \nu_i \tag{3.68}$$

should yield an $\hat{\theta}_1$ close to 1 and a $\hat{\theta}_2$ close to 0.

3.4.3 Goodness-of-Fit Tests

In contrast to continuous modeling, discrete data allow to calculate probabilities of single outcomes after the model has been estimated. In the domain of binary variables, this fact has been recognized for a long time, and a comparison between actual and predicted outcomes is usually contained in the available statistical software. Prediction tables have been criticized for being uninformative, since the fitted model can be outperformed by a naive model predicting all outcomes to be equal to the most frequent outcome in the sample (See, for instance, Veall and Zimmermann, 1992). For count data models, however, the situation is more favorable, although most of the applied literature has ignored the possibility of using the predictions to evaluate the goodness-of-fit. Notable exceptions are Gilbert (1979) and Dionne and Vanasse (1992). See also Alvarez and Delgado (2002).

Gilbert (1979) considers the Poisson model and measures the goodness-of-fit by the proportion of correct in-sample predictions. He suggests to predict the count for individual i (with given attributes x_i) either by its modal value, or the integer nearest to its expected value. This procedure thus basically rests on the goodness-of-fit of the mean function.

A related procedure is based on the Pearson statistic

$$P = \sum_{i=1}^{n} \frac{(y_i - \hat{\lambda}_i)^2}{\hat{\lambda}_i} \tag{3.69}$$

If the Poisson model is correctly specified, $\mathrm{E}[(y_i - \lambda_i)^2/\lambda_i] = 1$, and thus $\mathrm{E}[\sum_{i=1}^{n}(y_i - \lambda_i)^2/\lambda_i] = n$. In practice, P is compared to $(n-k)$ in order to adjust for lost degrees of freedom due to estimation of λ_i. $P \neq n-k$ indicates a misspecification of the conditional mean or the distributional assumption.

An alternative goodness-of-fit statistic is the deviance

$$D = \sum_{i-1}^{n} \left\{ y_i \ln\left(\frac{y_i}{\hat{\lambda}_i}\right) - (y_i - \hat{\lambda}_i) \right\} \tag{3.70}$$

(see McCullagh and Nelder, 1989). For the exponential Poisson model with intercept included, the sum over the second term on the right is zero, so that we can write

$$D = \sum_{i-1}^{n} y_i \ln\left(\frac{y_i}{\hat{\lambda}_i}\right)$$

(where $y_i \ln y_i = 0$ for $y_i = 0$). The deviance measures the difference between the maximum log-likelihood achievable and the log-likelihood achieved by the model under scrutiny. The deviance has an approximate chi-squared distribution with $n-k$ degrees of freedom.

Cameron and Windmeijer (1996) discuss the use of pseudo R-squared measures for determining goodness-of-fit within classes of count data regression models. They list desirable properties of pseudo R-squared measures:

1. $0 \leq R^2 \leq 1$.
2. R^2 does not decrease as regressors are added.
3. R^2 based on residual sum of squares coincides with R^2 based on explained sum of squares.
4. There is a correspondence between R^2 and a significance test on all slope parameters or on incremental slope parameters.

The preferred measure identified by Cameron and Windmeijer is one based on deviance residuals. For the Poisson regression model, it is given by

$$R^2 = \frac{\sum_{i=1}^{n} y_i \log(\hat{\lambda}_i/\bar{y})}{\sum_{i=1}^{n} y_i \log(y_i/\bar{y})} \tag{3.71}$$

Cameron and Windmeijer derive similar pseudo R-squared measures for the negative binomial regression model.

A final descriptive goodness-of-fit measure is used by Dionne and Vanasse (1992). They suggest to sum the individual predicted probabilities over all possible outcomes $0, 1, \ldots$, where for practical calculations some cutoff value has to be chosen (for instance, the maximum count observed in the sample, $m = \max_i(y_i)$). The summed probabilities for a specific outcome j

$$\hat{p}_j = \frac{1}{n}\sum_{i=1}^{n} \frac{e^{-\hat{\lambda}_i}\hat{\lambda}_i^j}{j!} \qquad j = 0, 1, \ldots, m$$

can then be compared to the relative frequencies in the sample $\bar{p}_j$. Discrepancies indicate a poor performance of the model. A comparison of two models can be based on the proximity between average predicted probabilities and relative frequencies in the sample. Andrews (1988) shows how these fitted distributions can be subject to a formal chi-squared test.

3.4.4 Tests for Non-Nested Models

Standard statistical theory provides a wide range of tools for the testing of hypotheses which come in the form of parametric restrictions. A restriction transforms a general model G into a restricted model F. We say that F is nested in G. The restricted model can never be "better" than the general model, measured in terms of likelihood or coefficient of determination. Examples for two nested count data models are the Poisson and the negative binomial model, or the Poisson and the Poisson-log-normal model.

In count data modelling, one is often confronted with two models that are not nested in the above sense (for a general discussion of tests for non-nested models, see Pesaran, 1974). Two cases can be distinguished. In the first case, two models are *partially nested* or *overlapping* (Vuong, 1989). In this case neither of the two can be derived from the other through a parametric restriction on either model, and at the same time the models are identical for some joint restrictions.

This case is indeed very common for count data. For example, any two generalizations of the Poisson regression model (i.e., models that can be transformed to a Poisson model through a suitable parametric restriction) are pairwise overlapping. Since each of them can by definition be restricted to a Poisson model, they are equivalent if this restriction is imposed. Examples are the negative binomial model, the Poisson-log-normal model and the generalized Poisson distribution model. Similarly, hurdle Poisson and negative binomial models overlap with zero inflated Poisson and negative binomial models. The equivalence of hurdle and zero inflated models in the case where there are constants only (i.e., all slope coefficients are restricted to zero) has been noted already by Mullahy (1986). A final example for two overlapping models is the Poisson model with log-linear mean function $\exp(x_i'\beta)$ and the Poisson model with linear mean function $x_i'\gamma$. Again, the two models are clearly distinct in the presence of genuine regressors. Yet, when slopes are zero and the constants obey the restriction $\exp(\beta_0) = \gamma_0$, the two models are the same.

Second, two models can be strictly non-nested. In this case, no set of restrictions is available that would render the two specifications formally equivalent. It is relatively hard to find an example for two strictly non-nested count data models. The only established example is the zero-inflated Poisson model with logit (or probit) parameterization of the probability of an excess zero, and the standard Poisson model. Here, the problem arises that the specific parameterization precludes that the extra probability takes the value 0 for any finite value of the model parameters. Hence, the two models cannot be equivalent.

There are two different ways of looking at models that are non-nested (strictly or overlapping). One is *hypothesis testing* and one is *model selection*. An hypothesis test addresses the issue whether the true conditional density $f_0(y|x)$ belongs to F or to G. By its very nature, it introduces an asymmetry between the null hypothesis and the alternative. To treat the models symmetrically, both models are considered consecutively under the null hypothesis. Combining the two tests, four outcomes are possible:

1. $H_0 = F$ is accepted and $H_0 = G$ is rejected.
2. $H_0 = F$ is rejected and $H_0 = G$ is accepted.
3. $H_0 = F$ is rejected and $H_0 = G$ is rejected.
4. $H_0 = F$ is accepted and $H_0 = G$ is accepted.

In situations (1) and (2), a coherent decision can be made. (1) leads to a decision in favor of F and (2) to a decision in favor of G. For (3) and (4), the results are conflicting. In (3) both models are rejected and in (4) the evidence cannot discriminate between the two models. For this reason an alternative approach, the *model selection* approach, considers situations where a decision in favor of one model has to be made. Also, model selection criteria are more suitable in situations where more than two models are considered. There, the hypothesis testing framework provides little guidance how to proceed.

The focus of this section, however, will be on hypothesis testing for two non-nested models. The existing results can be roughly divided into three approaches. The first generalizes the classical asymptotic tests, the second uses Monte Carlo simulations, and the third approach specifies and estimates a *hyper*-model.

Vuong Test

The extension of the likelihood ratio tests to situations of not nested models uses results on pseudo-true values minimizing the Kullback distance to the true conditional law to establish the asymptotic distribution of the test statistic. The corresponding results for the Wald and the Lagrange multiplier tests can be found in Gourieroux and Monfort (1989, Chap. 12). The likelihood ratio approach originates in Cox (1961) and has been extended by Vuong (1989). It is based on the observed difference

$$d_{\text{obs}} = \ell_f(\hat{\alpha}) - \ell_g(\hat{\beta}) \tag{3.72}$$

between the log-likelihood under model F and G evaluated at the maximum, respectively. Note, that in contrast to nested models, this difference can be either negative or positive. The distribution of d_{obs} under any of the two models is, however, unknown. Cox (1961) derived a modified test statistic for a test of $H_0 = F_\alpha$ against $H_1 = G_\beta$:

$$T_f = \{\ell_f(\hat{\alpha}) - \ell_g(\hat{\beta})\} - \mathrm{E}_{\hat{\alpha}}[\ell_f(\hat{\alpha}) - \ell_g(\hat{\beta})] \tag{3.73}$$

where $\mathrm{E}_{\hat{\alpha}}[\cdot]$ is the expectation with respect to $F_{\hat{\alpha}}$. Due to the difficulty in calculating this expectation, the use of the Cox test is unrealistic in many practical situations.

Vuong (1989) has developed a considerably simpler test. His test statistic for non-nested models is

$$\mathrm{LR}_{\mathrm{NN}} = \frac{1}{\sqrt{n}}(\ell_f(\hat{\alpha}) - \ell_g(\hat{\beta}))/\omega \tag{3.74}$$

where

$$\begin{aligned}\omega^2 &= \frac{1}{n}\sum_{i=1}^{n}(\ell_f(y_i|x_i,\hat{\alpha}) - \ell_g(y_i|x_i,\hat{\beta}))^2 \\ &\quad - \left(\frac{1}{n}\sum_{i=1}^{n}\ell_f(y_i|x_i,\hat{\alpha}) - \ell_g(y_i|x_i,\hat{\beta})\right)^2\end{aligned}$$

This test statistic is thus very simple to compute. All one needs to do is to divide the average difference of the log-likelihood functions (or the difference of the average), evaluated at the respective maximum likelihood estimates, by the estimated standard error of the average difference, which is just the standard deviation of the individual differences in log-likelihood divided by the square root of n.

None of the two models needs to be true. The test is aimed at selecting the model that is closest to the true conditional distribution. The null hypothesis is that the two models are equivalent:

$$H_0 : \mathrm{E}_0[\ell_f(\hat{\alpha}) - \ell_g(\hat{\beta})] = 0$$

Under the null hypothesis, the $\mathrm{LR_{NN}}$ statistic converges in distribution to a standard normal distribution. To implement the test, let c denote the critical value for some significance level. If the statistic exceeds c, one rejects the null hypothesis (of equivalence) in favor of model f being better than model g. If the statistic is smaller that $-c$, one rejects the null hypothesis (of equivalence) in favor of model g being better than model f. Finally, if $|\mathrm{LR_{NN}}| < c$ the null hypothesis is not rejected and one cannot discriminate between the two models given the data. In this case, both models are rejected. The possibility of rejecting both models is a conceptual weakness of this test, in which neither of the two models is true under H_0.

If the two models are overlapping – the standard situation in count data applications – the above test needs to be slightly modified. In particular, rather than applying it directly, a pre-test is required. In this pre-test, one needs to establish that the two models are not equivalent. In practice it is sufficient to separately perform t- or F-tests to see whether the parameter vectors violate the overlapping constraint, and the overlapping constraint can thus be rejected (see Vuong, 1989, particularly footnote 6).

For example, when testing the negative binomial model against the Poisson-log-normal model, this would amount to testing whether the two dispersion parameters are individually significantly greater than zero. When testing the hurdle Poisson model against the zero inflated Poisson model, it is sufficient that the slope vectors in both models are significantly different from zero, so that one rejects the null hypothesis of two constant index functions, the requirement for the overlap (Mullahy, 1986).

It should be noted that, in practice, many researchers ignore this issue, assuming right away that the overlapping models are not equivalent, and conducting the (nonnested) likelihood ratio test directly without pretesting.

Simulation-Based Tests

A second approach for testing non-nested hypotheses is simulation-based. This test has been put forward by Williams (1970). Essentially, a large number of data sets is generated under each of the two models. The models are re-estimated and a likelihood ratio is calculated. Denote by d_f (d_g) the distribution of the (log of the) likelihood ratio under F (G). Comparing the observed likelihood ratio with d_f (d_g) then provides evidence in favor of F_α or in favor of G_β. The following four steps can be distinguished:

- Obtain estimates $\hat{\alpha}$ and $\hat{\beta}$ and calculate the observed log-likelihood difference d_{obs}

- Simulate R sets of endogenous variables y_r under F_α with $\alpha = \hat{\alpha}$ and identical x. Then re-estimate each conditional model with y_r $i = 1, \dots R$ and x to obtain $\hat{\alpha}_{fr}$ and $\hat{\beta}_{fr}$ and calculate $d_{fr} = \ell_f(\hat{\alpha}_{fr}) - \ell_g(\hat{\beta}_{fr})$.
- Simulate R sets of endogenous variables y_r under G_β with $\beta = \hat{\beta}$ and identical x. Then re-estimate each conditional model with y_r $i = 1, \dots R$ and x to obtain $\hat{\alpha}_{gr}$ and $\hat{\beta}_{gr}$ and calculate $d_{gr} = \ell_f(\hat{\alpha}_{gr}) - \ell_g(\hat{\beta}_{gr})$.
- Compare the value d_{obs} with the empirical distribution of d_{fr} and of d_{gr} to provide evidence whether the observed log-likelihood difference is more compatible with model F or with model G.

As a result, none of the simulations may generate values close to the observed ones, the simulations may support a particular model, or they may not be able to discriminate between the two models.

The question arises, how the y_r can be simulated. Williams (1970) proposed *parametric simulation*. In case of the Poisson models with different mean functions, call them λ_f and λ_g, this amounts to repeated draws from Poisson distributions with means fixed at $\bar{\lambda}_f = \lambda_f(\hat{\alpha})$ $\bar{\lambda}_g = \lambda_g(\hat{\beta})$, respectively. Alternatively, Hinde (1992) suggests a non-parametric bootstrap simulation, i.e., re-sampling from the observed residuals.

Artificial Nesting

A third method for testing non-nested models is the construction of a hyper-model (See for instance Gourieroux and Monfort, 1989, Chap. 12). Ingeneral, hyper-models contain an additional parameter, and a test between two models comes down to a test of a restriction on the hyper-parameter. One important example in the count data literature, the Negbin$_k$ model, is presented in Chap. 4.2.2. It can be used to discriminate between the two common specifications of the negative binomial model, Negbin I and Negbin II. The general remarks about the coherence of the test for non-nested hypotheses apply here as well: The test may produce a conclusive result, or the evidence may be inconclusive, either rejecting both or none of the models.

Hausman Test

In certain situations, two non-nested models can be tested by way of a Hausman test (Hausman, 1978). The underlying test idea is quite general and not restricted to maximum likelihood estimation, nor to count data models. Assume that under the null hypothesis, model 1 gives an estimator that is both consistent and efficient. The alternative model 2 gives an estimator that is consistent but inefficient. Further, assume that under the alternative, model 1 is inconsistent but model 2 remains consistent. The Hausman test is based on the distance

$$HT = (\hat{\theta}_1 - \hat{\theta}_2)'[\text{Var}(\hat{\theta}_1 - \hat{\theta}_2)]^{-1}(\hat{\theta}_1 - \hat{\theta}_2) \tag{3.75}$$

Under H_0, both estimators are consistent. Hence, $\hat{\theta}_1$ and $\hat{\theta}_2$ should be similar. Under the alternative, $\hat{\theta}_1$ is inconsistent. Thus, "large" values of $\hat{\theta}_1 - \hat{\theta}_2$ lead to a rejection of H_0. The computation of the test statistic requires estimation of $\text{Var}(\hat{\theta}_1 - \hat{\theta}_2)$. Since $\hat{\theta}_1$ and $\hat{\theta}_2$ use the same data they could be correlated which would cause considerable difficulties.

However, recall that model 1 is by assumption efficient under the null hypothesis. It can be shown (Hausman, 1978) that the asymptotic covariance of an efficient estimator with its difference from an inefficient estimator must be zero:

$$\text{Cov}(\hat{\theta}_1, \hat{\theta}_1 - \hat{\theta}_2) = \text{Var}(\hat{\theta}_1) - \text{Cov}(\hat{\theta}_1, \hat{\theta}_2) = 0$$

and hence $\text{Cov}(\hat{\theta}_1, \hat{\theta}_2) = -\text{Var}(\hat{\theta}_1)$. Thus, under H_0,

$$\begin{aligned} \text{Var}(\hat{\theta}_1 - \hat{\theta}_2) &= \text{Var}(\hat{\theta}_1) - 2\text{Cov}(\hat{\theta}_1, \hat{\theta}_2) + \text{Var}(\hat{\theta}_2) \\ &= \text{Var}(\hat{\theta}_2) - \text{Var}(\hat{\theta}_1) \end{aligned} \tag{3.76}$$

The resulting test statistic has an asymptotic chi-squared distribution with q degrees of freedom, where q is the dimension of $\hat{\theta}_1$.

In count data modeling, an application of this test occurs with panel data, where the random effects Poisson model is tested against the fixed effects Poisson model. Implicitly, one tests the assumption that the individual specific unobserved error component is uncorrelated with the regressors. A related application arises in tests for endogeneity of regressors.

4

Econometric Modeling – Extensions

4.1 Introduction

The previous chapter identified various sources of misspecification. This chapter introduces alternative count data models that overcome the restrictiveness of the Poisson specification. Most of these models can be classified depending on the particular problem they address as models for unobserved heterogeneity or as models for selectivity. Unobserved heterogeneity reflects a lack of observability of independent variables, whereas selectivity reflects limits in the observability of the dependent variable. Hybrid models exist as well. A third category of models addresses endogeneity of the regressors.

One possible organizing principle for the variety of models and estimation procedures, and the one adopted here, is the distinction between models and estimation procedures that address a particular source of misspecification, and models and estimation procedures that address the consequences only. The idea behind this distinction can be best illustrated by making reference to the well known case of heteroskedasticity in the (normal) linear model. Here, one can estimate a covariance matrix based on observed residuals, and proceed by (feasible) generalized least squares. Alternatively, one can stay agnostic with respect to the exact nature of the heteroskedasticity and instead compute a heteroskedasticity-consistent covariance matrix of the ordinary least squares estimator (See, for instance, Greene, 2000).

The first method provides an asymptotically efficient estimator if the nature of heteroskedasticity is correctly defined. However, biased inference may arise if this is not the case. The second method provides an unbiased, though inefficient, estimator and valid inference. There is a trade-off between robustness and efficiency.

The analogy to the linear model seems appropriate since in count data models the various misspecifications in general lead to either over- or underdispersion, which may be interpreted as "heteroskedasticity" of the Poisson regression model. And in fact, two different possibilities for dealing with, say, overdispersion exist: one can use a modified estimator that includes the information on the form of the variance function, or, alternatively, one can proceed with Poisson maximum likelihood estimation as before and compute a modified estimator of the asymptotic covariance matrix. The former approach leads to an estimator with smaller variance compared to the latter, more "robust", approach. The two modeling strategies are compared in Chap. 4.10.

The analogy to the linear model fails in one important aspect. For count data models there is no equivalent to the Gauss-Markov theorem that would allow to establish certain optimality properties without making reference to distributional assumptions for the dependent variable. The particular structure of the linear model allows to derive the Gauss Markov result without assumptions on higher order moments of the error distribution. However, even in the linear model, a distributional assumption, like a normally distributed error term, is required in order to establish asymptotical efficiency among the class of all possible estimators. For normally distributed errors, the variance of the regression coefficients reaches the Cramer-Rao efficiency bound. If the errors were not normally distributed, a different (necessarily) non-linear estimator could improve on efficiency.

For count data models the issue of heteroskedasticity is intrinsically linked to the underlying data generating process. For instance, any model for unobserved heterogeneity that is characterized by a multiplicative error with constant variance (examples are the gamma, log-normal, or inverse Gaussian distributions) has a variance function that is quadratic in the conditional mean. Similarly, a quadratic variance function as embodied in the negative binomial model can reflect occurrence dependence. Hence, a particular type of heteroskedasticity, like a quadratic variance function, is in general compatible with a number of underlying causes. Ignoring more detailed information, that is, information on higher order moments of the conditional distribution, leads to inefficient estimators.

The above discussion suggests three basic approaches, or estimation strategies, in the case where the Poisson model is misspecified. Firstly, one can estimate an alternative parametric model. Secondly, one can identify the consequences of misspecification on the variance function, and incorporate a modified variance function. Thirdly, one can ignore the misspecification altogether for estimation purposes and rather just compute the correct covariance matrix of the Poisson maximum likelihood estimator.

Most of this chapter is devoted to the first option, to estimate a fully parametric alternative model. Only parametric models, if correctly specified, are asymptotically efficient, i.e., parametric estimators have a smaller asymptotic variance than semiparametric estimators. The emphasis on parametric models is supported by two further observations. First, such models can reveal important features of the underlying structural data generating process (beyond conditional mean and variance) which in turn might be of intrinsic interest. Second, only parametric models can be used to provide predictions for the probabilities of single outcomes if the dependent variable takes non-negative integer values. For these reasons, much attention has been devoted to the construction of refined parametric models for count data. Part of these developments have been previously reviewed in Cameron and Trivedi (1986) and in Winkelmann and Zimmermann (1995). In particular in the areas of selectivity and endogeneity, substantial developments have taken place since and these developments are presented in some detail.

4.2 Unobserved Heterogeneity

This chapter discusses mixture models for unobserved heterogeneity in the mean function. In Chap. 3.3.2 the consequences of unobserved heterogeneity were worked out. Unobserved heterogeneity was represented by a modified mean function with multiplicative error:

$$\tilde{\lambda} = \lambda u\,,$$

where $\lambda = \exp(x'\beta)$. By definition u and thus $\tilde{\lambda}$ are unobserved. Assume, however, that it is known that u has density function $g(u)$. As $\tilde{\lambda}$ must be non-negative to qualify as a mean parameter of a count data distribution, we require that the support of $g(u)$ be the positive real numbers. Moreover, by applying the basic change of variable technique where $\tilde{\lambda} = r(u) = \lambda u$ and $u = f(\tilde{\lambda}) = \tilde{\lambda}/\lambda$, we obtain that

$$\begin{aligned} h(\tilde{\lambda}) &= g[f(\tilde{\lambda})] \left| \frac{\mathrm{d}f(\tilde{\lambda})}{\mathrm{d}\tilde{\lambda}} \right| \\ &= g(\tilde{\lambda}/\lambda)/\lambda \end{aligned}$$

It is an arbitrary choice whether unobserved heterogeneity is introduced directly via $\tilde{\lambda}$ or indirectly via u. The latter option is slightly more common. Knowing the marginal distribution $g(u)$, we can express the joint density of y and u as

$$f(y,u) = f(y|u)g(u) \tag{4.1}$$

Finally, the marginal distribution for y is obtained by integrating the joint distribution $f(y,u)$ over u:

$$f(y) = \int_0^\infty f(y,u)\,du = \int_0^\infty f(y,\tilde{\lambda})\,d\tilde{\lambda} \tag{4.2}$$

For instance, if $f(y|u)$ is of the Poisson form, we get that

$$f(y) = \frac{\lambda^y}{y!} \int e^{-\lambda u} u^y g(u)\,du \tag{4.3}$$

Two cases can be distinguished. Firstly, in fully parametric mixture models, a parametric density function $g(u)$ is assumed. For instance, inspection of the first two terms after the integral in (4.3) reveals that, as a function of u, these resemble the core of a gamma distribution. As a consequence, if $g(u)$ is gamma distributed as well, we obtain the conjugate case and the integral has a closed form solution.

The gain of introducing a parametric assumption can be an increase in efficiency. The downside is a potential loss of consistency if the specific parametric assumption is wrong. More robust methods can be obtained by semiparametric methods that approximate the distribution of u, either using Laguerre polynomials and moment generating functions (Gurmu, Rilstone, and Stern, 1998), or discrete approximations (Brännäs and Rosenqvist, 1994). In either case, some efficiency is lost but substantial robustness may be gained from weaker assumptions. Both methods are discussed later on.

A third, from an estimation standpoint entirely different, situation is encountered if repeated measurements are available for the same individual, that is, if panel data are available. Methods to control for unobserved heterogeneity in panel count data models are presented in Chap. 5.2.

4.2.1 Parametric Mixture Models

Different Poisson mixture models can be distinguished depending on the specific assumptions made on the distribution of u. Johnson and Kotz (1969, Chap. 8) discuss a variety of mixing distributions. The choice is determined by the requirement that u be non-negative, and applications in regression analysis so far have been restricted to three distributions: the gamma distribution, the inverse Gaussian distribution, and the log-normal distribution.

Gamma Mixture (Greenwood and Yule (1920))

A random variable u is gamma distributed $\Gamma(\alpha, \beta)$ if the density takes the form

$$g(u; \alpha, \beta) = \frac{\alpha^{\beta}}{\Gamma(\alpha)} u^{\alpha-1} e^{-\beta u} \tag{4.4}$$

Mean and variance are $\mathrm{E}(u) = \alpha/\beta$ and $\mathrm{Var}(u) = \alpha/\beta^2$, respectively. Let $\alpha = \beta$, a reparameterization that reduces the number of free parameters from two to one. Then $\mathrm{E}(u) = 1$ and $\mathrm{Var}(u) = \alpha^{-1}$. Recall that $\tilde{\lambda} = \lambda u$. Applying the change of variable technique, $\tilde{\lambda}$ has a gamma distribution

$$\begin{aligned} h(\tilde{\lambda}; \lambda, \alpha) &= \frac{\alpha^{\alpha}}{\Gamma(\alpha)} \left(\frac{\tilde{\lambda}}{\lambda}\right)^{\alpha-1} e^{-\frac{\tilde{\lambda}\alpha}{\lambda}} \frac{1}{\lambda} \\ &= \frac{(\alpha/\lambda)^{\alpha}}{\Gamma(\alpha)} \tilde{\lambda}^{\alpha-1} e^{-\tilde{\lambda}\frac{\alpha}{\lambda}} \end{aligned} \tag{4.5}$$

with mean λ and variance $\alpha^{-1}\lambda^2$. The gamma distribution is a *scale* family, i.e., it is closed under scale transformations.

As demonstrated in Chap. 2.5.1, integration (4.2) of a Poisson-gamma mixture leads to the negative binomial distribution for y:

$$f(y|\alpha,\lambda) = \frac{\Gamma(\alpha+y)}{\Gamma(\alpha)\Gamma(y+1)}\left(\frac{\alpha}{\lambda+\alpha}\right)^{\alpha}\left(\frac{\lambda}{\lambda+\alpha}\right)^{y} \tag{4.6}$$

with

$$\mathrm{E}(y|\alpha,\lambda) = \lambda$$

and

$$\mathrm{Var}(y|\alpha,\lambda) = \lambda + \alpha^{-1}\lambda^2 \,.$$

The regression model is completed by setting $\lambda_i = \exp(x_i'\beta)$. An alternative common specification uses $\alpha = \sigma^{-2}\lambda_i$, resulting in a negative binomial model with linear variance function. Because of its importance and widespread use in applied work, this model and its various parameterizations are discussed in greater detail below.

Inverse Gaussian Mixture

The Poisson-inverse Gaussian mixture model was discussed, among others, by Dean, Lawless, and Willmot (1989). Early applications tended to focus on the univariate case. A regression parameterization that is directly comparable to the negative binomial regression model was used by Guo and Trivedi (2002) who included this model in a comparative performance study in an application to patent data. As before, the marginal probabilities are obtained through integration

$$f(y) = \int_0^{\infty} f(y|\tilde{\lambda})g(\tilde{\lambda})\,d\tilde{\lambda} \tag{4.7}$$

Let $\tilde{\lambda}$ be *inverse Gaussian* distributed with density

$$g(\tilde{\lambda}) = \sqrt{\frac{\alpha}{2\pi\tilde{\lambda}^3}}\exp\left(-\frac{\alpha(\tilde{\lambda}-\lambda)^2}{2\tilde{\lambda}\lambda^2}\right) \tag{4.8}$$

where $\alpha > 0$, $\tilde{\lambda} > 0$ and $\lambda > 0$. This inverse Gaussian distribution has mean λ and variance λ^3/α. With $\lambda_i = \exp(x_i'\beta)$, the mean and variance of the Poisson-inverse Gaussian distribution are

$$\mathrm{E}(y_i|x_i) = \exp(x_i'\beta)$$

and

$$\mathrm{E}(y_i|x_i) = \mathrm{E}(y_i|x_i) + \alpha^{-1}[\mathrm{E}(y_i|x_i)]^3$$

(see Guo and Trivedi, 2002).

Though the integration (4.7) does not yield a closed form for this choice of g, the probability generating function can be calculated using the methods introduced in Appendix A. Dean, Lawless, and Willmot (1989) note that the probabilities of the mixture distribution can be calculated recursively using a second order difference equation. They also derive analytical first and second derivatives of the count data log-likelihood.

Log-Normal Mixture (Hinde (1982))

First, recall some basic properties of the log-normal distribution: if x has a normal distribution with mean μ and variance σ^2, then $\exp(x)$ has a log-normal distribution with mean equal to $e^{\mu+1/2\sigma^2}$ and variance equal to $e^{2\mu+\sigma^2}(e^{\sigma^2}-1)$. In the present context, we can distinguish between an additive error ε_i in the index function: $\tilde{\lambda} = \exp(x_i'\beta + \varepsilon_i)$ and a multiplicative error $u_i = \exp(\varepsilon_i)$. If the additive error ε_i is normally distributed with mean $\mu = -1/2\sigma^2$ (a simplifying normalization that is inconsequential as long as the model has an overall constant) and variance σ^2, then $\mathrm{E}(u_i) = e^0 = 1$ and $\mathrm{Var}(u_i) = e^{\sigma^2} - 1$.

Putting things together, the probability function of the Poisson-log-normal model can be written as

$$f(y|x,\varepsilon) = \frac{\exp(-\exp(x'\beta+\varepsilon))\exp(x'\beta+\varepsilon)^y}{y!}$$

where $\varepsilon \sim N(0,\sigma^2)$, i.e.,

$$f(\varepsilon) = \frac{1}{\sqrt{2\pi}\sigma} e^{-\frac{1}{2}\left(\frac{\varepsilon}{\sigma}\right)^2}$$

Therefore, the observed data distribution function is obtained after integration

$$f(y|x) = \int_{-\infty}^{\infty} \frac{\exp(-\exp(x'\beta+\varepsilon))\exp(x'\beta+\varepsilon)^y}{y!} \frac{1}{\sqrt{2\pi}\sigma} e^{-\frac{1}{2}\left(\frac{\varepsilon}{\sigma}\right)^2} d\varepsilon \qquad (4.9)$$

No closed form solution of this integral is available for the Poisson-log-normal model. Hinde (1982) shows how maximum likelihood estimates can be obtained using a combination of numerical integration, the EM algorithm and iteratively reweighted least squares. With modern computing power, direct computation by Gauss-Hermite quadrature is quite straightforward, and maximum likelihood estimation in most situations is as fast as estimation of the negative binomial model. Details of the Gauss-Hermite procedure for this model are provided in Appendix B.

Relative to the Poisson model, the three models presented here share the common feature of one additional parameter. This parameter can be specified such as to preserve identical mean and variances for the distribution of u, and thus, for the mixture models for y. The three models have the same variance function as long as $e^{\sigma^2}-1 = \alpha^{-1}$, or, equivalently, $\sigma_u^2 = \ln(1+\alpha) - \ln(\alpha)$, where

α^{-1} is the variance of the multiplicative heterogeneity factor in the inverse gamma and gamma mixture models. The models are thus distinguished by their higher moments. Fig. 4.1 gives three density functions with $\mathrm{E}(u) = 1$ and $\mathrm{Var}(u) = 0.5$.

Fig. 4.1. Probability Density Functions of Gamma, Inverse Gaussian, and Log-Normal Distributions

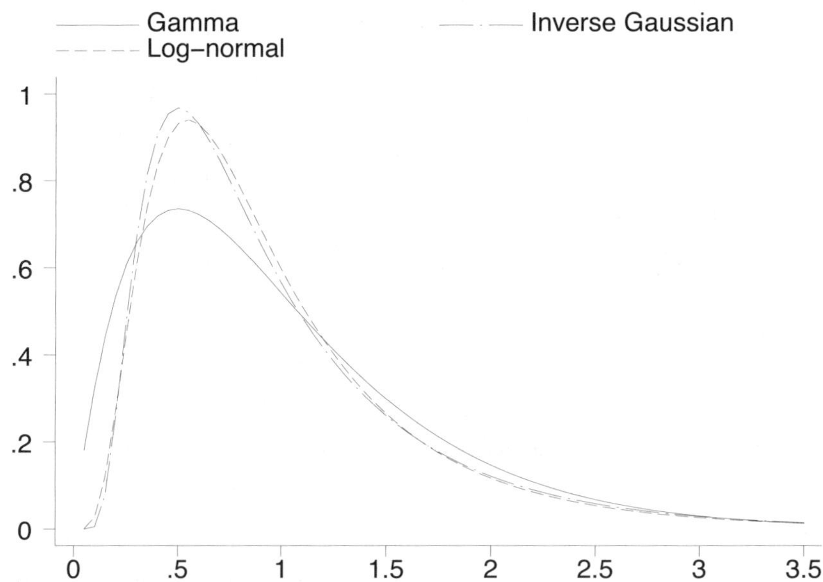

One of the practical advantages of the Poisson-log-normal model is that it is readily extended to the multivariate case (Aitchison and Ho, 1989; Chib and Winkelmann, 2001). Moreover, it has a natural interpretation. Assume that, as in the linear model, the error $\varepsilon_i = \ln u_i$ captures the effect of several additive omitted variables. If there are many omitted factors, and if these factors are independent, then central limit theorems can be invoked in order to establish normality of ε_i. This model is not only appealing from a theoretical perspective; results in the application section of this book show that it fits the data often much better than the negative binomial model. These result strongly suggest that the previous neglect of the Poisson-log-normal model in the literature should be reconsidered in future applied work.

4.2.2 Negative Binomial Models

The negative binomial distribution is the most commonly used alternative to the Poisson model when it is doubtful whether the strict requirements of independence of the underlying process, and inclusion of all relevant regressors, are satisfied. In particular, the negative binomial (Negbin) model is appropriate when the conditional distribution of $y|\tilde{\lambda}$ is Poisson distributed and $\tilde{\lambda}$ is independently gamma distributed. Thus, the Negbin model has the interpretation of a Poisson mixture model that accounts in a specific way for the randomness of the Poisson parameter $\tilde{\lambda}$. Alternatively, the Negbin model arises when the underlying count process is not independent, and when the dependence can be described through a specific type of *true contagion* (See Chap. 2.2.5). Further references on the Negbin model include Cameron and Trivedi (1986), Lawless (1987b) and Hausman, Hall and Griliches (1984).

The probability function of the negative binomial model has been given in (4.6). To make the step to the Negbin regression model, the parameters α and λ are specified in terms of exogenous variables.

The Negbin II model is obtained for $\alpha = \sigma^{-2}$ and $\lambda_i = \exp(x_i'\beta)$. The first two conditional moments for such a parameterization have been derived in Chap. 3.3.1 and Chap. 3.3.9. In particular, the relationship between conditional variance and mean, i.e., the variance function, is given by

$$\mathrm{Var}(y_i|x_i) = \exp(x_i'\beta) + \sigma^2[\exp(x_i'\beta)]^2 \tag{4.10}$$

The conditional variance is always greater than the conditional mean: the negative binomial model is a model for *overdispersion*. The Negbin I model is obtained by letting α_i vary across individuals such that $\alpha_i = \sigma^{-2}\exp(x_i'\beta)$ and $\lambda_i = \exp(x_i'\beta)$. This parameterization produces a variance that is a linear function of the mean:

$$\mathrm{Var}(y_i|x_i) = (1+\sigma^2)\exp(x_i'\beta) \tag{4.11}$$

Another way of characterizing the difference between the Negbin I and Negbin II models is in terms of a dispersion function ϕ_i, such that $\mathrm{Var}(y_i|x_i) = \phi_i \mathrm{E}(y_i|x_i)$. For the Negbin I model, $\phi_i = (1+\sigma^2)$, a constant function, whereas for the Negbin II model, $\phi_i = 1 + \sigma^2\exp(x_i'\beta)$.

Negbin II Log-Likelihood Function

The (conditional) probability function of the Negbin II model can be written as

$$f(y_i|\cdot) = \frac{\Gamma(\sigma^{-2}+y_i)}{\Gamma(\sigma^{-2})\Gamma(y_i+1)}\left(\frac{\sigma^{-2}}{\exp(x_i'\beta)+\sigma^{-2}}\right)^{\sigma^{-2}}\left(\frac{\exp(x_i'\beta)}{\exp(x_i'\beta)+\sigma^{-2}}\right)^{y_i}$$

For $\sigma^2 \to 0$, this model converges to the Poisson regression model (See Chap. 2.3.1). Since $\sigma^2 \geq 0$ the Poisson model is obtained at the boundary of the parameter space. This has to be kept in mind when evaluating the model: a modified likelihood ratio test has to be used to test H_0 : *f is Poisson* against H_1 : *f is negative binomial*. The problem of testing for restrictions at the boundary of the parameter space was discussed in Chap. 3.4.1.

Assuming an independent sample, the log-likelihood function of the Negbin II model is given by

$$\ell(\beta,\sigma^2) = \sum_{i=1}^{n}\left[\left(\sum_{j=1}^{y_i}\ln(\sigma^{-2}+j-1)\right) - \ln y_i! \right.$$
$$\left. -(y_i+\sigma^{-2})\ln(1+\sigma^2\exp(x_i'\beta)) + y_i\ln\sigma^2 + y_i x_i'\beta\right] \tag{4.12}$$

where the ratio of gamma functions in the first line was simplified with the help of equation (2.46). The Negbin II maximum likelihood estimators $\hat{\beta}$ and $\hat{\sigma}^2$ are obtained as solutions to the first-order conditions

$$\sum_{i=1}^{n}\frac{y_i-\exp(x_i'\beta)}{1+\sigma^2\exp(x_i'\beta)}x_i = 0 \tag{4.13}$$

and

$$\sum_{i=1}^{n}\left[\frac{1}{\sigma^4}\left(\ln(1+\sigma^2\exp(x_i'\beta)) - \sum_{j=1}^{y_i}\frac{1}{\sigma^{-2}+j-1}\right)\right.$$
$$\left. -\frac{(y_i+\sigma^{-2})\exp(x_i'\beta)}{1+\sigma^{-2}\exp(x_i'\beta)} + \frac{y_i}{\sigma^2}\right] = 0 \tag{4.14}$$

Moreover, it can be shown (See Lawless, 1987b) that the information matrix is block-diagonal. Therefore, $\hat{\sigma}^2$ and $\hat{\beta}$ are asymptotically independent. The variance of $\hat{\beta}$ is given by

$$\mathrm{Var}(\hat{\beta}) = \left(\sum_{i=1}^{n}\frac{\exp(x_i'\beta)}{1+\sigma^2\exp(x_i'\beta)}x_i x_i'\right)^{-1} \tag{4.15}$$

Negbin I Log-Likelihood Function

The Negbin I model has log-likelihood function

$$\ell(\beta,\sigma^2) = \sum_{i=1}^{n}\left[\left(\sum_{j=1}^{y_i}\ln(\sigma^{-2}\exp(x_i'\beta)+j-1)\right) - \ln y_i! \right.$$
$$\left. -(y_i+\sigma^{-2}\exp(x_i'\beta))\ln(1+\sigma^2) + y_i\ln\sigma^2\right] \tag{4.16}$$

with first-order conditions for $\hat{\beta}$:

$$\sum_{i=1}^{n}\left[\left(\sum_{j=1}^{y_i}\frac{\sigma^{-2}\exp(x_i'\beta)}{\sigma^{-2}\exp(x_i'\beta)+j-1}\right)x_i + \sigma^{-2}\exp(x_i'\beta)x_i\right] = 0 \tag{4.17}$$

In contrast to the Negbin II model, the first-order conditions of the Negin I model are not of the form $\sum(y_i-\mu_i)f(\mu_i)=0$. The Negbin I model does not fall within the class of linear exponential families, and the robustness results derived in Chap. 3.2.2 therefore do not apply in this case. In fact, the Negbin II model is the only model in that family. Relatedly, it is also the only Negbin model with block-diagonal information matrix.

Negbin$_k$ Model

Despite these advantages of the Negbin II model, one might nevertheless wish to embark on a search for alternative estimators that are asymptotically efficient if correctly specified. One such model is the generalized negative binomial model of Winkelmann and Zimmermann (1991, 1995). A similar model has been employed independently by Ruser (1991). This model was re-discovered by Saha and Dong (1997) who apparently were unaware of the previous literature. Let $\alpha_i = \sigma^{-2}\lambda_i^{1-k}$ and $\lambda_i = \exp(x_i'\beta)$. k is a continuous *non-linearity* parameter. Compared to the Poisson model, two additional parameters have to be estimated and this model has been called Negbin$_k$.

The Negbin$_k$ can be interpreted as a hyper-model for the non-nested Negbin I and Negbin II models. In particular, the Negbin$_k$ nests the Negbin II and Negbin I through the parametric restrictions $k = 1$ and $k = 0$, respectively. Thus, a test between the two non-nested sub-models can proceed as described in Chap. 3.4.4. (See Ozuna and Gomez (1995) for a number of other methods to test between the Negbin I and Negbin II models.)

One possible representation of the probability function of the Negbin$_k$ model makes use of the following notation. First,

$$\begin{aligned}\left(\frac{\alpha_i}{\lambda_i+\alpha_i}\right)^{\alpha_i} &= (\alpha_i^{-1}\lambda_i + 1)^{-\alpha_i} \\ &= (\sigma^2\lambda_i^k + 1)^{-\lambda_i^{1-k}/\sigma^2}\end{aligned}$$

Moreover,

$$\begin{aligned}\left(\frac{\lambda_i}{\lambda_i+\alpha_i}\right)^{y_i} &= (1 + \alpha_i\lambda_i^{-1})^{-y_i} \\ &= \prod_{i=1}^{y_i} \frac{1}{1+\sigma^{-2}\lambda_i^{-k}}\end{aligned}$$

Finally, using the recursive property of the gamma function,

$$\frac{\Gamma(\alpha_i + y_i)}{\Gamma(\alpha_i)\Gamma(y_i)} = \prod_{j=1}^{y_i} \frac{\sigma^{-2}\lambda_i^{1-k} + j - 1}{j}$$

Putting everything together, the probability function of the Negbin$_k$ can be expressed as

$$f(y_i|\lambda_i, \sigma^2, k) = C_i \times \begin{cases} \prod_{j=1}^{y_i} \frac{\sigma^{-2}\lambda_i^{1-k}+j-1}{(1+\sigma^{-2}\lambda_i^{-k})j} & \text{for } y_i = 1, 2, \ldots \\ 1 & \text{for } y_i = 0 \end{cases} \tag{4.18}$$

with

$$C_i = (\sigma^2 \lambda_i^k + 1)^{-\lambda_i^{1-k}/\sigma^2}$$
$$\lambda_i = \exp(x_i'\beta)\,, \sigma^2 \geq 0.$$

Given an independent sample of observations, the log-likelihood of the sample is the sum of the logarithm of the probabilities $f(y_i)$.

Negbin$_X$ Model

Yet another parameterization of the negative binomial distribution is proposed by Santos Silva and Windmeijer (2001). Remember from Chap. 2.5.2 that the negative binomial model can be represented as a stopped sum (or compound) distribution, where

$$Y = R_1 + R_2 + \ldots + R_S = \sum_{i=j}^{S} R_j$$

where $S = 0, 1, 2, \ldots$ is Poisson distributed, and the components $R_j = 1, 2, \ldots$ are identically logarithmically distributed, all independently. The logarithmic distribution has a single parameter θ with $0 < \theta < 1$. So it is natural to allow for covariates by letting

$$\theta_i = \frac{\exp(x_i'\gamma)}{1 + \exp(x_i'\gamma)}$$

Accordingly,

$$\frac{\theta_i}{1 - \theta_i} = \exp(x_i'\gamma)$$

From the properties of the logarithmic distribution, it follows that the expected number of counts in each component is

$$\mathrm{E}(R_{ij}) = \frac{\exp(x_i'\gamma)}{\ln[1 + \exp(x_i'\gamma)]}$$

If one lets for the Poisson part in addition

$$\lambda_i = \mathrm{E}(S_i) = \exp(x_i'\beta)$$

as usual, then it follows that Y is negative binomial distributed with parameters $\exp(x_i'\beta)/\ln[1 + \exp(x_i'\gamma)]$ and $\exp(x_i'\gamma)$. Substituting these expressions into the negative binomial probability function (2.37), and after some further simplifications, one obtains the Negbin$_X$ probability function

$$f(y_i|x_i) = \frac{\Gamma\left(y_i + \frac{\exp(x_i'\beta)}{\ln[1+\exp(x_i'\gamma)]}\right) \exp(-\exp(x_i'\beta))}{\Gamma(y_i + 1)\Gamma\left(\frac{\exp(x_i'\beta)}{\ln[1+\exp(x_i'\gamma)]}\right) (1 + \exp(-x_i'\gamma))^{y_i}} \tag{4.19}$$

with

$$\mathrm{E}(y_i|x_i) = \frac{\exp(x_i'\beta + x_i'\gamma)}{\ln[1+\exp(x_i'\gamma)]} .$$

Of course, one can modify the model further by including different sets of regressors z_i and x_i in the different parts of the model. Usually, there will be little a-priori reason to justify such a selection, however, and the model will include two coefficients for each available covariate. The interesting aspect of the model is the interpretation of the underlying data generating process. The overall effect of a regressor on the total number of counts is the sum of two separate effects. First, a variable may affect the number of components S. Second, a variable may affect the number of counts in each component R_j. This separation may have analogies in real life processes. Santos Silva and Windmeier motivate their model by the demand for doctor visits. Here, the total number of visits may depend on the total number of sickness spell over a period plus the number of visits within each spell.

Estimation

A remarkable result, due to Holgate and restated in Guo and Trivedi (2002) is that all continuous mixtures based on the Poisson distibution – this thus includes all three models discussed here, Negbin, Poisson inverse gamma and Poisson log-normal – have unimodal likelihood functions. Hence, applications of standard Newton-Raphson or BFGS algorithms will find the global maximum of the log-likelihood function.

4.2.3 Semiparametric Mixture Models

Gurmu, Rilstone, and Stern (1998) develop a semiparametric estimation approach for overdispersed count regression models based on series expansions for the unknown density of the unobserved heterogeneity component. They notice that while conventional approaches to unobserved heterogeneity impose *ad-hoc* restrictions on the functional form of the mixing distribution whose violation causes the estimator to be inconsistent, quasi-likelihood methods do not use information on higher order moments and hence are inefficient. Furthermore, quasi likelihood methods are in general not applicable if the count data are censored or truncated.

To illustrate the idea behind the semiparametric estimator for the Poisson mixture model, rewrite the marginal probability function (4.3) as

$$f(y) = \frac{\lambda^y}{y!}\mathrm{E}_u\left(e^{-\lambda u}u^y\right) \tag{4.20}$$

where E_u denotes the expectation with respect to the mixing distribution $g(u)$ which is left unspecfied. Recall the definition of a moment generating function

$$M(s) = \int e^{sx}f(x)dx$$

Taking y-th order derivatives with respect to s, we get

$$\begin{aligned} M^{(y)}(s) &= \int e^{sx} x^y f(x) dx \\ &= \mathrm{E}(e^{sx} x^y) \end{aligned}$$

For $s = -\lambda$ and $x = u$, this is precisely the expectation on the right side of (4.20) so that we can write the Poisson-mixture probability function as

$$f(y|x) = \frac{\lambda^y}{y!} \cdot M_u^{(y)}(-\lambda) \tag{4.21}$$

where $M_u^{(y)}$ is the y-th order derivative of the moment generating function of u. The log-likelihood for a sample of n independent observations is

$$\ell = \sum_{i=1}^{n} \left[y_i \ln \lambda_i - \ln y_i! + \ln M_u^{(y)}(-\lambda) \right] \tag{4.22}$$

Gurmu, Rilstone and Stern (1998) approximate $g(u)$ by Laguerre polynomials, derive the corresponding moment generating function, and use this function to estimate β together with additional parameters of the approximation by maximum likelihood, hence effectively avoiding the *a-priori* specification of a density function for the unobserved heterogeneity component. They show that the resulting estimator is consistent.

4.2.4 Finite Mixture Models

An alternative semiparametric approach of modeling unobserved heterogeneity has been popularized in econometrics by Heckman and Singer (1984). Earlier references in the statistical literature are Simar (1976) and Laird (1978). The semiparametric maximum likelihood estimator is based on a finite mixture specification in which the underlying distribution of u (that is, of the intercept) is approximated by a finite number of support points. This is a straightforward application of the mixture concept introduced in Chap. 2.5.1 where

$$F = \sum_{j=1}^{s} a_j F^j \tag{4.23}$$

and $a_j > 0$ and $\sum_j a_j = 1$. a_j gives the probability of belonging to class j. Estimation of the Poisson model with intercept heterogeneity is discussed in Brännäs and Rosenqvist (1994). Maximum likelihood estimation is based on

$$f(y_i) = \sum_{j=1}^{s} a_j \frac{\exp(-\exp(\beta_{0j} + x_i'\beta)) \exp(y_i(\beta_{0j} + x_i'\beta))}{y_i!} \tag{4.24}$$

where, since maximum likelihood estimators are invariant with respect to monotonic transformations, the simplex constraints on a_j may be imposed using the logit transformation

$$a_j = (1 + \exp(-q_j))^{-1}\,,\; j = 1, \ldots, s-1$$

and

$$a_s = 1 - \sum_{j=1}^{s-1} a_j$$

after which the log-likelihood function is maximized with respect to q_j. Maximum likelihood estimation yields a consistent estimator for β as shown in Simar (1976). Brännäs and Rosenqvist present an estimation algorithm due to Lindsay (1983) that accounts for the fact that s is unknown and has to be determined together with a_j and β. The algorithm is initiated with $s = 1$, i.e., $a_1 = 1$. Next, for given s, maximize the likelihood function with respect to the parameters β, a_j and β_{0j} and evaluate the function

$$D(\beta_{0j}) = \sum_{i=1}^{n} \frac{f(y_i|\beta_{0j})}{f(y_i)} - n \tag{4.25}$$

If $D(\beta_{0j}) \leq 0 \,\forall\, \beta_{0j}$ stop the algorithm and report the semiparametric maximum likelihood estimates. Otherwise, set $s = s + 1$ and return to maximization step. After convergence standard errors can be computed free of second derivatives using the BHHH algorithm. An alternative way of determining s is derived in Leroux (1992) who shows that the number of groups may be consistently estimated using the Akaike information criterion.

In a related development, Wedel et al. (1993) show that this semiparametric estimator is readily extended to the case where heterogeneity not only affects the intercept but the regression coefficients as well (See also Wang et al., 1996). The model then takes the form

$$f(y_i) = \sum_{j=1}^{s} a_j \frac{\exp(-\exp(x_i'\beta_j))\exp(y_i x_i'\beta_j)}{y_i!} \tag{4.26}$$

where the intercept is part of x_i, with likelihood function given by

$$L = \prod_{i=1}^{n} f(y_i|\beta_j) = \prod_{i=1}^{n}\sum_{j=1}^{s} a_j \frac{\exp(-\exp(x_i'\beta_j))\exp(y_i x_i'\beta_j)}{y_i!} \tag{4.27}$$

Formally, the model may be characterized as follows: Observations are independent. For observation i, class j is chosen according to a multinomial distribution with probability p_j. Subsequently, y_i is generated from a Poisson distribution with mean λ_{ij}.

As Wedel et al. point out this model has an interesting interpretation in terms of sample segmentation. In particular, the estimated proportions $\hat{a}_1$, $\hat{a}_2$, ... $\hat{a}_s$ give the fraction of the sample that is of type 1, type 2 ... type s, respectively. For a particular individual, the posterior probability of being in a specific segment j, conditional on the observed count y_i, can be computed using Bayes rule as

$$\hat{\theta}_{ij} = \frac{\hat{\alpha}_j f(y_i|\hat{\beta}_j)}{\sum_{k=1}^{s} \hat{\alpha}_k f(y_i|\hat{\beta}_k)} \tag{4.28}$$

In order to classify individuals into particular segments, one plausible rule is to select the segment for which the posterior probability (4.28) is greatest. For instance, an observation with a large count y_i and a large value for a particular x_{ik} will tend to be classified in segment j, where $\hat{\beta}_j$ is positive and large, ceteris paribus. In this finite mixture model, heterogeneity, rather than being a nuisance factor that leads to overdispersion and complicates statistical inference, provides useful information on features of the population.

Wedel at al. (1993) discuss an application in marketing research where the dependent variable is the number of customer purchases of books offered through direct mail. A study of market segmentation shows that different groups of individuals react differently to various marketing techniques (such as mail offers, mail promotions, or sweepstakes) and the ability of identifying market segments provides valuable information for targeted marketing.

More recently, this finite mixture model has been extended in two directions. First, van Duijn and Böckenholt (1995) and Deb and Trivedi (1997) consider a finite mixture negative binomial model. Second, Wang, Cockburn, and Puterman (1998) parameterize the mixture probabilities as a function of further explanatory variables.

4.3 Dependent Count Process

In Chap. 2.6.4 a distribution for a dependent count process, the gamma count distribution was derived. Winkelmann (1995) embeds this distribution in a regression model that nests the Poisson regression through a parametric restriction. In the gamma count distribution,

$$f(y) = G(\alpha y, \beta) - G(\alpha(y+1), \beta) \tag{4.29}$$

where

$$G(\alpha y, \beta) = \frac{1}{\Gamma(y\alpha)} \int_0^{\beta} u^{y\alpha-1} e^{-u} \mathrm{d}u \qquad n = 1, 2, \ldots \tag{4.30}$$

The expected value of this distribution is given by

$$\mathrm{E}(y) = \sum_{i=1}^{\infty} G(\alpha i, \beta) \tag{4.31}$$

Assume that

$$\frac{\beta}{\alpha} = \exp(x_i'\gamma)\,, \tag{4.32}$$

This parameterization yields the regression

$$\mathrm{E}(\tau_i|x_i) = \exp(-x_i'\gamma) \tag{4.33}$$

where τ is the waiting time between events. Hence, the estimated parameters $\hat{\gamma}$ measures minus the percentage change in the expected waiting time caused by a unit increase in x_i. If x_i is in logarithms, $-\gamma$ is a partial elasticity.

To express the estimated coefficients in terms of their effects on the event counts y, and hence make them comparable to those obtained from the negative binomial or Poisson regression models, a simple strategy is to hold all explanatory variables constant at their means, and to compute $\Delta\hat{Y}/\Delta x$ where x is the remaining explanatory variable, and the change is defined by a unit increase at the mean value (in the case of continuous variables), or by a change from 0 to 1 in the case of dummy variables. This ratio literally measures the estimated effect of an explanatory variable on the dependent count variable, holding everything else constant.

4.4 Censoring and Truncation

Models so far have been mostly concerned with the data generating process and unobserved heterogeneity. Unobserved heterogeneity can be interpreted as a lack of observability of *independent* variables. This section deals with limits in observability of the *dependent* variable which may stem from selective sampling or selective recording. The most common forms of sample selection arise from censoring and truncation. In the former case, the count dependent variable is observed only over a limited range, whereas in the latter case, certain observations are omitted entirely from the sample.

A common way to present such models is to distinguish between a latent count y_i^*, an observed count y, and a selection variable c_i. The binary variable c_i may indicate censoring, truncation, or non-reporting. We adopt the convention that truncation, censoring, or non-reporting occurs if $c_i = 0$. Models of "exogenous" censoring or truncation are based on the following mechanism:

$$c_i = \begin{cases} 1 \text{ if } y_i^* \in A \\ 0 \text{ if } y_i^* \notin A \end{cases} \tag{4.34}$$

that is, c_i is uniquely determined through the count dependent variable y_i^*. The two most commonly encountered situations are:

1. A is the set of positive integers (truncation/censoring at zero).
2. A is the set $\{0, \ldots, a\}$ where a is some positive integer (right truncation/censoring).

A common example of censored observations is given in Terza (1985). Here, the dependent variable is obtained from a survey question "How many times have you been to shopping area Q in the past thirty days?," with responses "zero", "one", "two", or "three or more". An example for truncated count data is the number of unemployment spells in the population of unemployed workers.

4.4.1 Truncated Count Data Models

The most common form of truncation is (left) truncation at zero. Truncated Poisson and negative binomial models have been discussed, among others, by Creel and Loomis (1990) and Grogger and Carson (1991). Gurmu (1991) refers to the truncated at zero Poisson model as "positive Poisson regression".

In the Poisson case, the observed data distribution is given by

$$
\begin{aligned}
f(y_i|x_i, c_i = 1) &= \frac{f(y_i^*, c_i = 1|x_i)}{f(c_i = 1|x_i)} \\
&= \frac{\exp(-\lambda_i)\lambda_i^{y_i}}{y_i!(1 - \exp(-\lambda_i))} \qquad y_i = 1, 2, \ldots
\end{aligned}
$$

where, as before, $\lambda_i = \exp(x_i'\beta)$. The truncated negative binomial model can be constructed in a similar way. Mean and variance of the truncated-at-zero Poisson model are given by

$$
\mathrm{E}_{\mathrm{tz}}(y_i|\lambda_i, y_i > 0) = \frac{\lambda_i}{1 - \exp(-\lambda_i)} \tag{4.35}
$$

and

$$
\mathrm{Var}_{\mathrm{tz}}(y_i|\lambda_i, y_i > 0) = E(y_i|\lambda_i, y_i > 0)\left(1 - \frac{\lambda_i}{\exp(\lambda_i) - 1}\right). \tag{4.36}
$$

Since λ_i (the mean of the untruncated distribution) is greater than zero, $0 < \exp(-\lambda_i) < 1$ and the truncated mean is shifted to the right. Moreover, the truncated-at-zero model displays underdispersion since $0 < 1 - \lambda_i(\exp(\lambda_i) - 1) < 1$.

Grogger and Carson (1991) apply the truncated model to the number of recreational fishing trips taken by a sample of Alaskan fishermen. Gurmu (1991) applies it to the Kennan data set on contract strikes.

4.4.2 Endogenous Sampling

A problem of truncated count data models is that they neglect an important consequence of the endogenous sampling scheme. As a result, it is not clear at all whether the parameters have the interpretation that is usually given to them. Grogger and Carson (1991) argue that truncated models can be used in sampling schemes where "inclusion in such a database is predicated on having engaged in the activity of interest" (p. 225). A typical example is a survey of shopping mall visitors, asking them on the number of shopping trips taken . Clearly, in such a case, the minimum response must be a count of one if the time frame of the response includes the present time, and zeros are unobserved. However, as first pointed out by Shaw (1988), this is only part of the story. The inclusion in the sample is based on the outcome of the dependent variable, and more frequent users are more likely part of the sample than less frequent users. Hence, large counts are overrepresented in the sample (relative to the population).

In other words, the distribution of positive counts in the sample cannot be proportional to the distribution of positive counts in the population. But the truncated models precisely postulate proportionality. Hence, the sample estimates do not estimate any interesting population parameter. In fact, the only practical use of truncated count data models is as building block for hurdle models, to be discussed below.

Shaw (1988) derives the likelihood function for an endogenous, or "on-site" sampling scheme. An alternative derivation is given by Santos Silva (1997a) who, in turn, applies results from Manski and Lerman (1977). To provide the intuition why high counts are overrepresented in an on-site sample, consider two individuals, the first with a count of one and the second with a count of two. The sample is taken "on-site" at a random point during the time interval. But then, an individual with a count of two is twice as likely to be included in the sample as an individual with a count of one. If $P(y=1)$ and $P(y=2)$ are the probabilities of outcomes one and two in the population, respectively, then the probabilities of outcomes one and two in the sample are given by $1 \times P(y=1)$ and $2 \times P(y=2)$, respectively. This reasoning can be generalized and, with appropriate normalization, we get

$$\begin{aligned} f^*(y_i|x_i) &= \frac{f(y_i|x_i)i}{\sum_{k=1}^{\infty} f(y_k|x_k)k} \\ &= \frac{f(y_i|x_i)i}{\mathrm{E}(y_i|x_i)} \end{aligned} \tag{4.37}$$

where $f^*(y_i|x_i)$ is the probability function of the sample. From here, it is clear that zeros are not observed and that individuals for which $y_i > \mathrm{E}(y_i|x_i)$ are overrepresented in the sample.

If the expressions for the Poisson probability function are substituted into (4.37) we obtain the on-site sample probability function

$$f_{\mathrm{os}}(y_i|\lambda_i) = \frac{\exp(-\lambda_i)\lambda_i^{y_i-1}}{(y_i-1)!}, \qquad y_i = 1, 2, 3 \ldots \tag{4.38}$$

Interestingly, this is exactly the probability function of a *shifted* or *displaced* Poisson distribution that is obtained by shifting the sample space by one count (Johnson and Kotz, 1969). Expected value and variance are given by

$$\begin{aligned} \mathrm{E}_{\mathrm{os}}(y_i|\lambda_i) &= \sum_{k=0}^{\infty} (k+1)\frac{\exp(-\lambda_i)\lambda_i^k}{k!} \\ &= \lambda_i + 1 \end{aligned} \tag{4.39}$$

and

$$\begin{aligned}\mathrm{Var}_{\mathrm{os}}(y_i|\lambda_i) &= \sum_{k=0}^{\infty}(k+1)^2\frac{\exp(-\lambda_i)\lambda_i^k}{k!} - (\lambda_i+1)^2 \\ &= \lambda_i\,, \end{aligned} \tag{4.40}$$

respectively. For λ_i close to zero, underdispersion is substantial but it vanishes asymptotically for $\lambda_i \to \infty$. Santos Silva (1997a) discusses estimation of the on-site sample Poisson model with unobserved heterogeneity. See also Santos Silva (2003) and the discussion in Chap. 4.6.5.

4.4.3 Censored Count Data Models

Censored count data models have been studied by Terza (1985) and Brännäs (1992b). Unlike for continuous data Tobit models, the type of censoring that is typically encountered in count data models is right-censoring. It arises in survey questionnaires where the highest category is “x or more” counts The standard definition of a censored count data model is then based on the observation mechanism

$$c_i = \begin{cases} 1 \text{ for } y_i^* \in A = \{0,\ldots,a\} \\ 0 \text{ for } y_i^* \in A = \{a+1, a+2, \ldots\} \end{cases} \tag{4.41}$$

Thus, $P(c_i = 1) = F(a)$ where $F = \sum_{j=0}^{a} f(j)$, and $P(c_i = 0) = 1 - F(a)$. The probability function of observed counts y can be written as

$$g(y_i|x_i, c_i) = f(y_i)^{c_i}[1 - F(a)]^{1-c_i} \tag{4.42}$$

and the log-likelihood function for an random sample of size n has now two components

$$\ell(\beta) = \sum_{i=1}^{n} c_i \ln f(y_i) + (1 - c_i)\ln(1 - F(a)) \tag{4.43}$$

The first term on the right gives the likelihood contribution of the non-censored observations, while the second term on the right gives the contribution of the censored observations. Terza (1985) provides details on implementing a Newton-Raphson algorithm for the maximum likelihood estimator $\hat{\beta}$.

Another type of censoring is considered in Caudill and Mixon (1995). They are concerned with estimating the determinants of completed fertility using survey data of women, some of whom are still in their childbearing years, defined to be the age of forty or less. Let y_i^* denote completed fertility, measured by the (final) number of children in the family, and y_i denote current fertility. Then $y_i^* = y_i$ if age ≥ 40 and $y_i^* \geq y_i$ if age < 40. Define a variable

$$c_i = \begin{cases} 1 \text{ if age} \geq 40 \\ 0 \text{ if age} < 40 \end{cases} \tag{4.44}$$

The log-likelihood function can then be written as

$$\ell(\beta) = \sum_{i=1}^{n} c_i \ln f(y_i) + (1 - c_i) \ln(1 - F(y_i - 1)) \tag{4.45}$$

In contrast to the standard censoring model (4.43), the censoring threshold varies now from observation to observation. An interesting modification of this idea was recently introduced by McIntosh (1999). Rather than assuming that childbearing years end for each person at the age of 40, McIntosh derived the censoring status from an additional survey question on the desirability of further children (independently of age). In this case, the selection variable in (4.44) simply needs to be re-defined such that $c_i = 1$ if no further children are desired, and $c_i = 0$ else.

4.4.4 Grouped Poisson Regression Model

Closely related to censoring is the concept of grouping. The grouped Poisson regression model was discussed by Moffatt (1995) who showed that the resulting log-likelihood function is globally concave and the maximum likelihood estimator $\hat{\beta}$ thus unique (See also Moffatt, 1997b, and Moffatt and Peters, 2000). In particular, consider a mutually exclusive and exhaustive partition of $I\!N_0$ into J subsets $A_1, \ldots, A_J$. Also, assume that the subsets are ordered, such that each set consists of an uninterrupted sequence of natural numbers, and one plus the largest number in set A_j is equal to the smallest number in set A_{j+1}. Clearly,

$$P(y_i \in A_j) = \sum_{k \in A_j} f(k) \tag{4.46}$$

where $f(k)$ is the Poisson probability function, and the log-likelihood function for a sample of independent observations can be written as

$$\ell(\beta) = \sum_{i=1}^{n} \sum_{j=1}^{J} d_{ij} \ln P(d_{ij}|x_i) \tag{4.47}$$

where the variable d_{ij} indicates membership of individual i in group j. Maximization of the log-likelihood function can make use of the Newton-Raphson algorithm in the usual way.

4.5 Incidental Censoring and Truncation

So far, all models were based on a single population model. This underlying model at the same time describes population distribution and the censoring or truncation process. Generalized models can be obtained by allowing for separate processes for the counts and for the observation mechanism. Among all possible generalizations, there is a particular structure that has attracted most attention. This structure is characterized by a correlation between the two processes. Thus, the observation mechanism and the count process are not independent. Estimating the outcome equation without taking into account of the observation mechanism would lead to inconsistent parameter estimates.

In order to implement count data models with correlated, or incidental selection (either truncation or censoring), one can adapt the basic structure of sample selection models for continuous data (Heckman, 1979) to the count data situation. Accordingly, the essential elements of count models with incidental (or endogenous) censoring and truncation are:

1. a mean function
$$\mathrm{E}(y_i^*|x_i,\varepsilon_i) = \exp(x_i'\beta + \varepsilon_i) \tag{4.48}$$
with unobserved heterogeneity, that is, omitted regressors, and consequently a conditional mean that varies more than can be explained by the observed covariates.
2. a latent process
$$c_i^* = z_i'\gamma + v_i \tag{4.49}$$
that generates a binary indicator variable c_i where
$$c_i = \begin{cases} 1 \text{ if } c_i^* \geq 0 \\ 0 \text{ if } c_i^* < 0 \end{cases} \tag{4.50}$$
3. a bivariate normal distribution for ε_i and v_i with mean vector zero and covariance matrix
$$\Sigma = \begin{bmatrix} \sigma^2 & \sigma\rho \\ \sigma\rho & 1 \end{bmatrix} \tag{4.51}$$
where ρ is the coefficient of correlation and σ^2 the variance of ε_i.

The main pay-off to this generalized framework is that correlation between y_i^* and c_i can be modeled through the joint distribution for $f(\varepsilon_i, v_i)$ (see Terza, 1998, Winkelmann, 1998). The variance of v_i is normalized to one, since the selection equation (4.50) identifies γ only up to a scale factor. This model defines a conditional expectation function $\mathrm{E}(y_i^*|x_i,u_i) = \exp(x_i'\beta)u_i$, where u_i has a log-normal distribution with mean $\exp(0.5\sigma^2)$ and variance $\exp(2\sigma^2) - \exp(\sigma^2)$. Int his model, u_i and x_i are correlated as long as x_i and z_i are correlated. A sufficient condition for such a correlation would be that x_i and z_i shaer in common some explanatory variables. In this case, direct estimation of (4.48) using observed data would yield an inconsistent estimator.

4.5.1 Bivariate Normal Distribution

The following derivations make use of results for conditional distributions involving two jointly normally distributed random variables (here: v_i and ε_i). In particular, we are interested in the following two conditional distributions

a) $f(v_i|\varepsilon_i)$
b) $f(\varepsilon_i|v_i > -z_i'\gamma) = f(\varepsilon_i|c_i = 1)$

The general result related to a) is that if x_1 and x_2 are bivariate normal with means μ_1 and μ_2, variances σ_1^2 and σ_2^2, and correlation ρ, then

$$x_2|x_1 \sim N\left[\mu_2 + \rho\sigma_2\left(\frac{x_1-\mu_1}{\sigma_1}\right), \sigma_2^2(1-\rho^2)\right] \tag{4.52}$$

Thus, under the assumptions for v_i and ε_i, it follows that

$$v_i|\varepsilon_i \sim N\left[\rho\varepsilon_i/\sigma, 1-\rho^2\right] \tag{4.53}$$

and we can write in explicit form

$$f(v_i|\varepsilon_i) = \phi\left(\frac{v_i - \rho\varepsilon_i/\sigma}{\sqrt{1-\rho^2}}\right) \tag{4.54}$$

The required distribution for b) can be written in general terms as

$$f(\varepsilon_i|v_i > -z_i'\gamma) = \int_{-z_i'\gamma}^{\infty} \frac{f(\varepsilon_i, v_i)}{1-\Phi(-z_i'\gamma)} dv_i$$

No closed form solution is available. To derive the conditional expectation, note first that in the univariate normal case

$$\mathrm{E}(v_i|v_i > -z_i'\gamma) = \frac{\phi(-z_i'\gamma)}{1-\Phi(-z_i'\gamma)}$$

(see Maddala, 1983). Moreover, from (4.52), $\mathrm{E}(\varepsilon_i|v_i) = \rho\sigma v_i$. Thus,

$$\begin{aligned}
\mathrm{E}(\varepsilon_i|v_i > -z_i'\gamma) &= \rho\sigma\mathrm{E}(v_i|v_i > -z_i'\gamma) \\
&= \rho\sigma\frac{\phi(-z_i'\gamma)}{1-\Phi(-z_i'\gamma)} \\
&= \rho\sigma\frac{\phi(z_i'\gamma)}{\Phi(z_i'\gamma)}
\end{aligned} \tag{4.55}$$

Similarly

$$\mathrm{E}(\varepsilon_i|v_i < -z_i'\gamma) = -\rho\sigma\frac{\phi(z_i'\gamma)}{1-\Phi(z_i'\gamma)} \tag{4.56}$$

The Poisson-bivariate-normal set-up allows for incidental selection in the sense that unobserved factors affecting c_i also affect y_i^*. Ignoring this correlation will lead to a misspecified model with the possibility of inconsistent parameter estimates. The effects of selectivity in count data models are similar to those found for continuous data (See e.g. Heckman, 1979), and the corresponding models will be presented in the next sections. Existing models for incidental truncation and censoring are presented here. Other models with similar structure (endogenous switching and endogenous reporting) follow in later part of the book.

Non-Normal Errors

The assumption of a bivariate normal distribution for ε_i and v_i can be relaxed (Lee, 1983). Weiss (1999) considers a class of models where $\exp(\varepsilon_i)$ has a gamma distribution so that the marginal distribution of $y_i|x_i$ is negative binomial. Consider the transformation

$$h(\varepsilon_i) = \Phi^{-1}[G(\exp \varepsilon_i)]$$

where Φ is the cumulative density function of the standard normal distribution and G is the cumulative density function of the gamma distribution. As a result, $h(\varepsilon_i)$ has a standard normal distribution. Moreover, Weiss (1999) lets

$$v_i = \rho h(\varepsilon_i) + \eta_i$$

where η_i is independently normally distributed with mean zero and variance σ_η^2. Thus, $h(\varepsilon_i)$ and v_i are bivariate normal and the methods to be presented below can be modified accordingly.

Discussion

Before proceeding, it is useful to consider the implications of alternative distributional choices for y_i^* in this class of models. It has been emphasized earlier that models can be specified with either Poisson or negative binomial distribution. By contrast, the negative binomial distribution is no longer suitable in the context of endogenous selectivity. The resulting model suffers from overparameterization. To illustrate this point, consider the Negbin II model with $\mathrm{E}(y_i|x_i, u_i) = \lambda_i u_i$, where $\lambda_i = \exp(x_i'\beta)$, $u_i = \exp(\varepsilon_i)$ and $\mathrm{E}(u_i) = 1$. Conditional on u_i, this model has the standard variance function of the Negbin II model:

$$\mathrm{Var}(y_i|x_i, u_i) = \lambda_i u_i + \alpha(\lambda_i u_i)^2$$

where α is the dispersion parameter. Marginalizing with respect to u_i yields

$$\begin{aligned}\mathrm{E}(y_i|x_i) &= \mathrm{E}_u[\mathrm{E}(y_i|u_i)] \\ &= \lambda_i \end{aligned} \tag{4.57}$$

and

$$\begin{aligned}\mathrm{Var}(y_i|x_i) &= \mathrm{E}_u[\mathrm{Var}(y_i|u_i)] + \mathrm{Var}_u[\mathrm{E}(y_i|u_i)] \\ &= \lambda_i + \lambda_i^2(\sigma_u^2 + \alpha\sigma_u^2 + \alpha) \\ &= \lambda_i + c\lambda_i^2 \end{aligned} \tag{4.58}$$

Let $\sigma_u^2 = (c-\alpha)/(1+\alpha)$ and it is clear that α and σ_u^2 are not identified in the variance function. While higher order moments can solve this problem, this approach is usually deemed unsatisfactory and may cause numerical instability in practice.

4.5.2 Incidental Censoring

A model for incidental censoring was introduced by Crépon and Duguet (1997b). In this model, which is formally equivalent to the class of zero-inflated count data models to be discussed in Chap. 4.7,

$$y_i = \begin{cases} 0 & \text{if } c_i = 1 \\ y_i^* & \text{if } c_i = 0 \end{cases} \tag{4.59}$$

and $c_i = 1$ for $v_i > -z_i'\gamma$. Recall from (4.53) that

$$f(v_i|\varepsilon_i) = \phi\left(\frac{v_i - \rho\varepsilon_i/\sigma}{\sqrt{1-\rho^2}}\right)$$

Thus

$$\begin{aligned} P(c_i = 1|\varepsilon_i) &= f(v_i > -z_i'\gamma|\varepsilon_i) \\ &= 1 - \Phi\left(\frac{-z_i'\gamma - \rho\varepsilon_i/\sigma}{\sqrt{1-\rho^2}}\right) \\ &= \Phi\left(\frac{z_i'\gamma + \rho\varepsilon_i/\sigma}{\sqrt{1-\rho^2}}\right) \\ &\equiv \Phi^*(\varepsilon_i) \end{aligned} \tag{4.60}$$

where Φ is the cumulative density function of the standard normal distribution. Furthermore, for $y_i^*|\varepsilon_i$ ∼Poisson with $\tilde{\lambda}_i = \exp(x_i'\beta + \varepsilon_i)$ the probability function of y_i, conditional on ε_i, is given by

$$f(y_i|\varepsilon_i, x_i, z_i) = \Phi^*(\varepsilon_i)d_i + [1 - \Phi^*(\varepsilon_i)]\frac{\exp(-\tilde{\lambda}_i)y_i^{\tilde{\lambda}_i}}{y_i!} \tag{4.61}$$

where $d_i = 1 - \min\{y_i, 1\}$. This probability function depends on the unobserved ε_i, and the observed data distribution is obtained after integration as

$$f(y_i|x_i, z_i) = \tag{4.62}$$
$$\int_{-\infty}^{\infty}\left\{\Phi^*(\varepsilon_i)d_i + [1 - \Phi^*(\varepsilon_i)]\frac{\exp(-\tilde{\lambda}_i)y_i^{\tilde{\lambda}_i}}{y_i!}\right\} f_\varepsilon(\varepsilon_i)\mathrm{d}\varepsilon_i$$

where f_ε is a normal distribution with mean 0 and variance σ^2. While marginalizing with respect to ε_i does not lead to a closed form solution, Crépon and Duguet suggest a feasible simulation method due to Gourieroux and Monfort (1993). Alternatively, Gauss-Hermite quadrature can be used for an exact evaluation of the integral (See Appendix B). Crépon and Duguet apply their model to a study of R&D productivity, where y_i^* gives the number of discoveries, y_i the number of patents applied for, and $c_i = 1$ if the firm decided to apply for patents in general.

4.5.3 Incidental Truncation

A model for endogenous truncation is proposed in Greene (1994). He considers the following model:

$$y_i = \begin{cases} y_i^* & \text{if } c_i = 1 \\ \text{unobserved} & \text{if } c_i = 0 \end{cases} \tag{4.63}$$

where c_i and y_i^* are determined as above, i.e., $y_i^* \sim \text{Poisson}(\tilde{\lambda}_i)$ with $\tilde{\lambda}_i = \exp(x_i'\beta + \varepsilon_i)$, $c_i = 1$ for $v_i > -z_i'\gamma$, and v_i and ε_i are bivariate normal distributed with correlation ρ.

In the standard model for unobserved heterogeneity, it was routinely assumed that ε_i and x_i are uncorrelated. This assumption is likely to be violated under endogenous truncation. In particular, applying (4.55), we know that

$$\begin{aligned} \text{E}(\varepsilon_i|c_i = 1) &= \rho\sigma\frac{\phi(z_i'\gamma)}{\Phi(z_i'\gamma)} \\ &\equiv \rho\sigma m_i \end{aligned}$$

where m_i is the inverse Mill's ratio. Clearly, $\text{E}(\varepsilon_i|c_i = 1) > 0$ for $\rho > 0$ and $\text{E}(\varepsilon_i|c_i = 1) < 0$ for $\rho < 0$. But endogenous truncation can affect more than the overall constant. Assume that x_i and z_i contain overlapping variables. As a consequence, m_i and x_i are correlated. To establish the sign of the correlation, note that $\text{d}m_i/\text{d}(z_i'\gamma) < 0$. Thus, for $x_{ik} = z_{ik}$, the sign of the correlation is given by $-\text{sign}(\rho)\text{sign}(\beta_k)$. In either case, Poisson regression using the truncated sample will cease to be consistent.

In order to obtain consistent estimators of β and γ we can use the maximum likelihood estimator based on the probability function (Greene, 1997)

$$\begin{aligned} f(y_i, c_i = 1) &= \int_{-\infty}^{\infty} f(y_i, c_i = 1|\varepsilon_i)f_\varepsilon(\varepsilon_i)\text{d}\varepsilon_i \\ &= \int_{-\infty}^{\infty} f(y_i|\varepsilon_i)\Phi^*(\varepsilon_i)f_\varepsilon(\varepsilon_i)\text{d}\varepsilon_i \end{aligned} \tag{4.64}$$

where $f(y_i|\varepsilon_i)$ is a Poisson distribution, Φ^* is defined as in (4.60), and Gauss-Hermite quadrature is required.

Greene (1994, 1997, 1998) discusses an alternative, moment based estimator, that follows the Heckman (1979) two-step procedure for the linear model, where sample selection is interpreted as an omitted variable problem. Greene suggests to estimate a "mean corrected" Poisson regression model, where

$$\lambda_i^* = \exp(x_i'\beta + \tau\hat{m}_i) \tag{4.65}$$

which can be estimated by a two-step procedure: a probit regression provides a consistent estimator $\hat{\gamma}$. The predicted selectivity term $\hat{m}_i = \phi(z_i'\hat{\gamma})/\Phi(z_i'\hat{\gamma})$ is then used as a regressor in a second step Poisson regression. As the Poisson model is member of the linear exponential family, consistency follows.

The functional from of the mean correction (4.65) can be justified as a linear approximation to the true mean (Greene, 1997). The difficulty arises since $E(\exp(\varepsilon_i|c_i = 1))$, the term we are really interested in, is not equal to $\exp(E(\varepsilon_i|c_i = 1))$. What is needed, then, is a result for the mean of a truncated log-normal distribution. Greene (1997), based on Johnson, Kotz and Balakrishnan (1994, p. 241), states that

$$E(\exp(\varepsilon_i)|c_i = 1) = \exp(1/2\sigma^2)\frac{\Phi(z_i'\gamma + \rho\sigma)}{\Phi(z_i'\gamma)} \tag{4.66}$$

Therefore

$$\begin{aligned} E(y_i|c_i = 1) &= \exp(x_i'\beta + 1/2\sigma^2)\frac{\Phi(z_i'\gamma + \rho\sigma)}{\Phi(z_i'\gamma)} \\ &= \exp(x_i'\beta^*)Q(\theta) \end{aligned} \tag{4.67}$$

where β^* is the same as β except that the intercept is shifted by $1/2\sigma^2$, and $\theta = \rho\sigma$. This model could be estimated directly by non-linear least squares. Alternatively, a first order Taylor-series expansion of $\ln Q(\theta)$ around $\theta = 0$ yields

$$\begin{aligned} \ln Q &\approx \rho\sigma\frac{\phi(z_i'\gamma)}{\Phi(z_i'\gamma)} \\ &= \rho\sigma m_i \end{aligned} \tag{4.68}$$

Thus, two-step estimation of a Poisson model with endogenous truncation can be justifiably based on the mean function (4.65). Greene applies his model to a study of the determinants of credit card default, where c_i indicates credit card approval.

4.6 Hurdle Count Data Models

Hurdle count data models were first discussed by Mullahy (1986). See also Creel and Loomis, 1990. Hurdle models allow for a systematic difference in the statistical process governing individuals (observations) below the hurdle and individuals above the hurdle. In particular, a hurdle model combines a dichotomous model for the binary outcome of the count being below or above the hurdle (the selection variable), with a truncated model for outcomes above the hurdle. For this reason, hurdle models sometimes are also referred to as two-part models.

The most widely used hurdle count data model sets the hurdle at zero. Only this model is discussed in this chapter. From a statistical point of view, the hurdle at zero formulation can account for a frequently observed empirical phenomenon, namely "excess zeros", i.e., more zero outcomes than can possibly be explained by a Poisson or negative binomial model. From an economic point of view, an intuitive appeal arises from its interpretation as a two-part decision process, a plausible feature of individual behavior in many situations.

The hurdle model is flexible and allows for both under- and overdispersion in a way to be discussed below. Applications include Pohlmeier and Ulrich

(1995), Arulampalam and Booth (1997), and Booth, Arulampalam and Elias (1997) who estimate hurdle negative binomial models for the determinants of visits to a physician and the incidence of training, respectively. Gurmu and Trivedi (1996) apply a hurdle model to the annual number of recreational boating trips by a family. Wilson (1992) proposed a hurdle Poisson model with endogenously determined position of the hurdle.

For a general formulation, assume that $g_1(0)$ is the probability of a zero outcome, and that $g_2(k)$, $k = 1, 2, \ldots$ is a probability function for positive integers. The probability function of the hurdle-at-zero model is then given by:

$$
\begin{aligned}
f(y = 0) &= g_1(0) \\
f(y = k) &= (1 - g_1(0))g_2(k) \; k = 1, 2, \ldots
\end{aligned}
$$

Mullahy (1986) advocates an approach where both parts of the hurdle model are based on probability functions for nonnegative integers, call them f_1 and f_2, respectively. In terms of the general model above, let $g_1(0) = f_1(0)$ and $g_2(k) = f_2(k)/(1 - f_2(0))$. In the case of g_2, a normalization is required since f_2 has support over the nonnegative integers whereas the support of g_2 must be over the positive integers. Formally, this corresponds to truncation of f_2. However, there is no truncation of the population here. All that is needed is a distribution with positive support, and the second part of a hurdle model can use a displaced distribution, or any distribution with positive support as well.

Under the Mullahy (1986) assumptions, the probability distribution of the hurdle-at-zero model is given by

$$
\begin{aligned}
f(y = 0) &= f_1(0) \\
f(y = k) &= \frac{1 - f_1(0)}{1 - f_2(0)} f_2(k) \\
&= \Phi f_2(k) , \qquad k = 1, 2, \ldots
\end{aligned}
\tag{4.69}
$$

where f_2 is referred to as *parent*-process. The numerator of Φ gives the probability of crossing the hurdle and the denominator is a normalization that accounts for the (purely technical) truncation of f_2. It follows that the hurdle model collapses to the parent model if $f_1 = f_2$ or, equivalently, $\Phi = 1$. The expected value of the hurdle model is given by

$$
\mathrm{E}_h(y) = \Phi \sum_{k=1}^{\infty} k f_2(k) = \Phi \mathrm{E}_2(y)
$$

It differs from the expected value of the parent model by a factor Φ. If the probability of crossing the hurdle is greater than the sum of the probabilities of positive outcomes in the parent model, Φ will exceed 1, thus increasing the expected value of the hurdle model relatively to the expected value of the parent model. Alternatively, if the probability of not crossing the hurdle is greater than the probability of a zero in the parent model – the usual case in an application with excess zeros – Φ will be less than 1, thus decreasing the expected value of the hurdle model relatively to the expected value of the parent model. This model thus provides a new interpretation of excess zeros as being a feature of the mean function rather than a feature of the variance function. The mean function of the hurdle model introduces additional non-linearities relative to the standard model in order to account for the corner solution outcome, much as in other corner solution models, such as for instance the Tobit model.

In addition, the hurdle model leads to a modified variance to mean ratio. The variance is

$$\mathrm{Var}_h(y) = \sum_{k=1}^{\infty} k^2 f_2(k)\Phi - \left[\Phi \sum_{k=1}^{\infty} k f_2(k)\right]^2$$

and the variance-mean ratio can be written as

$$\frac{\mathrm{Var}_h(y)}{\mathrm{E}_h(y)} = \frac{\sum_{k=1}^{\infty} k^2 f_2(k) - \Phi[\sum_{k=1}^{\infty} k f_2(k)]^2}{\sum_{k=1}^{\infty} k f_2(k)} \tag{4.70}$$

For $\Phi = 1$, (4.70) reduces to the variance-mean ratio of the parent model. If f_2 is a Poisson distribution function, this is equidispersion with $\mathrm{Var}(y)/\mathrm{E}(y) = 1$. For f_2 Poisson and $\Phi \neq 1$, (4.69) defines a *hurdle Poisson* model. $0 < \Phi < 1$ yields overdispersion, $1 < \Phi < c$ underdispersion. Mullahy (1986) sets $c = \infty$, but this does not hold in general since there is an upper limit to keep the variance positive. E.g., for the Poisson case

$$\sum_{k=1}^{\infty} k^2 f_2(k) = \lambda_2(\lambda_2 + 1)$$

where λ_2 is the expected value of the (untruncated) parent distribution. Hence

$$\mathrm{Var}_h(y) = \Phi\lambda_2(\lambda_2 + 1) - \Phi^2\lambda_2^2$$

with roots

$$\Phi_1 = 0\,, \qquad \Phi_2 = \frac{\lambda_2 + 1}{\lambda_2}$$

Thus, for the hurdle Poisson model underdispersion is obtained for $1 < \Phi < (\lambda_2 + 1)/\lambda_2$. For $\lambda_2 \to \infty$, underdispersion becomes impossible. This reflects the fact that underdispersion occurs if zeros are less frequent than the parent distribution would predict. The higher the expected value of the Poisson distribution, the lower the predicted probability of zero outcome and the lower the scope for underdispersion.

Double Hurdle Model

Hurdle models can be generalized to include more than a single hurdle. For example, let f_1, f_2, and f_3 be arbitrary probability distribution functions for non-negative integers. A double-hurdle model, or a model with two hurdles, in this example a first one at zero and a second one at one, has probability function

$$
\begin{aligned}
f(0) &= f_1(0) \\
f(1) &= \frac{1-f_1(0)}{1-f_2(0)} f_2(1) \\
f(k) &= \left[1 - f_1(0) - \frac{1-f_1(0)}{1-f_2(0)} f_2(1)\right] \frac{f_3(k)}{1-f_3(0)-f_3(1)}, \quad k = 2, 3, \ldots
\end{aligned}
$$

This is a straightforward generalization of (4.69). No applications of such a model are known so far.

Likelihood Function

The generic likelihood function for the hurdle-at-zero model with independent sampling is given by

$$
L = \prod_{i=1}^{n} f_1(0;\theta_1)^{d_i} [1 - f_1(0;\theta_1)]^{1-d_i} [f_2(y_i;\theta_2)/(1 - f_2(0;\theta_2))]^{1-d_i} \tag{4.71}
$$

or, in logarithmic form,

$$
\begin{aligned}
\ell = \sum_{i=1}^{n} & d_i \ln f_1(0;\theta_1) + (1-d_i)\ln[1 - f_1(0;\theta_1)] \\
& + (1-d_i)\ln[f_2(y_i;\theta_2)/(1 - f_2(0;\theta_2))]
\end{aligned}
$$

where $d_i = 1 - \min\{y_i, 1\}$. The first two terms on the right-hand side refer to the likelihood of the hurdle step, while the third term is the likelihood for positive counts. The log-likelihood of this parameterization is therefore separable, and maximization can be simplified by first maximizing a binary model log-likelihood using all observations, and then separately maximizing the log-likelihood for a truncated variable using the subset of observations for which the counts are positive.

4.6.1 Hurdle Poisson Model

Clearly, the hurdle model can be specified in a variety of ways by choosing different probability distributions f_1 and f_2 and specific parameterizations, like for instance Poisson, geometric, or negative binomial. Mullahy (1986) proposes the use of two Poisson distributions with $\lambda_{1i} = \exp(x_i'\beta_1)$ and $\lambda_{2i} =$

$\exp(x_i'\beta_2)$. This approach is convenient since the standard Poisson model can be tested via the parametric restriction $H_0 : \beta_1 = \beta_2$ using Wald or likelihood ratio test statistics.

The hurdle Poisson model has conditional expectation

$$\mathrm{E}(y_i|x_i) = \frac{(1 - e^{-\lambda_{1i}})}{(1 - e^{-\lambda_{2i}})}\lambda_{2i} \tag{4.72}$$

and variance function

$$\mathrm{Var}(y_i|x_i) = \mathrm{E}(y_i|x_i) + \frac{1 - \Phi_i}{\Phi_i}[\mathrm{E}(y_i|x_i)]^2 \tag{4.73}$$

where $\Phi_i = (1 - e^{-\lambda_{1i}})/(1 - e^{-\lambda_{2i}})$. It closely resembles the variance function (3.57) of the negative binomial model for $k = 1$, with the difference that the coefficient $\sigma_i^2 = (1 - \Phi_i)/\Phi_i$ now varies between individuals. The likelihood function of the hurdle Poisson model is given by

$$L = \prod_{i=1}^{n} \exp(-\exp(x_i'\beta_1))^{d_i}[1 - \exp(-\exp(x_i'\beta_1))]^{1-d_i}$$
$$\times \left[\frac{\exp(-\exp(x_i'\beta_2))\exp(y_i x_i'\beta_2))}{y_i!\exp(-\exp(x_i'\beta_2))}\right]^{1-d_i}$$

where $d_i = 1 - \min\{y_i, 1\}$.

4.6.2 Marginal Effects

Consider the marginal mean effect first. The conditional expectation function of the hurdle model was given in (4.72). Taking first derivatives, we obtain

$$\frac{d\mathrm{E}(y|x)}{dx} = \frac{\lambda_1 e^{-\lambda_1}}{1 - e^{-\lambda_2}}\lambda_2\beta_1 - \frac{1 - e^{-\lambda_1}}{(1 - e^{-\lambda_2})^2}\lambda_2^2 e^{-\lambda_2}\beta_2 + \frac{1 - e^{-\lambda_1}}{1 - e^{-\lambda_2}}\lambda_2\beta_2 \tag{4.74}$$

Note that this partial effect reduces to $\lambda\beta$ for $\beta_1 = \beta_2$, as it should.

It was mentioned before, in the context of the interpretation of parameters of the Poisson regression model, that sometimes the interest may be in marginal probability effects rather than marginal mean effects. In the case of the hurdle Poisson model, we can write

$$\frac{df(k;x)}{dx} = f(k;x)\left[\frac{\lambda_1 e^{-\lambda_1}}{1 - e^{-\lambda_1}}\beta_1 - \frac{\lambda_2 e^{-\lambda_2}}{1 - e^{-\lambda_2}}\beta_2 + (k - \lambda_2)\beta_2\right] \tag{4.75}$$

The single crossing restriction of the Poisson model does not apply here. It is softened by the introduction of a second parameter vector. In fact, one can show that the marginal probability effects of the hurdle Poisson model can switch signs twice. So it is still not perfectly flexible, although more so than the simple Poisson model.

An alternative representation of the effect of a marginal change in an explanatory variable can be obtained by decomposing the conditional mean explicitly into a hurdle part and a part for the positives. In particular, we can always write (this is true for any count data model, not only the hurdle Poisson model)

$$\mathrm{E}(y|x) = P(y > 0|x)\mathrm{E}(y|y > 0, x)$$

Therefore

$$\frac{\partial \mathrm{E}(y|x)}{\partial x} = \frac{\partial P(y > 0, x)}{\partial x}\mathrm{E}(y|y > 0, x) + \frac{\partial \mathrm{E}(y|y > 0, x)}{\partial x}P(y > 0|x)$$

or

$$\begin{aligned}\frac{\partial \mathrm{E}(y|x)/\mathrm{E}(y|x)}{\partial x} &= \frac{\partial P(y > 0, x)/P(y > 0, x)}{\partial x} \\ &+ \frac{\partial \mathrm{E}(y|y > 0, x)/\mathrm{E}(y|y > 0, x)}{\partial x}\end{aligned}$$

As stated, such a decomposition is possible for any count data model. In standard models, however, it remains purely tautological. Both effects $\partial P(y > 0|x)/\partial x$ and $\partial \mathrm{E}(y|y > 0, x)/\partial x$ are functions of the same single index and the same parameter β.

An interesting aspect of hurdle models is that this strict link is severed, since $P(y > 0|x)$ is a function of β_1 and $\mathrm{E}(y|y > 0, x)$ is a function of β_2. In this way, the hurdle model allows the marginal effects to be different in different parts of the distribution, relative to a single index model. For example, a given change in a regressor x can decrease the probability of a zero, but leave the conditional expectation $\mathrm{E}(y|y > 0, x)$ unchanged. The standard Poisson model would not allow for such a distinction to be made.

4.6.3 Hurdle Negative Binomial Model

By far the most popular hurdle model in practice is the hurdle-at-zero negative binomial model (Pohlmeier and Ulrich, 1995). In this case, $f_1 \sim NB(\beta_1, \sigma_1^2)$ and $f_2 \sim NB(\beta_2, \sigma_2^2)$. Estimation of the model can be based on the general likelihood factorization described above. This specification can give rise to an identification problem, first noted by Pohlmeier and Ulrich (1995).

Consider estimation of the hurdle part of the model, i.e., the parameters β_1 and σ_1. This estimation is based on the dichotomous model $f_1(0; \beta_1, \sigma_1^2)$ versus $1 - f_1(0; \beta_1, \sigma_1^2)$. From (4.6), we see that for the generic negative binomial model

$$P(y_i = 0) = \left(\frac{\alpha}{\alpha + \lambda_i}\right)^{\alpha}$$

The Negbin II model is obtained directly by letting $\lambda_i = \exp(x_i'\beta)$. For the Negbin I model, $\alpha = \lambda_i/\sigma$, such that

$$P(y_i = 0) = \left(\frac{1}{1 + \sigma^2}\right)^{\lambda_i/\sigma^2}$$

In the Negbin I model, we can thus write

$$\begin{aligned}\ln P(y_i = 0) &= \frac{\lambda_i}{\sigma^2}\ln(1+\sigma^2)\\ &= \exp(x_i'\beta + \ln\phi)\end{aligned}$$

where $\phi = \ln(1+\sigma^2)/\sigma^2 = f(\sigma^2)$. Hence, σ^2 is not identified, as long as the regression part of the model contains a constant. In the Negbin II model, by contrast, two overdispersion parameters can be estimated, since

$$\ln P(y_i = 0) = -\alpha \ln(1 + \alpha^{-1}\exp(x_i'\beta))$$

and functional form guarantees identification.

4.6.4 Nonnested Hurdle Models

All models considered so far did nest a standard count data model, typically through a restriction of the sort $\theta_1 = \theta_2$. This was achieved since the hurdle process f_1 and the process for the positives, f_2, were based on the same distribution. If one sees the hurdle step as an altogether different process, one can as well use standard models for binary dependent variables at this stage, such as the probit or the logit model. Such models were estimated by Grootendorst (1995). Gurmu (1998) discusses a model in which the hurdle step is parameterized by a generalized (asymmetric) logit model. Winkelmann (2003) combines a probit hurdle model with a truncated-at-zero Poisson-log-normal model. The probability function of this probit-Poisson-log-normal model can be written as

$$P(y_i = 0) = \Phi(x_i'\gamma) \tag{4.76}$$

$$P(y_i = k) = [1 - \Phi(x_i'\gamma)] \times \int_{-\infty}^{\infty} \frac{\exp(-\exp(x_i'\beta+\varepsilon))\exp(x_i'\beta+\varepsilon)^{y_i}}{[1-\exp(-\exp(x_i'\beta+\varepsilon))]y_i!}\frac{1}{\sqrt{2\pi}\sigma}e^{-\frac{1}{2}\left(\frac{\varepsilon}{\sigma}\right)^2}d\varepsilon \tag{4.77}$$

for $k = 1, 2, \ldots$, where Φ denotes the cumulative distribution function of the standard normal distribution. Note that the mixing in the second part of the distribution is done over the truncated Poisson model. As will be shown in the next chapter, this order is preferable over the alternative, to do the mixing first, and then form a truncated version of the resulting Poisson-log-normal model. Also, it is possible to generalize the model further by allowing correlation between the hurdle step and the process for the positives. A probit-Poisson-log-normal model with correlated hurdle is presented in Chap. 4.6.7.

While these hurdle models do not nest the Poisson or negative binomial model, they are also not strictly non-nested with the Poisson or negative binomial model, or among each other. In the terminology of Vuong (1989) these models are overlapping. The reason is that for certain parameter restrictions on both models, the two become equivalent. Take as an example the comparison between the logit-Poisson hurdle model and the simple Poisson model. In the logit-Poisson hurdle model

$$f_1(0; z_i, \gamma) = \frac{\exp(z_i'\gamma)}{1 + \exp(z_i'\gamma)}$$

whereas in the simple Poisson model

$$f_2(0; x_i, \beta) = \exp(-\exp(x_i'\beta))$$

Now, assume that all slope parameters are set to zero and only non-zero intercepts, γ_0 and β_0 say, are left. The two probabilities, and with them the full distributions, are the same as long as as

$$\beta_0 = -\exp(\gamma_0) - \ln(1 - \exp(-\exp(\gamma_0)))$$

By the same technique, one can establish that the probit-Poisson-log-normal model and the Poisson model are overlapping, as are for instance the probit-Poisson-log-normal model and the standard negative binomial model.

In these cases, to discriminate between the two models following the methods proposed by Vuong (1989), one needs to follow the procedure for overlapping models. In general, this requires a pre-test before the usual statistic is computed. However, in practice it is sufficient to establish that the condition for overlap, i.e., the restriction that all slope coefficients are zero, can be rejected in each model (see Vuong 1989, footnote 6).

4.6.5 Unobserved Heterogeneity in Hurdle Models

The standard approach to unobserved heterogeneity in hurdle models has introduced heterogeneity at the level of the parent distribution. Integration takes then place prior to the conversion into a hurdle model, the latter being based on modified densities $\tilde{f}_1$ and $\tilde{f}_2$ where

$$\tilde{f}_1 = \int f_1(y|u) g_1(u) du$$

and

$$\tilde{f}_2 = \int f_2(y|u) g_2(u) du$$

The hurdle negative binomial model is an example for this approach. In this case, f_1 and f_2 are Poisson probability functions, and g_1 and g_2 are gamma density functions.

The truncated probability function for the positives can then be written as

$$P(y|y > 0) = \frac{\tilde{f}_2(y)}{1 - \tilde{f}_2(0)} = \frac{\int f_2(y|u) g_2(u) du}{1 - \int f_2(0|u) g_2(u) du} \tag{4.78}$$

As pointed out by Santos Silva (2003), there is an alternative way of thinking of this problem, namely to define a distribution over the positive integers first (any truncated distribution will fulfill this requirement), and then do the mixing in a second step, over the positive part of the distribution only. In this case,

$$P(y|y>0) = \int \frac{f_2(y|u)}{1-f_2(0|u)}\tilde{g}_2(u)du \tag{4.79}$$

Clearly, these are not the same models, depending on how $g_2(u)$ and $\tilde{g}_2(u)$ are defined. In the regression context, where we consider conditional models, we typically assume independence between unobserved heterogeneity and the explanatory variables x. If we assume, for example, that $g_2(u|x)$ is a gamma density independent of x then (4.78) is a truncated at zero negative binomial distribution. If, however, f_2 is the Poisson distribution and $\tilde{g}_2(u|x)$ is a gamma density independent of x, the resulting probability function (4.79) is not that of a truncated negative binomial distribution. The question then becomes which of the two assumptions, and thus the two models, is more meaningful. Santos Silva (2003) argues that the population of interest is the actual population. In the case of the positive part of the hurdle models, this would favor an approach where the assumption is made that the unobservables in the (truncated) population of interest are independent of the x's, i.e., model (4.79). In this way, one avoids the akward step of needing to compute the integral $\int f_2(0|u)g_2(u)du$ although the zeros are generated by an altogether different process.

While hurdle models based on the negative binomial distribution assume that the unobservables are independent of the covariates in an hypothetical population, the probit-Poisson-log-normal model discussed above rather assumes that unobservables are independent of the covariates in the observed population. From this point of view, this makes it a more meaningful hurdle model.

4.6.6 Finite Mixture Versus Hurdle Models

A problem in health economics – how to model the demand for physician services – has prompted a controversy, whether finite mixture models or hurdle models are more appropriate for such data. The initial advocates of the hurdle model in this context, Pohlmeier und Ulrich (1995), maintained that the hurdle model may describe well the agency problem in the demand for doctor consultations, where the initial contact decision is made by the individual whereas further referrals are influenced by the physician's objectives. Arguably, then, two different parameterizations may be needed to capture this two-part decision process.

Proponents of finite mixture models take a less strict view. According to this view, every individual is a potential user but the population is composed of different types, or classes, of users. If there are two types, for instance, one could label them 'light users' and 'heavy users'. The econometrician does not observe which class an individual belongs to. Finite mixture models therefore are also called "latent class" models. The sample is a mixture of the two groups, and estimation of the group specific parameters and the group proportions is possible.

In a number of applications, both types of models have been estimated in order to determine which of the two better fits the data. For that purpose, one usually compares the hurdle negbin model

$$f(y_i) = f_{NB}(0;\theta_1)^{d_i} \left(\frac{f_{NB}(0;\theta_1)}{f_{NB}(0;\theta_2)} f_{NB}(y_i;\theta_2) \right)^{1-d_i}$$

where $d_i = 1 - \min(y_i, 1)$ as before, with a two component latent class negative binomial model

$$f(y_i) = \alpha f_{NB}(y_i;\theta_1) + (1-\alpha) f_{NB}(y_i;\theta_2)$$

The evidence is mixed. Using various model selection criteria (accounting for the fact that the second model has on additional parameter) Deb and Trivedi (2002) and Winkelmann (2003) find that the finite mixture model is superior, although Winkelmann (2003) shows that this is only the case for the Negbin hurdle models and not necessarily for the hurdle models as a class. Jimenéz-Martin, Labeaga and Martinez-Granado (2002) report instances, where the hurdle model is better. See also Doorslaer, Koolman and Jones (2002). This question still awaits a final answer.

4.6.7 Correlated Hurdle Models

In the spirit of Chap. 4.5, it is straightforward to develop a generalized hurdle count data model where the hurdle process and the process for the positives are correlated. Such correlation may originate, for example, from common but unobserved variables. Such a model was proposed and estimated by Winkelmann (2003).

In that paper, a probit model for the hurdle is combined with a truncated Poisson-log-normal model for the positives. Accordingly, it can be referred to as the *probit-Poisson-log-normal model.* Let z_i be a latent indicator variable such that

$$z_i = x_i'\gamma + v_i$$

and

$$y_i = 0 \text{ iff } z_i \geq 0$$

Moreover, for the positive part of the distribution

$$y_i | y_i > 0 \sim \text{truncated Poisson}(\lambda_i)$$

where

$$\lambda_i = \exp(x_i'\beta + \varepsilon_i)$$

The model is completed by the assumption that v_i and ε_i are bivariate normal distributed with mean 0 and covariance matrix

$$\Sigma = \begin{bmatrix} 1 & \rho\sigma \\ \rho\sigma & \sigma^2 \end{bmatrix}$$

To derive the log-likelihood function, note first that

$$v_i|\varepsilon_i \sim N(\rho\varepsilon_i/\sigma, 1-\rho^2)$$

and

$$\begin{aligned} P(y_i = 0|\varepsilon_i) &= P(v_i \geq -x_i'\gamma|\varepsilon_i) \\ &= \Phi\left(\frac{x_i'\gamma + \rho\varepsilon_i/\sigma}{\sqrt{1-\rho^2}}\right) \\ &= \Phi_i^*(\varepsilon_i) \end{aligned}$$

Thus one obtains, with $d_i = 1 - \min(y_i, 1)$,

$$f(y_i|\varepsilon_i) = \Phi_i^*(\varepsilon_i)^{d_i} \times \left[(1-\Phi_i^*(\varepsilon_i))\frac{\exp(-\lambda_i(\varepsilon_i))(\lambda_i(\varepsilon_i))^{y_i}}{[1-\exp(-\lambda_i(\varepsilon_i))]y_i!}\right]^{1-d_i} \tag{4.80}$$

and

$$f(y_i) = \int_{-\infty}^{\infty} f(y_i|\varepsilon_i)\frac{1}{\sqrt{2\pi}\sigma}e^{-\frac{1}{2}\left(\frac{\varepsilon_i}{\sigma}\right)^2}\mathrm{d}\varepsilon_i \tag{4.81}$$

The parameters β, γ and ρ can be estimated by maximum likelihood, using Gauss-Hermite integration to evaluate the likelihood function.

4.7 Zero-Inflated Count Data Models

Zero-inflated Poisson or negative binomial models (ZIP, ZINB), like their hurdle-at-zero counterparts, address the problem that the data display a higher fraction of zeros, or non-occurrences, than is likely to be compatible with any fitted standard count regression model. The zero inflated model combines a binary variable c_i with a standard count variable y_i^* (with support over the nonnegative integers) such that the observed count y_i is given by

$$y_i = \begin{cases} 0 & \text{if } c_i = 1 \\ y_i^* & \text{if } c_i = 0 \end{cases} \tag{4.82}$$

If the probability that $c_i = 1$ is denoted by ω_i, the probability function of y_i can be written compactly as

$$f(y_i) = \omega_i d_i + (1-\omega_i)g(y_i)\,, \qquad y_i = 0, 1, 2, \ldots \tag{4.83}$$

where $d_i = 1 - \min\{y_i, 1\}$ and $g(y_i)$ is a regular count data probability function such as the Poisson or the negative binomial probability function.

The difference between the zero-inflated model and the hurdle model is that in the latter, there is a single type of zeros whereas in the former one obtains two types of zeros: zero outcomes can either arise from regime 1 ($c_i = 1$) or from regime 2 ($c_i = 0$ and $y_i^* = 0$).

Which of the two models is more appropriate will depend on the particular application. Tests for non-nested hypotheses can be used. These tests need to account for the fact that the two models are overlapping (see Chap. 3.4.4), since the two models are equivalent if the slope coefficients are zero (Mullahy, 1986). The Poisson model is nested in the ZIP model for $\omega = 0$. Note, however, that the two model become non-nested if one adopts the specification $\omega = \exp(\theta)$ (or similar reparameterizations, where $\omega \neq 0$ for all finite parameter values, see next paragraph).

Zero-Inflated Poisson Model

Mullahy (1986) discussed the zero-inflated Poisson model with constant ω_i. Lambert (1992) extended it by specifying a logit model for ω_i in order to capture the influence of covariates on the probability of extra zeros:

$$\omega_i = \frac{\exp(z_i'\gamma)}{1 + \exp(z_i'\gamma)}$$

The log-likelihood function of this model is

$$\ell = \sum_{y_i=0} \ln(\exp(z_i'\gamma) + \exp(-\exp(x_i'\beta)) \\ + \sum_{y_i>0} y_i x_i'\beta - \exp(x_i'\beta) - \ln(y!) - \sum_{i=1}^{n} \ln(1 + \exp(z_i'\gamma)) \tag{4.84}$$

There are a couple of points pertinent to this model. First, strictly speaking, the model allows for zero-deflation as well as zero-inflation. Zero-deflation occurs whenever ω is negative, subject to the constraint that $0 < f(y_i) < 1, \quad y_i = 0, 1, \ldots$. In particular, we require that $\omega > -g(0)/(1 - g(0))$ in order to keep $f(0)$ nonnegative. Of course, if ω is modeled as a logit function in order to introduce covariates, zero-deflation is excluded. Second, the "zero-altered" model has a structurally different mean function than the base model. In particular, if the base model has mean $\exp(x_i'\beta)$, then

$$\mathrm{E}(y_i) = (1 - \omega_i)\exp(x_i'\beta)$$

Hence, this type of extra zeros are important to account for. Direct estimation of the base model in general will lead to an inconsistent estimator if the zero inflated model is the true data generating process. Also note that the extension to a zero inflated negative binomial model is straightforward. For example, with a Negbin II specification, the log-likelihood function is given by

$$
\begin{aligned}
\ell = & \sum_{y_i=0} \ln(\exp(z_i'\gamma) + \alpha(\ln\alpha - \ln(\exp(x_i'\beta) + \alpha)) \\
& + \sum_{y_i>0} \ln(\Gamma(\alpha + y_i)/\Gamma(\alpha) + \alpha\ln(\alpha - \ln(\exp(x_i'\beta)) \\
& + y_i(x_i'\beta - \ln(\exp(x_i'\beta) + \alpha)) - \sum_{i=1}^{n} \ln(1 + \exp(z_i'\gamma))
\end{aligned}
$$

Marginal Effets

Typically, in this model, one would be interested in how $\mathrm{E}(y_i|x_i)$ changes, as x_{ij} increases by one unit. This effect will depend on whether or not z and x are the same, overlap, or are disjoint. In most applications one has little prior knowledge that would suggest to exclude a variable from one part of the model but leave it in the other. In this case $z_{ij} = x_{ij}$, and

$$
\begin{aligned}
\frac{\partial \mathrm{E}(y_i|x_i)}{\partial x_{ij}} & = -\frac{\partial \omega_i}{\partial x_{ij}} \exp(x_i'\beta) + (1 - \omega_i) \frac{\partial \exp(x_i'\beta)}{\partial x_{ij}} \\
& = \frac{\exp(x_i'\beta)}{(1 + \exp(x_i'\gamma))^2} \beta_j
\end{aligned}
$$

As usual, one needs to evaluate these marginal effects at some appropriate value of the explanatory variables.

Alternative Estimation

Santos Silva and Covas (2000) have pointed out that the conditional distribution $f(y_i|y_i > 0)$ does not depend on ω:

$$
\begin{aligned}
f(y_i|y_i > 0) & = \frac{(1 - \omega)g(y_i)}{1 - [\omega + (1 - \omega)g(0)]} \\
& = \frac{g(y_i)}{1 - g(0)} \quad y_i = 1, 2, \ldots
\end{aligned}
$$

Hence, one can use positive observations only and estimate the parameters of g using a truncated-at-zero count data model without the need to specify ω. In fact, this is the same estimator as the second part of the hurdle model. The advantage is that these estimators are robust to the misspecification of ω, and they can serve as a basis for specification tests.

The zero inflated model has become quite popular in the recent applied count data literature. Lambert (1992) developed the model for the occurrence of defects in manufacturing. Economic applications of zero-inflated models are given in Greene (1994) on the frequency of loan defaults, in Grootendorst (1995) on prescription drug utilization, in List (2002) on the number of job interviews secured by a job seeker and in Tomlin (2000) on the empirical connection between exchange rates an the number of foreign direct investment occurrences. Beckmann (2002) uses the ZINB model for the number of apprentices trained by a firm. Gameren and Woittiez (2002) estimate the determinants of the demand for home care by the elderly. Guo and Melkersson (2002) estimate a one-inflated model for the number of visits to a dentist, and Melkersson and Roth (2000) extend the zero inflated model in order to simultaneously account for extra zeros and extra twos in an application to fertility.

Finally, the paper by Crépon and Duguet (1997b) on patents was already discussed in Chap. 4.5.2. This paper can be interpreted as a generalization of the zero-inflated Poisson model, in that the process for the extra zeros (a probit model) and the count process (a Poisson-log-normal model) are correlated, conditional on covariates.

4.8 Underreporting

Consider the following data generating process: Events occur randomly over time according to a Poisson process. In contrast to the standard Poisson model, however, only a subset of events is reported. The number of reported events falls short of the total number of events. "Reporting" can be understood in a very broad sense. The basic distinction is whether events are "successful", or "unsuccessful". Observed counts give the number of successful events.

This model is applicable in many situations: In labor economics it can describe the frequency of job changes in a given period (Winkelmann and Zimmermann, 1993c, Winkelmann, 1996b). This frequency will depend on both the frequency at which outside job offers are received, and the probability that outside offers are accepted. In industrial organization, the model can be used for the number of firms entering an industry in a given period (Berglund and Brännäs, 1995). Here, the base entity is the number of potential entrants who decide to whether or not to enter. Finally, in modeling accident proneness one might distinguish between the total number of accidents (in a region, for instance) and the number of accidents involving fatalities.

4.8.1 Poisson-Logistic Regression

In the simplest case, the reporting decision is independent of both the count process and previous reporting decisions. The separation of successful and unsuccessful events is assumed to follow an independent binomial distribution. The model then corresponds to the finite Poisson mixture presented in Chap. 2.5.1, and the total number of successful events is again Poisson distributed with a modified mean function. (The strong independence assumption is relaxed in Chap. 4.8.3)

Formally, let the total number of events y_i^* in a given period of time be Poisson distributed with

$$f(y_i^*) = \exp(-\lambda_i)\lambda_i^{y_i^*}/y_i^*! \tag{4.85}$$

where

$$\lambda_i = \exp(z_i'\gamma) \tag{4.86}$$

and z_i is a vector of individual covariates. Also assume, following Winkelmann and Zimmermann (1993c), that the binomial model is of the logistic form

$$P(\text{Event is successful}) = \Lambda(x_i'\beta) \tag{4.87}$$

where x_i are individual covariates and Λ is the *logistic* cumulative density

$$\Lambda(x_i'\beta) = \frac{\exp(x_i'\beta)}{1+\exp(x_i'\beta)} \tag{4.88}$$

The set of explanatory variables z and x may be identical, different or overlapping. It will be assumed that consecutive decisions determining the success or non-success of an event are independent and identical. The number of successful events y_i is then given by the sum of i.i.d. distributed Bernoulli variables, where the number of summands is itself a random variable:

$$y_i = B_{1i} + \ldots + B_{y_i^*,i} \tag{4.89}$$

where $B_i \sim Bn(1, \Lambda(x_i'\beta))$ and $y_i^* \sim \text{Poisson}(\exp(z_i'\gamma))$. Under independence, the number of successful events y_i has a Poisson distribution with parameter

$$\lambda_i^S = \exp(z_i'\gamma)\Lambda(x_i'\beta) \tag{4.90}$$

where z_i is a $(k_1 \times 1)$ vector and x_i is a $(k_2 \times 1)$ vector (see Chap. 2.5.2 and Feller, 1969). Estimation of the parameters β and γ by maximum likelihood is straightforward. The log-likelihood function has the form

$$\ell(\beta,\gamma) = \sum_{i=1}^{n} -\frac{\exp(x_i'\beta + z_i'\gamma)}{1+\exp(x_i'\beta)} + y_i \ln\left[\frac{\exp(x_i'\beta + z_i'\gamma)}{1+\exp(x_i'\beta)}\right] - \ln y_i! \tag{4.91}$$

Collecting the coefficients β and γ in a parameter vector θ, the gradient vector can be written as

$$\frac{\partial \ell(\theta; y, z, x)}{\partial \theta} = \sum_{i=1}^{n} \left(\frac{y_i - \lambda_i^S}{\lambda_i^S}\right) \frac{\partial \lambda_i^S}{\partial \theta} \tag{4.92}$$

or

$$\frac{\partial \ell(\theta; y, z, x)}{\partial \theta} = \sum_{i=1}^{n} (y_i - \lambda_i^S) \begin{bmatrix} z_i' \\ x_i'(1-\Lambda(x_i'\beta)) \end{bmatrix} \tag{4.93}$$

If z contains an intercept, (4.93) states that the sum of the residuals $u_i = y_i - \lambda_i^S$ is equal to zero. The Hessian matrix has the form

$$\frac{\partial^2 \ell(\theta; y, z, x)}{\partial \theta \partial \theta'} = \sum_{i=1}^{n} -\lambda_i^S \times$$

$$\begin{bmatrix} z_i z_i' & z_i x_i'(1-\Lambda(x_i'\beta)) \\ \cdot & x_i x_i' \left[(1-\Lambda(x_i'\beta))^2 + \frac{(y-\lambda_i^S)}{\lambda_i^S}\Lambda(x_i'\beta)(1-\Lambda(x_i'\beta))\right] \end{bmatrix} \tag{4.94}$$

Hence, the information matrix of the model is given by

$$I_n(\theta) = \sum_{i=1}^{n} \lambda_i^S \begin{bmatrix} z_i z_i' & z_i x_i'(1-\Lambda(x_i'\beta)) \\ x_i z_i'(1-\Lambda(x_i'\beta)) & x_i x_i'(1-\Lambda(x_i'\beta))^2 \end{bmatrix} \tag{4.95}$$

Identification of the parameters requires that $I_n(\theta)$ is nonsingular for arbitrary values in the parameter space. For instance, the information matrix is singular for $x_i = z_i$. For example, assume that there is only one regressor. Then, at the point $\beta = 0$,

$$I_n(\gamma, 0) = \sum_{i=1}^{n} \lambda_i^S \begin{pmatrix} x_i^2 & x_i^2(1-0.5) \\ x_i^2(1-0.5) & x_i^2(1-0.5)^2 \end{pmatrix} = \begin{pmatrix} 1 & 0.5 \\ 0.5 & 0.25 \end{pmatrix} \sum_{i=1}^{n} \lambda_i^S x_i^2$$

which is a singular matrix. It can be shown more generally that identifiability requires that neither x nor z does belong to the column space of the other. Under the assumptions of the model, the maximum likelihood estimator $\hat{\theta}$ is consistent for θ and $\sqrt{n}(\hat{\theta}-\theta) \to N(0, n\,I_n(\theta)^{-1})$, where $I_n(\theta)^{-1}$ is the inverse of the information matrix.

4.8.2 Count Amount Model

In the Poisson-logistic model, the probability of reporting an event is modeled as a function of individual specific (but event-unspecific) covariates. In a related model, the probability of recording an event depends on the (partially observed) "magnitude" of the event. The structure of this model is best illustrated with an application to consumer purchase behavior for which it was originally developed by van Praag and Vermeulen (1993). In this application, purchases are reported, together with their money value, as long as they exceed a minimum threshold value a. Let y^* denote the total number of purchases over a given period of time, and let y denote the number of recorded purchase occurrences. Denote the purchase amount by c^*. A purchase is recorded whenever $c^* > a$. The probability for this event is given by $P(c^* > a) = 1 - F(a)$, where F is the cumulative density function of c^*. In this set-up, observed purchase values c have a truncated density function $f(c)/(1 - F(a))$.

The model is completed by imposing some distributional assumptions. First, c^* is assumed to be normally distributed, conditional on individual specific covariates: $c_i^* = z_i\gamma + \varepsilon_i$ where ε_i has a normal distribution $N(0, \sigma^2)$. Second, y_i^* is assumed to be Poisson (or negative binomial) distributed with parameter $\lambda_i = \exp(x_i'\beta)$.

Since both the number of events and the money values are observed, van Praag and Vermeulen estimate the parameter vector $\theta = [\beta, \gamma]$ from the joint likelihood function of $c_i = (c_{i1}, \ldots, c_{iy})$ and y_i which is given by

$$
\begin{aligned}
g(y_i, c_i) &= \prod_{i=1}^{y_i} \frac{f(c_i;\gamma)}{1-F(a)} \\
&\quad \times \sum_{y_i^*=y_i}^{\infty} f(y_i^*;\beta) \frac{y_i^*!}{y_i!(y_i^*-y_i)!}(1-F(a))^{y_i}(F(a))^{y_i^*-y_i} \\
&= \prod_{i=1}^{y_i} f(c_i;\gamma) \sum_{y_i^*=y_i}^{\infty} f(y_i^*;\beta) \frac{y_i^*!}{y_i!(y_i^*-y_i)!}(F(a))^{y_i^*-y_i}
\end{aligned}
\tag{4.96}
$$

where $f(y_i^*;\beta)$ is a Poisson or negative binomial probability function and $f(c_i;\gamma)$ is the normal density. The term under the summation sign is a Poisson-Binomial mixture. To understand the meaning of this expression, note that the event "y_i purchases are recorded" can arise in a multitude of ways:

1. there were $y_i^* = y_i$ purchases, all with amounts greater than a and therefore all recorded.
2. there were $y_i^* = y_i + 1$ purchases, of which y_i with amounts greater than a and one with amount less than a (and therefore unrecorded).
3. there were $y_i^* = y_i + 2$ purchases, of which y_i with amounts greater than a and two with amount less than a (and therefore unrecorded), and so forth.

The probability for each event in the above list is a joint probability $f(y_i^*, y_i)$ which can be written as a product of marginal and conditional distribution:

$$
f(y_i^*, y_i) = f(y_i^*) f(y_i | y_i^*) \tag{4.97}
$$

The first probability function on the right side is a Poisson (or negative binomial) distribution. Under the assumption that purchase amounts at subsequent purchase occasions are independent, the second (conditional) probability function is a binomial distribution with parameter $p = 1 - F(a)$, i.e. the probability of success, here recording an event, is equal to the probability that the purchase amount exceeds the threshold of a. Without the independence assumption, it would not be possible to obtain such simple probability expressions.

As van Praag and Vermeulen (1993) point out this model has a wide range of potential applications. Examples include the modeling of the number of insurance claims where the insurance includes a deductible amount, the modeling of crime statistics, where official authorities do not file formal reports for minor crimes, or the number of unemployment spells, where only spells exceeding a certain minimum duration are observed.

4.8.3 Endogenous Underreporting

A count data model with endogenous reporting was considered in Winkelmann (1997, 1998). The model is closely related to the models on incidental censoring and truncation of Chap. 4.5. A restriction of the standard model is the assumption of independence between the count process and the binary reporting outcome. Consider, for instance, the study by Winkelmann and Zimmermann (1993c), where the model is applied to data on labor mobility. y_i^* gives then the (unobserved) number of job offers, $\lambda_i = \exp(x_i'\beta)$ the offer arrival rate, p_i the acceptance probability and y_i the (observed) number of accepted offers. The explicit assumption is that

a) the offer arrival rate is a deterministic function of observed covariates, and
b) the offer arrival rate is independent of the acceptance probability.

Yet, it is unreasonable to assume that all relevant variables are observed in practice and that arrival rates and acceptance decisions are independent. For instance, economic models of efficient job search predict that the reservation wage depends on the offer arrival rate and hence a correlation between the two should exist (See Mortensen, 1986). Therefore, a more general model that allows for endogenous underreporting is desirable. Such a model is now introduced.

Let $y_i^*|\varepsilon_i$ have a count data distribution with mean

$$\mathrm{E}(y_i^*|x_i, \varepsilon_i) = \exp(x_i'\beta + \varepsilon_i) \tag{4.98}$$

As before, an event j is reported and $c_{ij} = 1$ if the net utility from doing so is positive, i.e.

$$c_{ij}^* = c_i^* = z_i'\gamma + v_i > 0 \tag{4.99}$$

where, by assumption, the net utility does not depend on the specific event. Furthermore, assume that ε_i and v_i are jointly normal distributed with correlation ρ. Note, that this model is based on a probit-type specification whereas the standard underreporting model was based on the logit model. This change is dictated by convenience as the probit model leads to a straightforward extension for the correlated case.

The number of reported counts is given by

$$y_i = \sum_{j=1}^{y_i^*} c_i \tag{4.100}$$

To derive the probability function of y_i, consider first the case where ε_i is given. As before

$$P(c_i = 1|\varepsilon_i) = \Phi^*(\varepsilon_i) \tag{4.101}$$

where Φ^* is defined as in (4.60). Moreover, conditional on ε_i, x_i and z_i, c_i and y_i^* are independent. Assume that $y_i^*|\varepsilon_i$ is Poisson distributed. It follows directly from results in Chap. 4.8 that $y_i|\varepsilon_i$ is Poisson distributed with mean

$$\tilde{\lambda}_i = \exp(x_i'\beta + \varepsilon_i) \times \Phi^*(\varepsilon_i) \tag{4.102}$$

while $y_i|x_i, z_i$ has distribution

$$g(y_i|x_i, z_i) = \int_{-\infty}^{\infty} \frac{\exp(-\tilde{\lambda}_i(\varepsilon_i))\tilde{\lambda}_i(\varepsilon_i)^{y_i}}{y_i!} f_\varepsilon(\varepsilon_i)\mathrm{d}\varepsilon_i \tag{4.103}$$

or, in explicit notation

$$g(y_i|x_i, z_i; \beta, \gamma, \rho, \sigma) = \tag{4.104}$$

$$\int_{-\infty}^{\infty} \exp\left[-\exp(x_i'\beta + \varepsilon_i)\Phi\left(\frac{z_i'\gamma + \rho\varepsilon_i/\sigma}{\sqrt{1-\rho^2}}\right)\right]$$

$$\times \left[\exp(x_i'\beta + \varepsilon_i)\Phi\left(\frac{z_i'\gamma + \rho\varepsilon_i/\sigma}{\sqrt{1-\rho^2}}\right)\right]^{y_i} \times \frac{1}{y_i!\sigma}\phi(\varepsilon_i/\sigma)\mathrm{d}\varepsilon_i$$

The parameters of the model, β, γ, ρ, and σ are estimated by maximum likelihood. The resulting log-likelihood function involves simple integrals that can be evaluated by Gauss-Hermite quadrature. Details are given in Appendix B.

The model is quite general and encompasses a variety of interesting special cases that can be tested using parametric restrictions. For $\rho = 0$ the selection and count equations are independent. For $\rho = 0$ and $\sigma = 0$, the model reduces to a version of the Poisson-logistic regression model in Winkelmann and Zimmermann (1993c) where the logit type expression for the reporting probability is replaced by a probit type expression. Positive values for σ indicate unobserved heterogeneity in the count regression. In particular, the implicit variance function for y_i^* is

$$\mathrm{Var}(y_i^*|x_i) = \lambda_i + \alpha\lambda_i^2 \tag{4.105}$$

where $\alpha = \exp(2\sigma_u^2) - \exp(\sigma_u^2)$.

4.9 Endogenous Regressors

As for the linear model, correlation between a regressor and the error term (endogeneity) leads to inconsistency of the estimated regression parameters in count data models. Correlation may arise due to omitted variables that are correlated with the included ones, or, more generally, due to a simultaneous determination of the regressor through a related model. The term "regressor" is used somewhat loosely in this chapter. Strictly speaking, if some variables are correlated with the errors they are not regressors because in this case the model is not a regression (i.e., a conditional expectation function).

An important example where the issue of regressor endogeneity is a major worry is the estimation of a treatment effect on a count. In experimental sciences, individuals are randomly assigned to treatment group and control group, and differences in outcomes will thus be a good estimator of the treatment effect. An example is the effect of a drug on the number of epileptic seizures (Diggle, Liang and Zeger, 1995).

In (non-experimental) observational data, treatment is not assigned randomly. In many instances, individuals self-select into treatment, i.e., treatment becomes a matter of choice. For example, the number of doctor consultations may depend on the health insurance status (the "treatment" variable in this case). But insurance coverage is a choice variable that can depend, among other things, on health status and the expected number of doctor visits itself. Moreover, third factors can exist that are unobserved by the econometrician and yet affect both insurance cover and the number of doctor consultations over a given period.

In order to consistently estimate the ceteris paribus, or causal, effect of an endogenous regressor on the dependent variable, three general methods are available: estimation using instrumental variables; identification by parametric assumptions and functional form; and repeated measurements. The first two methods are discussed in this chapter. The third approach, based on repeated measurements, or panel data, is postponed until Chap. 5.2.

4.9.1 Instrumental Variable Estimation

A count data model with endogenous regressors can be estimated using instrumental variables. A general exposition of the method is provided by Mullahy (1997a), who approached the issue in the context of the Poisson regression model. Since the starting point is the log-linear conditional expectation function rather than the full distribution, the method immediately generalizes to any exponential regression model. By the same token, this also means that the technique does not generalize to arbitrary alternative count data models, such as hurdle or zero-inflated models where the conditional expectation function is more complex.

Recall that the Poisson maximum likelihood estimator solves the first-order conditions

$$\sum_{i=1}^{n}[y_i - \exp(x_i'\beta)]x_i = 0$$

The solution is a consistent estimator of β as long as the conditional moment restriction $E(v_i|x_i) = 0$, where $v_i = y_i - \exp(x_i'\beta)$, holds in the population. In other words, it must be the case that $E(y_i|x_i) = \exp(x_i'\beta)$. In this framework, one can usefully think of two alternative ways, in which the moment condition fails. Either

$$E(y_i|x_i, \xi_i) = \exp(x_i'\beta) + \xi_i$$

and $E(\xi_i|x_i) \neq 0$ (endogeneity with additive errors). Or

$$E(y_i|x_i, u_i) = \exp(x_i'\beta)u_i$$

and $E(u_i|x_i) \neq 1$ (endogeneity with multiplicative errors)

Endogeneity With Additive Errors

Let

$$y_i = \exp(x_i'\beta) + \xi_i$$

Endogenous regressors imply that $\mathrm{E}(\xi_i|x_i) \neq 0$ for some elements of x_i. Thus

$$\mathrm{E}[(y_i - \exp(x_i'\beta))|x_i] = \mathrm{E}(\xi_i|x_i) \neq 0$$

and Poisson regression is inconsistent. However, suppose that instruments z_i are available such that $\mathrm{E}(\xi_i|z_i) = 0$, or

$$\mathrm{E}[(y_i - \exp(x_i'\beta))|z_i] = 0$$

Estimation can then be based on non-linear instrumental variable techniques (See, for instance, Davidson and MacKinnon, 1993). Grogger (1990b) discusses this approach with special reference to count data modeling. He points out that a comparison of the non-linear instrumental variable estimator with the Poisson maximum likelihood estimator provides a simple test for exogeneity of the regressors.

Endogeneity With Multiplicative Errors

Let

$$y_i = \exp(x_i'\beta)u_i \tag{4.106}$$

where $\mathrm{E}(u_i|x_i) \neq 1$. The required conditional moment restriction implicit in the first-order condition of the Poisson regression model can be written as

$$\begin{aligned}
\mathrm{E}[(y_i - \exp(x_i'\beta))|x_i] &= \mathrm{E}[\exp(x_i'\beta)(u_i - 1)|x_i] \\
&= \mathrm{E}_x[\exp(x_i'\beta)\mathrm{E}_u(u_i - 1)|x_i] \\
&\neq 0
\end{aligned}$$

since $\mathrm{E}_u(u_i - 1|x_i) = \mathrm{E}(u_i|x_i) - 1 \neq 0$.

Next, assume that an instrument z is available such that $\mathrm{E}(u_i|z_i) = 1$. From (4.106), $u_i = \exp(-x_i'\beta)y_i$ from where it follows that

$$\mathrm{E}[\exp(-x_i'\beta)y_i - 1|z_i] = 0$$

This moment condition can be used to estimate β consistently (Mullahy, 1997a). As pointed out by Windmeijer and Santos Silva (1997), the same instrument cannot be valid in the multiplicative and additive case. Rewrite the orthogonality condition in the multiplicative case as $\mathrm{E}[(y_i - \mu_i)/\mu_i|z_i] = 0$ where μ_i is the mean. If a variable is independent of $\xi_i = y_i - \mu_i$ then it usually is not also independent of ξ_i/μ_i, because ξ_i and μ_i are correlated.

4.9.2 Simultaneous Equations

In the linear model the main sources for endogeneity of a regressor are measurement error, omitted variables, and simultaneity. Consider the following system of equations

$$y_{1i} = \alpha_1 y_{2i} + x_i'\beta_1 + u_{1i} \tag{4.107}$$

$$y_{2i} = \alpha_2 y_{1i} + z_i'\beta_2 + u_{2i} \tag{4.108}$$

where $\mathrm{Cov}(x, u_1) = \mathrm{Cov}(z, u_1) = \mathrm{Cov}(x, u_2) = \mathrm{Cov}(z, u_2) = 0$ and thus x and z are exogenous. One type of endogeneity arises if $\alpha_1 \neq 0$, $\alpha_2 = 0$ and $\mathrm{Cov}(u_1, u_2) \neq 0$. For example, in the omitted variable case, $u_{1i} = z_i'\gamma + v_{1i}$, where y_{2i} and z_i are correlated. Simultaneity requires $\alpha_1 \neq 0$ and $\alpha_2 \neq 0$, regardless of whether the errors are correlated or not.

If endogeneity is due to omitted variables, there are in principle two remedies. One is to include proxies for these variables, the other is to use instruments. If endogeneity is due to simultaneity, estimation of structural parameters is not possible without the availability of instruments, i.e., variables in z that are not included in x (exclusion restrictions). It is well known that in the linear case, the instrumental variable estimator is equivalent to two-stage least squares, where the reduced form predictions $\hat{y}_{2i} = z_i'\hat{\pi}_1 + x_i'\hat{\pi}_2$ are used instead of y_2 in an OLS estimation of the first equation. This works because

$$\begin{aligned} y_{1i} &= \alpha_1 y_{2i} + x_i'\beta_1 + u_{1i} \\ &= \alpha_1 \hat{y}_{2i} + x_i'\beta_1 + \alpha_1 v_{1i} + u_{1i} \\ &= \alpha_1 \hat{y}_{2i} + x_i'\beta_1 + e_i \end{aligned}$$

where v_1 is the reduced form residual and $\hat{y}_2$ is thus uncorrelated with e by construction.

Unfortunately, the situation for count data is more complex, and the analogy with linear models does not go very far. Difficulties arise already if one considers the conditional expectation aspect of the model only, disregarding distributional issues for a moment (The following discussion follows Windmeijer and Santos Silva (1997)). Let

$$y_{1i} = \exp(\alpha_1 y_{2i} + x_i'\beta_1) + u_{1i} \tag{4.109}$$

(or $y_{1i} = \exp(\alpha_1 y_{2i} + x_i'\beta_1 + v_{1i})$ – whether the model is specified with additive or multiplicative error is inconsequential for this discussion). If $\mathrm{E}(u_{1i}|y_{2i}, x_i) = 0$ (or $\mathrm{E}(\exp(v_{1i})|y_{2i}, x_i) = 1$) this is just the standard model with log-linear conditional expectation function $\mathrm{E}(y_{1i}|y_{2i}, x_i) = \exp(\alpha_1 y_{2i} + x_i'\beta_1)$. If, however, y_{2i} and u_{1i} are correlated, as they must be in the presence of a feedback process

$$y_{2i} = \alpha_2 y_{1i} + z_i'\beta_2 + u_{2i} \tag{4.110}$$

then $\mathrm{E}(u_{1i}|y_{2i}, x_i) \neq 0$ and $\mathrm{E}(y_{1i}|y_{2i}, x_i) \neq \exp(\alpha_1 y_{2i} + x_i'\beta_1)$. The natural approach would be to replace y_{2i} by a prediction from a reduced form model such that $\mathrm{E}(y_{1i}|y_{2i}, x_i)$ can be consistently estimated by $\exp(\alpha_1 \hat{y}_{2i} + x_i'\beta_1)$.

This is not possible, however, since a simple, practical reduced form does not exist. To see this, substitute equation (4.109) into equation (4.110) to obtain

$$y_{2i} = \alpha_2 \exp(\alpha_1 y_{2i} + x_i'\beta_1) + z_i'\beta_2 + \alpha_2 u_{1i} + u_{2i}$$

which does not reduce to a simple reduced form equation for y_2. But the reduced form model is required to implement a two-stage least squares type estimator.

Next, consider the distributional implication of full simultaneity. For example, let $y_{1i} \sim \text{Poisson}(\lambda_i)$ where

$$\begin{aligned} \lambda_i &= \exp(\alpha_1 y_{2i} + x_i'\beta_1) \\ &= \exp[\alpha_1(\alpha_2 y_{1i} + z_i'\beta_2 + u_{2i}) + x_i'\beta_1] \end{aligned}$$

Again, it would be impossible to determine the reduced form distribution of y_{1i}. Therefore, realistically, we have to confine our attention to the class of recursive models

$$y_{1i} = \exp(\alpha_1 y_{2i} + x_i'\beta_1) + u_{1i} \tag{4.111}$$

$$y_{2i} = z_i'\beta_2 + u_{2i} \tag{4.112}$$

where endogeneity arises because $\text{Cov}(u_1, u_2) \neq 0$. Assume that u_2 is statistically independent of x and z. In this case, one can indeed proceed in close analogy to the linear model, namely first obtain a predicted regressor $\hat{y}_{2i} = z_i'\hat{\gamma}$ from the auxiliary regression. A second stage Poisson regression (or any other count data regression), in which y_{2i} is replaced by $\hat{y}_{2i}$ gives then a consistent estimator for α_1. In order to see why this works, write the model as

$$\begin{aligned} y_{1i} &= \exp(\alpha_1(z_i'\beta_2 + u_{2i}) + x_i'\beta_1) + u_{1i} \\ &= \exp(\alpha_1 z_i'\beta_2 + x_i'\beta_1)\exp(\alpha_1 u_{2i}) + u_{1i} \end{aligned}$$

By assumption, u_2, and therefore $\exp(\alpha_1 u_{2i})$, are statistically independent of x and z. Hence, the conditional expectation of the count variable can be written as

$$\begin{aligned} \text{E}(y_{1i}|z_i'\beta_2, x_i) &= \text{E}_{u_2}[\exp(\alpha_1 z_i'\beta_2 + x_i'\beta_1)\exp(\alpha_1 u_{2i})] \\ &= \exp(\alpha_1 z_i'\beta_2 + x_i'\beta_1^*) \end{aligned}$$

where the constant term in β_1^* has been shifted by $\ln \text{E}[\exp(\alpha_1 u_{2i})]$. The standard errors of the second stage estimates $\hat{\alpha}_1$ and $\hat{\beta}_1^*$ need to be adjusted to account for the sampling variation introduced from the estimation of β_2, which can be done using standard formulas for two-step estimation. Alternatively, one can bootstrap the standard errors.

However, there is still an important difference to the linear case. Here, the linear reduced form model for the right hand side endogenous variable actually amounts to an additional assumption that is required (Mullahy, 1997a, Wooldridge, 1997b). The second stage Poisson estimators using first stage fitted values is inconsistent unless the model for the first stage conditional expectation function is correct. On the other hand, conventional two-stage least squares estimators using a linear second stage model are consistent whether or not the first stage conditional expectation function is linear. In fact, two-stage least squares is equivalent to instrumental variable estimation then.

Thus, there is some pay-off to actually avoiding the exponential mean function. If one is willing to specify the Poisson regression model with linear mean function, consistency holds regardless of the true mean function of the endogenous regressor (See, for instance, Mullahy and Portney, 1990, and also Sander, 1992). Angrist (2001) uses a related argument in the context of a linear outcome equation combined with a binary endogenous regressor to argue for using a linear probability model at the first stage. Also note that the approach for count data is closely related to other limited dependent variable models with endogenous regressors. Corresponding results for probit models are for example discussed in Heckman (1978) and in Rivers and Vuong (1988) who advocate a conditional maximum likelihood estimator.

4.9.3 Binary Endogenous Variables

A special case of some practical importance arises if a count dependent variable depends on a binary endogenous variable. This setting comprises the important case of estimating the effect of a treatment on the treated. As in Heckman (1978), one can distinguish between the two cases of binary endogenous and latent endogenous variables. In the above notation, we have

$$y_{2i}^* = \alpha_2 y_{1i} + z_i' \beta_2 + u_{2i}$$

However, only the sign of y_2^* is observed and

$$d_i = \begin{cases} 1 & \text{if } y_{2i}^* \geq 0 \\ 0 & \text{else} \end{cases}$$

Moreover, in the binary endogenous variable case, we can write for the count variable

$$y_{1i} = \exp(\alpha_1 d_{2i} + x_i' \beta_1) + u_{1i}$$

whereas in the latent endogenous variable case

$$y_{1i} = \exp(\alpha_1 y_{2i}^* + x_i' \beta_1) + u_{1i}$$

The latent endogenous variable case is more straightforward, and it will be discussed first. There is no substantial difference between this case and the system of equations (4.109) and (4.110) in the previous chapter. Again, intractable reduced form equations suggest the imposition of a triangular structure, by letting $\alpha_2 = 0$. Again, the conditional expectation of the count variable can be written as

$$\mathrm{E}(y_{1i} | z_i' \beta_2, x_i) = \exp(\alpha_1 z_i' \beta_2 + x_i' \beta_1^*)$$

where the constant term in β_1^* has been shifted by $\ln \mathrm{E}[\exp(\alpha_1 u_{2i})]$. Initial estimates for β_2 can be obtained from a first stage logit or probit model, whereas α_1 and β_1^* are estimated in a second stage standard Poisson model. Again, the computation of standard errors requires adjustment to account for the two-step estimation.

Binary Variable

A somewhat different case arises if the count model includes an endogenous dummy variable such that

$$y_{1i} = \exp(\alpha_1 d_{2i} + x_i'\beta_1) + u_{1i}$$

The first thing to note is that in this case, the fully interdependent system has not only the problem of lacking a manageable reduced form. Full simultaneity is impossible on purely logical grounds. To understand this point, assume that we have a simultaneous model with

$$y_{2i}^* = \alpha_2 y_{1i} + z_i'\beta_2 + u_{2i}$$

such that $d_{2i} = 1$ if $y_{2i}^* > 0$ and $d_{2i} = 0$ if $y_{2i}^* \le 0$. It must be the case that

$$P(d_i = 0) + P(d_i = 1) = 1$$

where $P(d_i = 1) = F(\alpha_2 y_{1i} + z_i'\beta_2)$ and F is the cumulative distribution function of u_2 (assuming symmetry of the density of u_2). Upon substituting the count equation for y_1 we obtain

$$\begin{aligned} &F[\alpha_2(\exp(x_i'\beta_1 + \alpha_1) + u_{1i}) + z_i'\beta_2] \\ &\qquad + \{1 - F[\alpha_2(\exp(x_i'\beta_1) + u_{1i}) + z_i'\beta_2]\} = 1 \end{aligned}$$

or

$$F[\alpha_2(\exp(x_i'\beta_1 + \alpha_1) + u_{1i}) + z_i'\beta_2] = F[\alpha_2(\exp(x_i'\beta_1) + u_{1i}) + z_i'\beta_2]$$

This condition implies that either $\alpha_1 = 0$ or $\alpha_2 = 0$. Logical consistency prohibits full simultaneity and requires a recursive system. Note that the issue of logical consistency does not arise if either y_2^* itself is observed, or if the model is formulated with a latent endogenous variable.

In the following, consider the triangular model

$$y_{1i} = \exp(\alpha_1 d_{2i} + x_i'\beta_1) + u_{1i}$$

$$y_{2i}^* = z_i'\beta_2 + u_{2i}$$

where

$$d_{2i} = \begin{cases} 1 & \text{if } y_{2i}^* > 0 \\ 0 & \text{else} \end{cases}$$

and $\mathrm{E}(u_1|x,z) = \mathrm{E}(u_2|x,z) = 0$ and $\mathrm{Cov}(u_1,u_2) \neq 0$. One could be tempted to replace in the first equation as before d_{2i} by its probability $F(z_i'\beta_2)$. However, this will not work in this non-linear model. In this case,

$$\mathrm{E}(y_{1i}|z_i'\beta_2, x_i, d_i) = \exp(\alpha_1 F(z_i'\beta_2) + x_i'\beta_1)\exp(\alpha_1 w_i)$$

where $w_i = d_i - F(z_i'\beta_2)$ Now, although u_2 and z_i are independent, $\exp(w_i)$ and z_i are not. The reason is that the moments of w_i depend on z_i. For example, $\mathrm{E}(w_i^2|z_i) = F(z_i'\beta_2)[1-F(z_i'\beta_2)]$. But $\mathrm{E}(\exp(w_i))$, because of the convexity of the exponential transformation, is an increasing function of the variance, and therefore depends on both parameters and regressors. Windmeijer and Santos Silva (1997) propose instead an instrumental variables estimator, instrumenting d_i by $F(z_i'\hat{\beta}_2)$, where $\hat{\beta}_2$ is obtained from estimating a probit or logit model first.

Full Information Maximum Likelihood Estimation

Consider now a fully specified count data model with endogenous binary regressor. For example, let y_{1i} have a known count data distribution $f(y_{1i})$ with conditional expectation function

$$\mathrm{E}(y_{1i}|d_i, x_i, \varepsilon_i) = \exp(\alpha_1 d_i + x_i'\beta + \varepsilon_i) \tag{4.113}$$

As before, d_i is determined through a latent model

$$y_{2i}^* = z_i'\gamma + u_i$$

where

$$d_i = \begin{cases} 1 & \text{if } y_{2i}^* > 0 \\ 0 & \text{else} \end{cases}$$

Finally, ε_i and u_i have a bivariate normal distribution with correlation parameter ρ. Both y_{1i} and d_i are fully observed, and parameter estimation by maximum likelihood can be based on the joint probability function

$$\begin{aligned} f(y_{1i}, d_i) &= d_i f(y_{1i}, d_i = 1) + (1-d_i) f(y_{1i}, d_i = 0) \\ &= d_i[f(y_{1i}|d_i = 1)P(d_i = 1)] + (1-d_i)[f(y_{1i}|d_i = 0)P(d_i = 0)] \end{aligned}$$

Under independence, the expressions $f(y_{1i}|d_i = 1)P(d_i = 1)$ and $f(y_{1i}|d_i = 0)P(d_i = 0)$ could be readily determined by multiplication of standard distributions. Because of the correlated errors (the bivariate normal distribution for ε_i and u_i) the situation is somewhat more complex. In particular, one needs first to consider these expressions conditional on ε_i, and then obtain in a second step the desired quantities by integration of $f(y_{1i}, d_i, \varepsilon_i)$ over ε_i:

$$\begin{aligned} f(y_{1i}, d_i) &= \int_{-\infty}^{\infty} f(y_{1i}, d_i, \varepsilon_i)\mathrm{d}\varepsilon_i \\ &= \int_{-\infty}^{\infty} f(y_{1i}|d_i, \varepsilon_i) f(d_i|\varepsilon_i) g(\varepsilon_i)\mathrm{d}\varepsilon_i \end{aligned} \tag{4.114}$$

The first distribution under the integral is simply the specified count data probability function (Poisson or Negbin) with mean function (4.113). The second distribution under the integral is a Bernoulli distribution

$$f(d_i|\varepsilon_i) = P(d_i = 1|\varepsilon_i)^{d_i}[1 - P(d_i = 1|\varepsilon_i)]^{1-d_i}$$

where

$$\begin{aligned} P(d_i = 1|\varepsilon_i) &= P(v_i > -z_i'\gamma|\varepsilon_i) \\ &= \Phi^*(\varepsilon_i) \end{aligned} \tag{4.115}$$

and Φ^* is defined as in (4.60). Finally, $g(\varepsilon_i)$ is a normal distribution with mean 0 and variance σ^2.

Putting everything together, (4.114) can be rewritten as

$$f(y_{1i}, d_i) = \int_{-\infty}^{\infty} f(y_{1i}|d_i, \varepsilon_i)\Phi^*(\varepsilon_i)^{d_i}[1 - \Phi^*(\varepsilon_i)]^{1-d_i} g(\varepsilon_i)\mathrm{d}\varepsilon_i \tag{4.116}$$

Computation of the integral using quadrature or other simulation methods provides no major difficulty. The parameters can be estimated by maximizing the log-likelihood function of the sample

$$\ell = \sum_{i=1}^{n} \ln f(y_{1i}, d_i; x_i, \gamma_i)$$

with respect to β, γ and σ. Alternatively, Terza (1998) derives a two-stage moment estimators that does not require the specification of the full distribution of y_{1i}. A Bayesian analysis of this model is provided by Kozumi (1999).

What would be the consequences of ignoring endogeneity of d_i? From (4.113), we know that

$$\begin{aligned} \frac{\mathrm{E}(y_{1i}|x_i, d_i = 1)}{\mathrm{E}(y_{1i}|x_i, d_i = 0)} &= \frac{\mathrm{E}_\varepsilon \mathrm{E}(y_{1i}|x_i, \varepsilon_i, d_i = 1)}{\mathrm{E}_\varepsilon \mathrm{E}(y_{1i}|x_i, \varepsilon_i, d_i = 0)} \\ &= \frac{\exp(x_i'\beta + \alpha)\mathrm{E}(\varepsilon_i|d_i = 1)}{\exp(x_i'\beta)\mathrm{E}(\varepsilon_i|d_i = 0)} \end{aligned}$$

In order to evaluate the expectations, we need to make reference to the aforementioned results on truncation in the lognormal distribution (see equation (4.66)). In particular,

$$\mathrm{E}(\exp(\varepsilon_i)|d_i = 1) = \exp(1/2\sigma^2)\frac{\Phi(z_i'\gamma + \rho\sigma)}{\Phi(z_i'\gamma)}$$

and

$$\mathrm{E}(\exp(\varepsilon_i)|d_i = 0) = \exp(1/2\sigma^2)\frac{\Phi(-z_i'\gamma - \rho\sigma)}{\Phi(-z_i'\gamma)}$$

Therefore, under the assumption of this model

$$\frac{\mathrm{E}(y_{1i}|x_i, d_i = 1)}{\mathrm{E}(y_{1i}|x_i, d_i = 0)} = \exp(\alpha)\frac{\Phi(z_i'\gamma + \rho\sigma)}{\Phi(z_i'\gamma)}\frac{\Phi(-z_i'\gamma)}{\Phi(-z_i'\gamma - \rho\sigma)} \tag{4.117}$$

If $\rho > 0$, it is easily verified that the factor following $\exp(\alpha)$ is greater than 1. In other words, the overall relative difference between the two expected counts exceeds then $\exp(\alpha) - 1$, the causal difference that would be observed for two randomly selected, otherwise identical individuals, for one of whom $d_i = 1$ and for the other $d_i = 0$. Ignoring the endogeneity of d_i would lead to an upward bias in the estimated effect.

Switching Regression

The binary endogenous variable approach can be easily extended to a full switching regression model. Such a model has been considered by Terza (1998). Here, a selection variable c_i is an indicator for the regime j the model is in. In general terms, we can write the two-part data generating process as

$$f_1(y_i|x_i) \text{ if } c_i = 1$$

$$f_0(y_i|x_i) \text{ if } c_i = 0$$

If all regression coefficients are allowed to depend on the regime but the functional form is otherwise the same, we can write

$$f_j(y_i|x_i) = f(y_i|x_i; \beta_j)$$

van Ophem (2000) discusses this general case.

4.9.4 Mixed Discrete-Continuous Models

A number of models discussed so far involved both discrete and continuous random variables. Examples are count data models with incidental censoring or truncation, as well as the simultaneous equation models presented before. Another type of mixed discrete-continuous data model arises if one observes jointly two outcome variables, one being a count and the other being a continuous variable, without having a direct interaction between the two. For example, let y_{1i} have a Poisson (or negative binomial) distribution $f_1(y_{1i})$ with conditional expectation function

$$\mathrm{E}(y_{1i}|d_i, x_i, \varepsilon_i) = \exp(x_i'\beta + \varepsilon_i) \tag{4.118}$$

and let y_{2i} be determined by

$$y_{2i} = z_i'\gamma + u_i$$

Thus, there is no direct feedback. In the taxonomy of the simultaneous equation system (4.109) and (4.110), both $\alpha_1 = 0$ and $\alpha_2 = 0$. On the face of it, one would think that the models can just be estimated independently. Now assume, however, that ε_i and u_i are correlated. For example, they might be bivariate normal distributed with correlation parameter $\rho \neq 0$. In this case, separate estimation of the two equations is no longer efficient. Rather, we are operating now in the context of *seemingly unrelated regression models*. If we are willing to specifiy a joint distribution for the two error terms, estimation can proceed by maximum likelihood. The applicable techniques are very similar to those discussed before in the context of incidental censoring or a binary dependent variable. Again, one dimensional numerical integration will be required, and the Gauss-Hermite algorithm can be used.

A model with such a structure has been suggested and estimated by Prieger (2002). Here, the two outcome variables are the number of innovations in the telecommunication markets (a discrete variable modelled by a count data model) and the time until adoption in the market, or regulatory delay (a continuous variable modelled by a duration model, here a Weibull distribution). In such an application, the correlation ρ may have a substantive interpretation, providing evidence on possible congestion effects, as an increased number of innovations may for instance increase the regulatory delay due to congestion. Other interesting uses of such a model are conceivable, such as the joint modelling of the individual number of unemployment spells and their durations.

4.10 Generalized Variance Models

A common feature of the generalized count data models presented in this chapter is their agnosticism with respect to the origins of the Poisson failure. Models for unobserved heterogeneity or models for selectivity were each tailored to a specific violation of the standard Poisson model which could be described in terms of the underlying data generating process. It was noticed that all such violations cause, among other things, a break-down of the Poisson requirement of equal mean and variance.

This observation is the starting point for a different approach for overcoming the restrictiveness of the Poisson model, namely to allow directly for a generalized variance function that does not impose the Poisson restriction of equal mean and variance, and to embed such a variance function in a parametric or non-parametric count data model. The four discussed approaches are the generalized event count model, generalized Poisson regression, robust Poisson regression, and a non-parametric variance estimator. An important aspect of the models discussed in this chapter is that they allow for both over- and underdispersion relative to the Poisson model.

4.10.1 Generalized Event Count Model

The introduction of the Negbin$_k$ generated substantial flexibility in the relationship between variance and mean. Still, the model is restricted to situations where overdispersion prevails. On the other hand, there are several reasons why the data may display underdispersion: negative contagion; positive duration dependence; spell specific heterogeneity; and zero-deflation, to name but a few. The following model, the *generalized event count model* (GEC_k), allows for both over- and underdispersion.

The derivation follows King (1989b). It is based on the Katz family of distributions, which has been introduced in Chap. 2.5.4. The Katz family is defined by a recursive formula for the probabilities $f(y)$:

$$\frac{f(y+1)}{f(y)} = \frac{\theta + \gamma y}{1+y} \quad \text{for } y = 0, 1, 2, \ldots \text{ and } \theta + \gamma y \geq 0\,. \tag{4.119}$$

Using recursive substitution, (4.119) can be rewritten as

$$f(y|\theta,\gamma) = f(0) \prod_{j=1}^{y_i} \left[\frac{\theta + \gamma(j-1)}{j} \right], y_i = 1, 2, \ldots \tag{4.120}$$

where $f(0)$ is determined by the fact that the probabilities have to sum to one. Mean and variance are given by

$$\mathrm{E}(y) = \frac{\theta}{(1-\gamma)}\,, \quad \mathrm{Var}(y) = \frac{\theta}{(1-\gamma)^2} \tag{4.121}$$

It is easily seen that this family produces equidispersion for $\gamma = 0$, overdispersion for $0 < \gamma < 1$, and underdispersion for $\gamma < 0$. The following parameterization has been suggested.

$$\gamma = \frac{(\sigma^2 - 1)\lambda_i^k}{(\sigma^2-1)\lambda_i^k + 1}\,, \quad \theta = \frac{\lambda_i}{(\sigma^2-1)\lambda_i^k + 1}\,, \lambda_i = \exp(x_i'\beta) \tag{4.122}$$

It differs slightly from the Negbin_k specification introduced before, replacing σ^2 by $\sigma^2 - 1$. This parameterization is more convenient to discuss underdispersion ($0 < \sigma^2 < 1$). In this case

$$\mathrm{E}(y_i|x_i) = \exp(x_i'\beta)$$

and

$$\mathrm{Var}(y_i|x_i) = \mathrm{E}(y_i|x_i) + (\sigma^2 - 1)[\mathrm{E}(y_i|x_i)]^{k+1}$$

The complete distribution is given by:

$$f_{\mathrm{geck}}(y_i|\cdot) = f(0|\lambda_i, \sigma^2, k) \times \begin{cases} \prod_{j=1}^{y_i} \left[\frac{\lambda_i + (\sigma^2-1)\lambda_i^k (j-1)}{[(\sigma^2-1)\lambda_i^k + 1]\,j} \right] & \text{for } y_i = 1, 2, \ldots \\ 1 & \text{for } y_i = 0 \end{cases} \tag{4.123}$$

where

$$f(0|\lambda_i,\sigma^2,k) = \begin{cases} (1+(\sigma^2-1)\,\lambda_i^k)^{\nu_i} & \text{for } \sigma^2 \geq 1 \\ (1+(\sigma^2-1)\,\lambda_i^k)^{\nu_i} D_i^{-1} & \text{for } 0<\sigma^2<1\,, \\ & \lambda^k \leq 1/(1-\sigma^2) \\ & \text{and } y_i \leq \text{int}^*(\nu_i) \\ 0 & \text{otherwise} \end{cases}$$

$$\nu_i = \lambda_i^{1-k}/(1-\sigma^2)\,,$$

$$D_i = \textstyle\sum_{m=0}^{\text{int}^*(\nu_i)} f_{\text{bn}}(m|\lambda_i,\sigma^2,k)\,,$$

$$\text{and int}^*(\text{y}) = \begin{cases} \text{int(y)+1} & \text{for int(y)<y} \\ \text{y} & \text{for int(y)=y} \end{cases}$$

The limit of $f(0|\lambda_i,\sigma^2,k)$ for $\sigma^2 \to 1$ is $e^{-\lambda_i}$ and the GEC_k converges to the Poisson model, as does the Negbin_k. The log-likelihood has the following form:

$$\ell(\beta,\sigma^2,k|y) = \sum_{i=1}^{n}\Big\{\ln(f(0|\lambda_i,\sigma^2,k)) + \sum_{j=1}^{y_i}\ln\left[\frac{\lambda_i+(\sigma^2-1)\lambda_i^k(j-1)}{[(\sigma^2-1)\lambda_i^k+1]\,j}\right]\Big\} \tag{4.124}$$

for $f(0|\lambda_i,\sigma^2,k) \neq 0$.

The maximizing values for β, σ^2 and k can be found by using a numerical optimization algorithm. The hypothesis that the data form a random sample from a Poisson distribution can be tested with $H_0 : \sigma^2 = 1$. The hypothesis that the data follow the Negbin II can be tested using a sequential testing strategy: First, test whether the Poisson hypothesis can be rejected; second, perform a Wald test for $H_0 : k = 1$ to test for the shape of the variance function.

4.10.2 Generalized Poisson Regression

A detailed presentation of the *generalized Poisson distribution* can be found in a monograph by Consul (1989). Further references are Consul and Famoye (1992), Famoye (1993), and Wang and Famoye (1997). The latter references explicitly introduce exogenous variables and thus a generalized Poisson regression model (GPR). Santos Silva (1997b) extended the model to truncated data.

Generalized Poisson regression is a genuine alternative to the generalized event count model as it allows for both over- and underdispersion and nests the

Poisson regression model as a special case. This is achieved by introducing one additional parameter θ. The probability distribution function can be written as (Consul 1989, p. 4)

$$f(y) = \begin{cases} \dfrac{\theta(\theta + y\gamma)^{y-1} e^{-\theta - y\gamma}}{y!}, & y = 0, 1, 2, \ldots \\ 0 \quad \text{for } y > m\,, \text{ when } \gamma < 0 \end{cases} \tag{4.125}$$

where $\theta > 0$, $\max[-1, -\theta/m] < \gamma \leq 1$ and $m(\geq 4)$ is the largest positive integer for which $\theta + m\gamma > 0$ when γ is negative.

The generalized Poisson distribution nests the Poisson distribution for $\gamma = 0$. Mean and variance are given by $\text{E}(y) = \theta(1 - \gamma)^{-1}$ and $\text{Var}(y) = \theta(1 - \gamma)^{-3}$, respectively. Thus, the generalized Poisson distribution displays overdispersion for $0 < \gamma < 1$, equidispersion for $\gamma = 0$ and underdispersion for $\max[-1, -\theta/m] < \gamma \leq 0$. Therefore, the parameter space is restricted in case of underdispersion.

The generalized Poisson regression model uses the following reparameterizations:

$$\theta_i = \frac{\lambda_i}{1 + a\lambda_i}$$

$$\gamma_i = \frac{a\lambda_i}{1 + a\lambda_i}$$

where $\lambda_i = \exp(x_i'\beta)$ as usual. Now, the probability function can be written as

$$f(y_i) = \left(\frac{\lambda_i}{1 + a\lambda_i}\right)^{y_i} \frac{(1 + ay_i)^{y_i - 1}}{y_i!} \exp\left(-\frac{\lambda_i(1 + ay_i)}{1 + a\lambda_i}\right) \tag{4.126}$$

and the conditional mean and variance of y_i are given by

$$\text{E}(y_i|x_i) = \lambda_i$$

$$\text{Var}(y_i|x_i) = \lambda_i(1 + a\lambda_i)^2$$

When $a = 0$, (4.126) reduces to the standard Poisson model. a acts like a dispersion parameter, with underdispersion for $a < 0$ and overdispersion for $a > 0$. The log-likelihood function can be written as

$$\ell(a, \beta; y_i) = \sum_{i=1}^{n} \left\{ y_i \ln\left(\frac{\lambda_i}{1 + a\lambda_i}\right) + (y_i - 1)\ln(1 + ay_i) - \frac{\lambda_i(1 + ay_i)}{1 + a\lambda_i} - \ln(y_i!) \right\}$$

and the first derivative with respect to β and a take relatively simple forms

$$\frac{\partial \ell}{\partial \beta} = \sum_{i=1}^{n} \frac{(y_i - \lambda_i)x_i}{(1 + a\lambda_i)^2}$$

and

$$\frac{\partial \ell}{\partial a} = \sum_{i=1}^{n} \left\{ \frac{-a\lambda_i}{1+a\lambda_i} + \frac{y_i(y_i-1)}{1+ay_i} - \frac{\lambda_i(y_i-\lambda_i)}{(1+a\lambda_i)^2} \right\}$$

These can be solved numerically using an iterative solution algorithm where for $a < 0$ the restrictions $1 + a\lambda_i > 0$ and $1 + ay_i > 0$ have to be observed.

4.10.3 Robust Poisson Regression

The Poisson regression model specifies that data generating process *up to some parameters*. This is an example of *parametric* modeling, leading to *parametric* estimation. *Robust* modeling and estimation, by contrast, does not assume that the exact distribution generating the observations be known (up to some parameters). Rather, possibilities for valid inferences under weaker parametric assumptions are explored. This kind of robust estimation is therefore often referred to as *semiparametric* estimation. It differs from alternative uses of the term 'robust' that refer to methods of reducing the effects of outliers on the estimator.

As an example, assume that observations form a random sample with a known, and correctly specified, mean function. Then, it is the case that maximum likelihood estimation based on any *linear exponential family* (which includes normal, Poisson, Negbin II for known α, and Gamma) yields a consistent estimator of the parameters of the mean function. This important result is due to Gourieroux, Monfort and Trognon (1984a,b). See also Wooldridge (1997b). The choice of a particular linear exponential family has an effect only insofar as the choice leads to different variance matrices of the estimator. Only the correctly specified full parametric model is asymptotically efficient.

These ideas were first dicussed in Chap. 3.2.2 in the context of *quasi-maximum likelihood estimation* (QML) and *pseudo-maximum likelihood estimation* (PML). QML refers to a situation in which a misspecified likelihood function is maximized. In contrast to QML, PML denotes a situation of QML under a particular type of misspecification: correctly specified mean function and estimation based on a linear exponential family. In this case the Gourieroux, Monfort and Trognon (1984a,b) consistency result applies.

In this section, the implications of the PML result are explored in the context of the Poisson regression model. PML estimation exploits the fact that, as the Poisson distribution is a linear exponential family, departure from the standard variance function do not affect consistency of the parameter estimates as long as the mean is correctly specified. The only effect of a misspecified variance function is then that the estimated variance matrix under the maximum likelihood assumption is "wrong" and has to be adjusted.

First, it is derived "how wrong" the variance estimates are. Second, three different methods for calculating an asymptotically valid variance matrix are presented. Finally, the performance of these methods in finite samples is compared using a small Monte Carlo study.

Under correct specification of the Poisson regression, the variance matrix of $\hat{\beta}$ can consistently be estimated by

$$\widehat{\mathrm{Var}}(\hat{\beta}) = -[H(\hat{\beta})]^{-1}$$

where H is the (expected) Hessian matrix of the log-likelihood function:

$$H(\hat{\beta}) = -\sum_{i=1}^{n} x_i x_i' \hat{\lambda}_i$$

The asymptotic distribution of a PMLE is given by (see Chap. 3.2.2)

$$\sqrt{n}(\hat{\beta} - \beta_0) \xrightarrow{d} N(0, n\mathcal{J}^{-1}\mathcal{I}\mathcal{J}^{-1}) \tag{4.127}$$

The matrices $\mathcal{J}(\beta)$ and $\mathcal{I}(\beta)$ are defined as minus the expected value of the Hessian and the expected value of the outer product of the score vector (which is equal to its variance), respectively, where the expectations are taken with respect to the true density:

$$\mathcal{J}(\beta) = \{-E_0(\partial^2 \ell(\beta; y, x)/\partial\beta\partial\beta')\}$$

$$\mathcal{I}(\beta) = \{E_0(\partial\ell(\beta; y, x)/\partial\beta \cdot \partial\ell(\beta; y, x)/\partial\beta')\}$$

The matrices are identical if the true probability function is chosen and equal to the Fisher information matrix. In case of misspecification, i.e., for a PMLE, $\mathcal{J}$ and $\mathcal{I}$ diverge.

Using the conditional moments of Y, which converge in probability to the marginal moments by the law of large numbers, the corresponding matrices $\mathcal{J}$ and $\mathcal{I}$ for the Poisson model are readily established:

$$\mathcal{J} = \sum_{i=1}^{n} x_i x_i' \mathrm{E}(y_i|x_i) \tag{4.128}$$

$$\mathcal{I} = \sum_{i=1}^{n} x_i x_i' \mathrm{Var}(y_i|x_i) \tag{4.129}$$

Wrongly assuming correct specification would mean estimating the variance matrix of $\hat{\beta}$ by $\mathcal{J}^{-1}$ instead of $\mathcal{J}^{-1}\mathcal{I}\mathcal{J}^{-1}$. The error is thus

$$\begin{aligned} \Delta &= \mathcal{J}^{-1} - \mathcal{J}^{-1}\mathcal{I}\mathcal{J}^{-1} \\ &= \mathcal{J}^{-1}(\mathcal{J} - \mathcal{I})\mathcal{J}^{-1} \end{aligned}$$

where

$$\mathcal{J} - \mathcal{I} = \sum_{i=1}^{n} x_i x_i' [\mathrm{E}(y_i|x_i) - \mathrm{Var}(y_i|x_i)]$$

In case of overdispersion $\text{Var}(y_i|x_i) > \text{E}(y_i|x_i)$ and Δ is negative definite. This implies that the standard variance estimator of the Poisson regression model $\mathcal{J}^{-1}$ underestimates the true standard errors, and that the t-values are too high. The opposite is true for the case of underdispersion. In both cases, the resulting inference based on the Poisson regression model is invalid.

These results suggest the use of the following strategy: rely on the consistency of the parameter estimates and calculate (asymptotically) valid standard errors. This is basically pseudo-maximum likelihood estimation and it will be referred to as robust Poisson regression. The central issue is how to calculate a valid variance matrix for $\hat{\beta}$. First, the matrices $\mathcal{J}$ and $\mathcal{I}$ are replaced by their sample moments:

$$\hat{\mathcal{J}}(\hat{\beta}) = \sum_{i=1}^{n} x_i x_i' \hat{\lambda}_i \tag{4.130}$$

$$\hat{\mathcal{I}}(\hat{\beta}) = \sum_{i=1}^{n} x_i x_i' \widehat{\text{Var}}(y_i|x_i;\hat{\beta}) \tag{4.131}$$

It follows that the estimated variance matrix of $\hat{\beta}$ is of the form:

$$\widehat{\text{Var}}(\hat{\beta}) = \left[\sum_{i=1}^{n} x_i x_i' \hat{\lambda}_i\right]^{-1} \sum_{i=1}^{n} x_i x_i' \widehat{\text{Var}}(y_i|x_i;\hat{\beta}) \left[\sum_{i=1}^{n} x_i x_i' \hat{\lambda}_i\right]^{-1}$$

The crucial step is the evaluation of the term $\widehat{\text{Var}}(y_i|x_i;\hat{\beta})$. The following three assumptions for the variance function can be distinguished:

a) <u>No Assumption</u> (White, 1982, Breslow, 1990)
This leads to the usual heteroskedasticity-consistent variance matrix, and the following estimate is used:

$$\widehat{\text{Var}}(y_i|x_i;\hat{\beta}) = (y_i - \hat{\lambda}_i)^2$$

b) <u>Linear Variance Function</u> (McCullagh and Nelder, 1989)
The estimated variance function is given by:

$$\widehat{\text{Var}}(y_i|x_i;\hat{\beta}) = \hat{\sigma}^2 \hat{\lambda}_i$$

In this case, the estimated variance matrix of $\hat{\beta}$ thus simplifies to $\widehat{\text{Var}}(\hat{\beta}) = \hat{\sigma}^2 \frac{1}{n} \hat{\mathcal{J}}^{-1}$.
This approach is common in generalized linear modeling. McCullagh and Nelder (1989) propose to estimate σ^2 by a moment estimator:

$$\hat{\sigma}^2 = \frac{1}{n-k} \sum_{i=1}^{n} \frac{(y_i - \hat{\lambda}_i)^2}{\hat{\lambda}_i}$$

c) Quadratic Variance Function (Gourieroux, Monfort and Trognon, 1984b) The estimated variance function is given by:

$$\widehat{\text{Var}}(y_i|x_i;\hat{\beta}) = \hat{\lambda}_i + \hat{\sigma}^2\hat{\lambda}_i^2$$

An estimate for σ^2 can be obtained by the auxiliary regression (See also Cameron and Trivedi, 1990):

$$(y_i - \hat{\lambda}_i)^2 - \hat{\lambda}_i = \sigma^2\hat{\lambda}_i^2 + v_i$$

This regression yields a strongly consistent estimator for σ^2.

A cautionary remark applies to the the third assumption of a quadratic variance function. While for the first two assumptions, PML estimation based on the Poisson distribution uses the *available* information efficiently, this is not the case for the third. Gourieroux, Monfort and Trognon (1984b) show that incorporating this information on the variance into the estimation provides a gain in efficiency, and they call this procedure *quasi-generalized pseudo maximum likelihood estimation* (QGPML). In the special case, where the true density is itself a member of a linear exponential family (which is fully characterized by its first two moments), QGPML estimation is asymptotically equivalent to ML estimation and hence fully efficient.

Monte Carlo Study

Given the three different estimators for a robust variance covariance matrix of $\hat{\beta}$, a Monte Carlo study might give some indication on whether they provide a substantial gain as compared to the use of the conventional Poisson-variance matrix. We also study, whether the three estimators lead to substantially different results and whether the validity of inference is robust with respect to the choice of a particular assumption. The latter finding would increase the overall confidence in Robust Poisson regression since otherwise one would need to rely, for example, on pre-tests to justify the particular variance assumption.

There have been a number of previous Monte Carlo studies to evaluate the finite sample properties of the Poisson regression model with adjusted covariance matrix estimator. Examples are Winkelmann and Zimmermann (1992a), Brännäs (1992a) and Bourlange and Doz (1988). The earlier results are extended by including an investigation of the performance of the covariance estimator that is based on the assumption of a quadratic variance function.

The design of the study is as follows: Firstly, random samples of count data with different degrees of overdispersion are drawn. Then, a PMLE are obtained based on the Poisson regression model. This procedure is repeated 1000 times. The sampling distribution of the PMLE over the 1000 replications can be investigated.

Simulating random counts with equidispersion proceeded as follows: A linear predictor is modeled as

$\eta = -1 + x$

where x is a $(n \times 1)$ vector of standard uniform random numbers. Thus, the true parameter vector is β_0 =(-1,1)'. The predictor is kept fixed over replications as we are interested in the *conditional* distribution of y_i given λ_i, where $\lambda_i = \exp(\eta_i)$. Pseudo-random Poisson numbers are obtained with an algorithm described in Knuth (1969).

To obtain random counts with overdispersion, a stochastic expected value is introduced: $\tilde{\lambda}_i = \exp(\eta_i + \varepsilon_i) = \lambda_i u_i$. ε_i are independent random normal numbers with constant variance σ_ε^2 and mean $\mu_\varepsilon = -0.5\,\sigma_\varepsilon^2$. Thus $u_i = \exp(\varepsilon_i)$ has mean $\mathrm{E}(u) = 1$ and variance $\sigma_u^2 = \exp(\sigma_\varepsilon^2) - 1$. The conditional variance of the Poisson-log-normal model is given by $\mathrm{Var}(y_i|x_i) = \lambda_i + \sigma_u^2 \lambda_i^2$. The degree of overdispersion depends on σ_ε^2 which is chosen in a way as to yield the values $0.2, 1.0$ and 5.0 for σ_u^2. In this way, the experiments cover a range from modest to substantial overdispersion.

In order to study the impact of increasing sample size on the quality of the approximation for three degrees of overdispersion, the experiments were conducted for samples of size 100 and 1000, respectively. The results are given in Tab. 4.1 and Tab. 4.2. For both sample sizes the mean and standard errors of the slope coefficient $\hat{\beta}_1$ are given. Furthermore, the empirical size for a two-sided asymptotic t-test under two alternative nominal significance levels are reported. The t-values were calculated in four different ways, following the different possible assumptions discussed above: $\widehat{\mathrm{Var}}_{\mathrm{psn}}$ is based on the assumption of a correctly specified model. $\widehat{\mathrm{Var}}_{\mathrm{White}}$ allows for all kinds of misspecification, while $\widehat{\mathrm{Var}}_{\mathrm{lvf}}$ and $\widehat{\mathrm{Var}}_{\mathrm{qvf}}$ are based on specific violations of the variance assumption. In the former case, the variance is a linear function of the expected value, while in the latter case, it is assumed to be a quadratic function (this, incidentally, is the 'true' model).

In interpreting the results, the focus is on consistency and valid inference. For $n = 100$, the deviation of the mean of the estimated coefficient $\hat{\beta}_1$ from its true value 1.0 is only in the second decimal place. For $n = 1000$, the deviation reduces to the third decimal place. For both sample sizes, this holds independently of the magnitude of overdispersion. The sample standard error increases with increasing overdispersion, and decreases strongly with sample size. These results confirm that the Poisson model yields consistent parameter estimates also in the presence of overdispersion.

A very different conclusion has to be drawn from the estimated standard errors. The empirical size of the test can be compared to its nominal size, i.e., the significance level of the test. If the empirical size exceeds the nominal size by an amount that is beyond the sampling variation expected from 1000 replications, it indicates an underestimation of the standard errors. And in fact, the test based on the Poisson variance estimator systematically overstates the nominal size. Not surprisingly, the underestimation of the true standard errors is the more severe, the larger the overdispersion. In the case of slight overdispersion, the asymptotic t-test for the smaller sample leads to type-I errors that are around 25% higher than the significance levels $\alpha = 0.05, 0.10$. For $\sigma_u^2 = 1$ and $n = 100$, the effective type-I error is already two times higher than the significance level $\alpha = 0.05$. For $\sigma_u^2 = 5$, the underestimation of the standard errors further increases. As expected, the bias does not improve with increased sample size.

Given this poor performance of the Poisson standard errors, the three alternative ways to calculate robust standard errors offer a clear improvement. For the larger sample, the nominal size of the test is closely realized even in the case of extreme overdispersion. The observed Poisson type-I error of 40.8% is reduced to 12.5% ($\widehat{\text{Var}}_{White}$), 10.9% ($\widehat{\text{Var}}_{lvf}$), and 11.4% ($\widehat{\text{Var}}_{qvf}$), respectively. Also in the small sample the performance of the three robust test procedures is much better than the one based on Poisson standard errors for $\sigma_u^2 = 1$ and $\sigma_u^2 = 5$ and is comparable to the performance of the test based on Poisson standard errors for $\sigma_u^2 = 0.2$.

These experiments demonstrate how misleading the use of Poisson standard errors can be in the presence of overdispersion, and how well the robust standard errors perform already in a medium sized sample. Not surprisingly, they also indicate a slight superiority of the t-test based on the assumption of a quadratic variance function (which is the correct one) which realizes closest the nominal size of the test in most, though not in all, of the experiments.

Table 4.1. Simulation Study for Poisson-PMLE: N=100

	$\sigma_u^2 = 0.2$	$\sigma_u^2 = 1$	$\sigma_u^2 = 5$	
Mean $\hat{\beta}_1$	1.0282	0.9900	1.0351	
Std $\hat{\beta}_1$	0.4997	0.5606	0.9045	
	$\widehat{\text{Var}}_{\text{psn}}$	$\widehat{\text{Var}}_{\text{White}}$	$\widehat{\text{Var}}_{\text{lvf}}$	$\widehat{\text{Var}}_{\text{qvf}}$
$\sigma_u^2 = 0.2$				
t-test ($\alpha = 0.10$)	0.135	0.118	0.110	0.108
t-test ($\alpha = 0.05$)	0.058	0.066	0.049	0.056
$\sigma_u^2 = 1$				
t-test ($\alpha = 0.10$)	0.184	0.103	0.085	0.126
t-test ($\alpha = 0.05$)	0.105	0.055	0.044	0.064
$\sigma_u^2 = 5$				
t-test ($\alpha = 0.10$)	0.342	0.150	0.125	0.121
t-test ($\alpha = 0.05$)	0.272	0.087	0.069	0.064

4.10.4 Non-Parametric Variance Estimation

Delgado and Kniesner (1997) propose an estimation method for a Poisson model with variance of unknown form that relies on generalized least squares

Table 4.2. Simulation Study for Poisson-PMLE: N=1000

	$\sigma_u^2 = 0.2$	$\sigma_u^2 = 1$	$\sigma_u^2 = 5$	
Mean $\hat{\beta}_1$	0.9947	0.9953	0.9975	
Std $\hat{\beta}_1$	0.1507	0.1754	0.2927	
	$\widehat{\text{Var}}_{\text{psn}}$	$\widehat{\text{Var}}_{\text{White}}$	$\widehat{\text{Var}}_{\text{lvf}}$	$\widehat{\text{Var}}_{\text{qvf}}$
$\sigma_u^2 = 0.2$				
t-test ($\alpha = 0.10$)	0.116	0.095	0.087	0.101
t-test ($\alpha = 0.05$)	0.057	0.043	0.045	0.048
$\sigma_u^2 = 1$				
t-test ($\alpha = 0.10$)	0.171	0.087	0.085	0.102
t-test ($\alpha = 0.05$)	0.103	0.045	0.039	0.048
$\sigma_u^2 = 5$				
t-test ($\alpha = 0.10$)	0.408	0.125	0.109	0.114
t-test ($\alpha = 0.05$)	0.323	0.065	0.055	0.058

using non-parametric estimation of the conditional variance. In particular, assume that y_i is distributed with mean $\mathrm{E}(y_i|x_i) = \exp(x_i'\beta)$ and conditional variance

$$\sigma_i^2 = \text{Var}(y_i|x_i) \tag{4.132}$$

of unspecified functional form. The conditional variances can be estimated non-parametrically as

$$\hat{\sigma}_i^2 = \sum_{j=1}^{k}(y_j - \exp(x_j'\tilde{\beta}))^2 w_{ij} \tag{4.133}$$

where $\tilde{\beta}$ is a root-n consistent estimator (for instance the Poisson PMLE), and w_{ij} are non-parametric k nearest neighbors probabilistic weights (see Delgado and Kniesner, 1997, for further details).

The semiparametric weighted least squares estimator β_0 is then obtained as a solution to

$$\sum_{i=1}^{n}\exp(x_i'\beta)x_i(y_i - \exp(x_i'\beta))\hat{\sigma}_i^{-2} = 0 \tag{4.134}$$

Delgado and Kniesner (1997) show that this estimator reaches the semiparametric efficiency bound, and state conditions for asymptotic normality. The model is applied to the number of work absence days in a sample of London Bus conductors and drivers during the early eighties. Robustness checks show that parameter estimates are sensitive to the choice of regressors but insensitive to the adopted econometric technique, including how the variance function is specified.

4.11 Quantile Regression for Count Data

Quantile regression has been in use in the context of linear regression models for a long time (see Koenker and Hallok, 2001, for a general introduction to the topic). Regression based on least absolute deviations, or median regression, offers an alternative to least squares. One of the advantages of quantile regression is that is robust to outliers. More importantly, though, quantile regressions can be performed for arbitrary quantiles of a distribution. Seen in this way, quantile regression becomes a tool for modelling the effect of regressors on the full distribution of the outcome variable, rather than modelling their effect on the first noncentral moment of the distribution only.

Quantile regression is not the only possibility to model the whole distribution. In fact, it has been pointed out before that all count data models proper are probability models and allow inferences to be drawn on all possible aspects of the outcome distibution, including the computation of marginal probability effects. Also, if the underlying model is specified with sufficient flexibility (for instance using hurdle type models) such inferences are not tautological but informed by data. Probability based models and quantile based models are two sides of the same coin: probability models are based on the representation of a random variable through its probability function, whereas quantile models are based on the representation of a random variable through its distribution function.

Fig. 4.2. Count Data Distribution Function Without Uniform Distribution Added

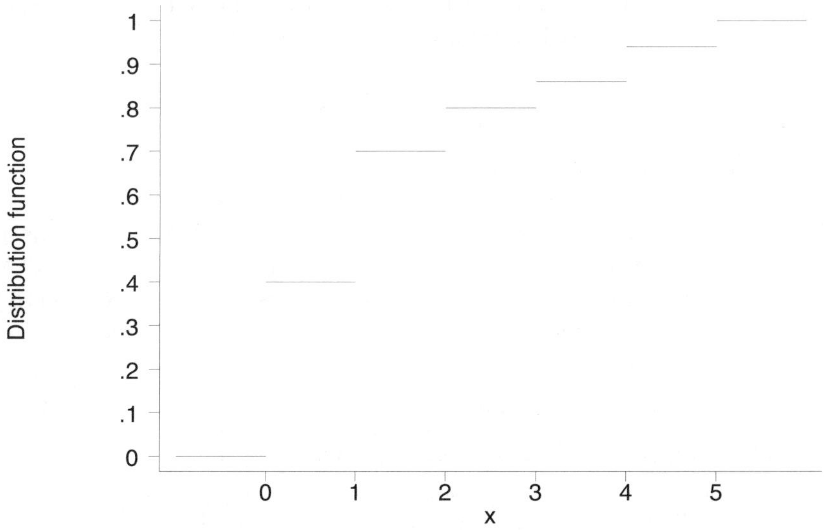

The main drawback of the latter is then of a more technical nature, namely the fact that the distibution function of a discrete random variable is not continuous. See Fig. 4.2 for a stylised distribution function with its typical jumps. However, this difficulty can be overcome, as shown by Machado and Santos Silva (2002). Let y be a count random variable. The α-quantile of y is defined by

$$Q_y(\alpha) = \min(\eta | P(y \leq \eta) \geq \alpha)$$

where $0 \leq \alpha < 1$. The object of interest is the conditional quantile $Q_y(\alpha|x)$. Since $Q_y(\alpha|x)$ has the same support as y, it is discrete and cannot be a continuous function of x (such as $\exp(x'\beta)$). Therefore, Machado and Santos Silva suggest to introduce "jittering": consider a new variable z, obtained by adding a uniform random variable to the count variable

$$z = y + u \text{ where } u \sim \text{ uniform } [0, 1)$$

where y and u are independent. Hence, z has density function

$$f(z) = \begin{cases} p_0 \text{ for } 0 \leq z < 1 \\ p_1 \text{ for } 1 \leq z < 2 \\ \text{and so forth} \end{cases}$$

(using notation $P(Y = k) = p_k$). Moreover, the distribution function of z can be written as

$$F(z) = \begin{cases} p_0 z \text{ for } 0 \leq z < 1 \\ p_0 + p_1(z-1) \text{ for } 1 \leq z < 2 \\ \text{and so forth} \end{cases}$$

The distribution function of z is also shown in Fig. 4.3. We see that the quantiles of z are continuous. For example,

$$z_\alpha = \frac{\alpha}{p_0} \text{ for } \alpha < p_0$$

$$z_\alpha = 1 + \frac{\alpha - p_0}{p_1} \text{ for } p_0 \leq \alpha < p_0 + p_1$$

Having established the relationship between the probability function of y and the quantiles of z, we can now turn to the estimation of the quantiles. The probabilities being unknown, the conditional z-quantiles are estimated directly by adopting a regression framework. Machado and Santos Silva (2002) specify the quantiles $Q_z(\alpha|x)$ as

$$Q_z(\alpha|x) = \alpha + \exp(x'\gamma(\alpha)), \quad \alpha \in (0, 1) \tag{4.135}$$

The reason for adding α on the right side of (4.135) is that the z-quantiles are bounded from below at α. The minimum is reached in the degenerate case where $p_0 = 1$. The dependent variable z (and with it the z-quantiles since quantiles are invariant under transformation, i.e., $Q_{g(z)} = g(Q_z)$) can

Fig. 4.3. Count Data Distribution Function With Uniform Distribution Added

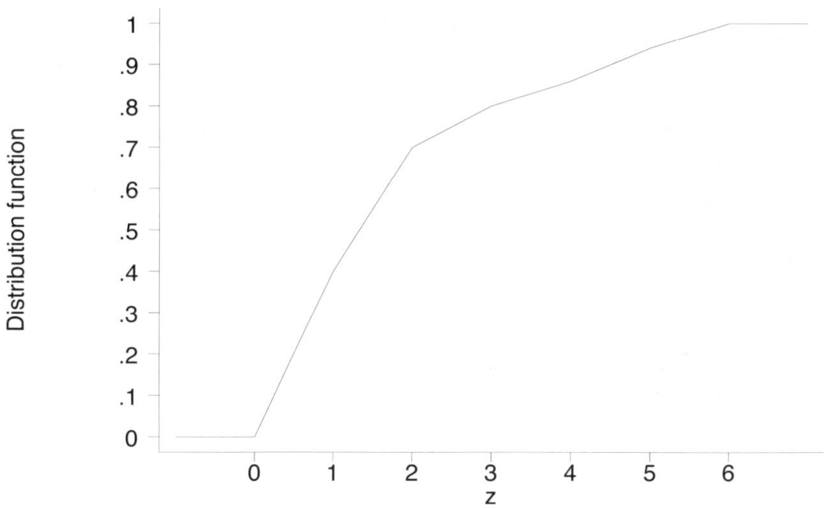

be transformed such that the transformed quantile function is linear in x and γ. Observe that

$$Q_{T(z;\alpha)}(\alpha|x) = x'\gamma(\alpha)$$

where

$$T(z;\alpha) = \begin{cases} \log(z-\alpha) & \text{for } z > \alpha \\ \log(\xi) & \text{for } z \le \alpha \end{cases} \tag{4.136}$$

and $0 < \xi < \alpha$.

The model suggest the following empirical implementation. First, one adds uniformly distributed pseudo random numbers to the observed counts. Second, one transforms the resulting data according to (4.136). Third and finally, the parameter estimates are obtained as solution to

$$\min \sum_{i=1}^{n} \rho_\alpha(T(z_i;\alpha) - x_i'\gamma)$$

where $\rho_\alpha(\nu) = \nu \times (\alpha - I(\nu < 0))$. For example, if $\alpha = 0.5$, the *rho* function returns simply the absolute value $0.5|T(z_i;\alpha) - x_i'\gamma|$.

Machado and Santos Silva (2002) prove consistency and asymptotic normality of this estimator. Although the quantile function is not differentiable everywhere (the distribution function has corners), these points do not affect the derivation as long as there is at least one continuous regressor, because in this case these corner points have measure zero.

The final question is how the parameters should be interpreted. In this approach, one estimates $Q_z(\alpha|x)$. The object of interest is $Q_y(\alpha|x)$, though. There is a unique correspondence between the two, since $Q_y(\alpha|x) = \text{int}^*[\mathrm{Q_Z(\alpha|x)}]$, where $\text{int}^*(a)$ is the ceiling function which returns the smallest integer greater than, or equal, to a. Hence, for example, testing the null hypothesis that a variable has no effect on $Q_Y(\alpha|x)$ is equivalent to testing that a variable has no effect on $Q_Z(\alpha|x)$.

5

Correlated Count Data

5.1 Multivariate Count Data

Multivariate count data are likely to have a non-trivial correlation structure. For instance, omitted variables may simultaneously affect more than one count. The modeling of the correlation structure is important for the efficiency of the estimator and the computation of correct standard errors, i.e., valid inference. Beyond that, the nature of the stochastic interaction between several counts may be of independent and intrinsic interest.

Following the usual notation for multivariate data let y_{ij}, $i = 1, \ldots, n$, $j = 1, \ldots, J$ denote the count for individual i and outcome j. Let $y_i = (y_{i1} \cdots y_{iJ})'$ denote the vector of counts for individual i over the different outcomes. Interest is in a class of models where observations are uncorrelated across individuals but correlated over outcomes:

$$\text{Cov}(y_{ij}, y_{kl}) \begin{cases} = 0 & \text{for } i \neq k \\ \neq 0 & \text{for } i = k,\ j \neq l \end{cases}$$

In matrix notation, correlated outcomes mean that $\text{Var}(y_i)$ is not a diagonal matrix. Five models for correlated counts are discussed in this chapter

- Multivariate Poisson Model
- Multivariate Negative Binomial Model
- Multivariate Poisson-Gamma Mixture Model
- Multivariate Poisson-Log-Normal Model
- Latent Poisson-Normal Model

The first three models have in common that they are based on a so-called one-factor structure: correlation is generated through an individual specific random factor u_i that does not vary over outcomes. A direct implication of this relatively simplistic assumption is that the covariance structure of these models is restricted to non-negative correlations. A more flexible covariance structure is provided by the multivariate Poisson-log-normal and the latent Poisson-normal models, two examples of multi-factor models. The distinction between one- and multi-factor models will become clearer once the models are presented.

The class of models pursued in this chapter encompasses a variety of interesting sub-cases and is more general than might appear at first. The initial interpretation is in terms of genuine multivariate outcomes. One example occurs when studying the provision of health services where a researcher might be interested in a joint analysis of the number of individual visits to a doctor and to a non-doctor health specialist (Gurmu and Elder, 1998). A further example is in the study of labor mobility, where one may be interested in the joint determination of voluntary and involuntary job changes (Jung and Winkelmann, 1993).

An alternative interpretation is in terms of seemingly unrelated regression (SURE). In this case, one considers a single outcome variable, for example the annual number of airtraffic incidents by major air carrier over a 30 year period as in Rose (1990). In a sense, these are also panel data, see below. However, they differ from usual panel data in at least two respects. First, in typical SURE applications, the time dimension is large relatively to the cross-section dimension. As a consequence, it becomes possible to estimate unit-specific constants and slope parameters for each cross-sectional unit. Second, because the number of cross sectional units is small and these units may interact (such as firms operating in the same market, or countries) and the random sampling assumption may be invalid. To be specific, under the assumption of the SURE model, observations are assumed to be (contemporaneously) correlated across observations (and hence not an independent random sample) but uncorrelated over time, given the unit-specific fixed effects. In order to make the SURE notation conform to the multivariate approach, let i denote time and j denote the cross-sectional observations. In this way, $\text{Cov}(y_{ij}, y_{kl})$ is a contemporaneous covariance for $i = k$, $j \neq l$.

A third area of application is to panel data. As mentioned before, both panel and SURE-type data have an individual and a time dimension. However, for panel data proper, the individual dimension is typically large relative to the time dimension. Often, such datasets includes measurements for thousands of individuals over a few years. Moreover, contemporaneous correlation is precluded by the assumption of random sampling, a reasonable assumption in such context. In this case, the interest is rather in the correlation of observations for a given individual over time. Such correlation may arise, for instance, from individual specific random coefficients. To adjust the multivariate set-up to the panel case i denotes the individual and j the period. Models for panel data are discussed in greater detail in Chap. 5.2.

A further difference between the panel view and the other two interpretation is that for multivariate and SURE data, one is interested in the correlation structure mainly in order to improve efficiency. For panel data, however, omitted variable bias, and hence the question whether the individual specific effects should be treated as fixed or random, is a major issue. Because T is small relative to n, estimating n separate constants is not a trivial issue. It reduces the degrees of freedom considerably. Typical multivariate and SURE data do not have this problem, and in these case, we can always let

$$\mathrm{E}(y_{ij}|x_{ij}) = \exp(x'_{ij}\beta_j)$$

Random coefficients models, though feasible (see Chib, Greenberg and Winkelmann, 1998, and Chap. 6.4) have been rarely used in practice so far. In order to keep notation simple, the following exposition does abstract from heterogeneity in slope parameters, be it observed or unobserved.

The classification of data into multivariate, SURE, or panel may be useful. It should not hide the fact, however, that the labels sometimes will be ambiguous. Also, hybrid cases clearly can exists, such as multivariate panel data, as discussed in Million (1998) and in Riphahn, Wambach and Million (2003).

5.1.1 Multivariate Poisson Distribution

In this chapter, the multivariate Poisson (MVP) distribution is derived and characterized. Its appeal stems both from its relative simple structure and its derivation based on a common additive error. The three main features of the multivariate Poisson distribution are

1. Two MVP distributed random variables are independent if and only if they are uncorrelated.
2. The marginal distributions of the MVP distribution are univariate Poisson.
3. The conditional distributions of the bivariate Poisson distribution are sums of an independent Poisson and binomial distribution. The conditional expectation functions are linear.

Kocherlakota and Kocherlakota (1992) provide a comprehensive discussion of the MVP distribution. They also point out that the adopted formulation is just one possible extension of the Poisson distribution to the multivariate setting (See Lakshminarayana, Pandit and Rao, 1999, for a recent example for a bivariate Poisson distribution based on a polynomial factor). So far, the MVP distribution seems to be the only one that has been put to practical use in econometrics, and we thus restrict the attention to this particular model.

The derivation, sometimes referred to as "trivariate reduction method", is based on a convolution structure: Let the random variables z_{ij}, $j = 1, \ldots, J$ and u_i have independent Poisson distributions with $z_{ij} \sim \text{Poisson}(\lambda_{ij})$, and $u_i \sim \text{Poisson}(\gamma)$ where $\lambda_{ij} = \exp(x'_{ij}\beta)$. New random variables y_{ij} can be constructed as

$$y_{ij} = z_{ij} + u_i \tag{5.1}$$

Using the convolution property of independent Poisson variables, $y_{ij} \sim \text{Poisson}(\lambda_{ij} + \gamma)$, $j = 1, \ldots, J$, that is, the bivariate Poisson distribution is characterized by Poisson *marginals*.

Furthermore,

$$\text{Cov}(y_{ij}, y_{kl}) = \begin{cases} 0 & \text{for } i \neq k \\ \lambda_{ij} + \gamma & \text{for } i = k,\ j = l \\ \gamma & \text{for } i = k,\ j \neq l \end{cases}$$

The last equality follows since for $j \neq l$

$$\begin{aligned} \text{Cov}(z_{ij} + u_i, z_{il} + u_i) &= \text{Var}(u_i) \\ &= \gamma \end{aligned}$$

Normalization by the standard deviations of the two random variables yields the correlation form

$$\text{Corr}(y_{ij}, y_{kl}) = \frac{\gamma}{\sqrt{(\lambda_{ij} + \gamma)(\lambda_{kl} + \gamma)}} \tag{5.2}$$

It can be shown that the correlation coefficient cannot exceed the square root of the ratio of the smaller to the larger of the means of the two marginal distributions. The covariance matrix of y_i can be written in more compact form. Let $\Lambda_i = \text{diag}(\lambda_{ij})$ and $\mathbf{1}$ a $(J \times 1)$ vector of ones. The covariance matrix of $y_i = (y_{i1} \cdots y_{iJ})'$ is then given by

$$\text{Var}(y_i) = \Lambda_i + \gamma \mathbf{1}\mathbf{1}'$$

Since $\gamma \geq 0$ the model allows only non-negative correlations. This property of the MVP, as deplorable as it might look at first glance, is in fact a direct result of its simple one-factor structure. The same result would be obtained if z_{ij} and u_i were independently normally distributed with variances σ_z^2 and σ_u^2, respectively, from where $\text{Cov}(y_{ij}, y_{kl}) = \sigma_u^2 \geq 0$. But such random effects models are widely used in econometrics, in particular for panel data. Thus, the MVP model should (at least) be useful in related situations involving panel count data.

The joint probability function of the MVP distribution for individual i is given by

$$\begin{aligned} f(y_{i1}, ..., y_{iJ}) &= \exp\left[-(\lambda_{i1} + \ldots + \lambda_{iJ} + \gamma)\right] \\ &\times \sum_{u_i=0}^{s_i} \frac{\gamma^{u_i}}{u_i!} \frac{\lambda_{i1}^{y_{i1}-u_i}}{(y_{i1} - u_i)!} \cdots \frac{\lambda_{iJ}^{y_{iJ}-u_i}}{(y_{iJ} - u_i)!} \end{aligned} \tag{5.3}$$

with $s_i = \min(y_{i1}, ..., y_{iJ})$. The intuition behind this joint probability function is as follows. First, u_i cannot exceed any of the y_{ij}'s because each count is the sum of u_i and a non-negative count z_{ij}. Hence, its upper bound is s_i. Secondly, the joint probability of observing $(y_{i1}, ..., y_{iJ})$ is the sum over the joint probabilities $f(u_i, y_{i1} - u_i, ..., y_{iJ} - u_i)$ where $u_i = 0, \ldots, s_i$. From independence, it follows that the joint probability can be factored such that

$$f(u_i, y_{i1} - u_i, ..., y_{iJ} - u_i) = f(u_i) f(y_{i1} - u_i) \cdots f(y_{iJ} - u_i)$$

In order to derive the conditional distributions of the MVP, consider for simplicity the bivariate case:

$$
\begin{aligned}
f(y_{i1}|y_{i2}) &= \frac{f(y_{i1},y_{i2})}{f(y_{i2})} \\
&= \frac{\exp(-\lambda_{i1}-\lambda_{i2}-\gamma)\sum_{u_i=0}^{s_i}\frac{\gamma^{u_i}}{u_i!}\frac{\lambda_{i1}^{y_{i1}-u_i}}{(y_{i1}-u_i)!}\frac{\lambda_{i2}^{y_{i2}-u_i}}{(y_{i2}-u_i)!}}{\exp(-\lambda_{i2}-\gamma)(\lambda_{i2}+\gamma)^{y_{i2}}/y_{i2}!} \\
&= \sum_{u_i=0}^{s_i}\left\{\frac{y_{i2}!}{u_i!(y_{i2}-u_i)!}\left(\frac{\gamma}{\lambda_{i2}+\gamma}\right)^{u_i}\left(\frac{\lambda_{i2}}{\lambda_{i2}+\gamma}\right)^{y_{i2}-u_i}\right. \\
&\qquad \left.\times\exp(-\lambda_{i1})\frac{\lambda_{i1}^{y_{i1}-u_i}}{(y_{i1}-u_i)!}\right\}
\end{aligned} \tag{5.4}
$$

This is the distribution of the sum of two independent variables (Recall the generic formula for a convolution: $f(z)=\sum_{i=0}^{z} f_x(z-i)f_y(i)$). Here, y_{i1} is Poisson distributed with parameter λ_{i1}, and $u_i|y_{i2}$ is binomial distributed with $n=y_{i2}$ and $p=\gamma/(\lambda_{i2}+\gamma)$. It follows that

$$
\begin{aligned}
\mathrm{E}(y_{i1}|y_{i2}) &= \mathrm{E}(y_{i1})+\mathrm{E}(u_i|y_{i2}) \\
&= \lambda_{i1}+\frac{\gamma}{\lambda_{i2}+\gamma}y_{i2}
\end{aligned}
$$

Thus, the bivariate Poisson distribution defines a linear regression between y_{i1} and y_{i2} (and conversely). This property could be used in order to test for correlated counts using OLS. If λ_{ij}, $j=1,2$ was specified as a non-linear function of additional parameters (such as $\lambda_{ij}=\exp(x'_{ij}\beta)$) the conditional expectation function would need to be estimated by non-linear least squares.

Interestingly, a slight modification of the conditional expectation function leads to a model that allows for both positive and negative correlations between y_{i1} and y_{i2} (and is thus unrelated to the BVP). Berkhout and Plug (1996) study the situation where y_{i1} is Poisson distributed and $y_{i2}|y_{i1}$ is also Poisson distributed with mean

$$
\begin{aligned}
\mathrm{E}(y_{i2}|y_{i1}) &= \lambda_{i2}\exp(\alpha y_{i1}) \\
&= \exp(x'_{i2}\beta_2+\alpha y_{i1})
\end{aligned}
$$

Thus, the conditioning variable enters multiplicatively rather than additively. They show that the sign of the correlation between y_{i1} and y_{i2} corresponds to the sign of α.

Probability Generating Function of the MVP

Joint distributions for non-negative integer random variables can be modeled using joint probability generating functions (See Appendix A). This method provides an alternative characterization of the multivariate Poisson distribution. For notational convenience the exposition is limited to the bivariate case.

The bivariate probability generating function of two random variable X and Y is defined as $\mathcal{P}(s_1, s_2) = \mathrm{E}(s_1^X s_2^Y)$. Thus, in the bivariate Poisson model, the probability generating function for the joint distribution of (y_{i1}, y_{i2}) is given by:

$$\begin{aligned} \mathcal{P}(s_1, s_2) &= \mathrm{E}(s_1^{y_{i1}} s_2^{y_{i2}}) \\ &= \mathrm{E}(s_1^{z_{i1}+u_i} s_2^{z_{i2}+u_i}) \\ &= \mathrm{E}((s_1 s_2)^{u_i} s_1^{z_{i1}} s_2^{z_{i2}}) \\ &\overset{(*)}{=} \mathrm{E}((s_1 s_2)^{u_i})\mathrm{E}(s_1^{z_{i1}})\mathrm{E}(s_2^{z_{i2}}) \\ &= \exp(-\gamma + s_1 s_2 \gamma) \exp(-\lambda_{i1} + \lambda_{i1} s_1) \exp(-\lambda_{i2} + \lambda_{i2} s_2) \\ &= \exp(-\lambda_{i1} - \lambda_{i2} - \gamma + \lambda_{i1} s_1 + \lambda_{i2} s_2 + \gamma s_1 s_2) \end{aligned} \tag{5.5}$$

where $(*)$ follows from the independence assumption.

The probability function can be derived from (5.5) using the relationship

$$f(y_{i1}, y_{i2}) = (y_{i1}! y_{i2}!)^{-1} \left. \frac{\partial^{y_{i1}+y_{i2}} \mathcal{P}}{(\partial s_1)^{y_{i1}} (\partial s_2)^{y_{i2}}} \right|_{s_1=s_2=0}$$

One can verify that this representation leads to the probability function (5.3). The marginal distributions are defined as (see Appendix A):

$$\begin{aligned} \mathcal{P}^{(y_{i1})}(s_1) &= \mathcal{P}(s_1, 1) \\ &= \exp[-\lambda_{i1} - \gamma + (\lambda_{i1} + \gamma) s_1] \end{aligned}$$

$$\begin{aligned} \mathcal{P}^{(y_{i2})}(s_2) &= \mathcal{P}(1, s_2) \\ &= \exp[-\lambda_{i2} - \gamma + (\lambda_{i2} + \gamma) s_2] \end{aligned}$$

The covariance of y_{i1} and y_{i2} can be calculated as

$$\begin{aligned} \mathrm{Cov}(y_{i1}, y_{i2}) &= \mathrm{E}(y_{i1} y_{i2}) - \mathrm{E}(y_{i1})\mathrm{E}(y_{i2}) \\ &= \gamma \end{aligned}$$

since

$$\begin{aligned} \mathrm{E}(y_{i1} y_{i2}) &= \sum_{y_{i1}=0}^{\infty} \sum_{y_{i2}=0}^{\infty} y_{i1} y_{i2} f(y_{i1}, y_{i2}) \\ &= \left[\frac{\partial^2 \mathcal{P}(s_1, s_2)}{\partial s_1 \partial s_2} \right]_{s_1=s_2=1} \\ &= \gamma + (\lambda_{i1} + \gamma)(\lambda_{i2} + \gamma) \end{aligned}$$

For $\gamma = 0$, the probability generating function can be factored:

$$\mathcal{P}(s_1, s_2) = \mathcal{P}(s_1, 1)\mathcal{P}(1, s_2)$$

and, therefore, y_{i1} and y_{i2} are independent (See Appendix A). Like for the bivariate normal distribution, independence and no correlation are equivalent notions. There are, however, two important differences: Whereas for the bivariate normal distribution both marginal *and* conditional distributions are again normal, here this holds only for the marginal distributions. Moreover, sums of Poisson random variables are again Poisson distributed if and only if they are independent: the probability generating function of the sum $y_{i1} + y_{i2}$ is obtained by setting $s_1 = s_2 = s$:

$$\mathcal{P}(s) = \exp[(\lambda_{i1} + \lambda_{i2})(s-1) + \gamma(s^2-1)]$$

For $\gamma = 0$, i.e. if the two Poisson variables are independent, this is the probability generating function of a Poisson distribution.

Bivariate Poisson Process

Yet another characterization of the bivariate Poisson distribution is based on the bivariate Poisson process: Let $y_1(t, t+\Delta), y_2(t, t+\Delta)$ be the number of events of two different types that occured between t and $t+\Delta$, $t, \Delta \in \mathbb{R}^+$. Assume that the probabilities of events y_1 or y_2 occuring in the interval $(t, t+\Delta)$ are independent of the previous process, and that

(i) The probability of one occurence of type 1 and no occurence of type 2 in the interval $(t, t+\Delta)$ is given by:

$$P(y_1 = 1, y_2 = 0) = \lambda_1 \Delta + \circ(\Delta)$$

(ii) The probability of one occurence of type 2 and no occurence of type 1 in the interval $(t, t+\Delta)$ is given by:

$$P(y_1 = 0, y_2 = 1) = \lambda_2 \Delta + \circ(\Delta)$$

(iii) The probability of one occurence of type 1 and one occurence of type 2 in the interval $(t, t+\Delta)$ is given by:

$$P(y_1 = 1, y_2 = 1) = \gamma \Delta + \circ(\Delta)$$

(iv) The probability of no event occuring is given by:

$$P(y_1 = 0, y_2 = 0) = 1 - \lambda_1 \Delta - \lambda_2 \Delta - \gamma \Delta + \circ(\Delta)$$

It can then be shown that the resulting probability generating function must be of the form

$$\mathcal{P}(t, s_1, s_2) = \exp[(-\lambda_1 - \lambda_2 - \gamma + \lambda_1 s_1 + \lambda_2 s_2 + \gamma s_1 s_2)t] \;. \tag{5.6}$$

Setting $t = 1$, the probability generating function for the bivariate Poisson distribution derived in (5.5) is obtained. This derivation of the bivariate Poisson distribution was proposed as early as 1926 by McKendrick (See the historical remarks in Kocherlakota and Kocherlakota, 1992). It can be given a spatial interpretation of moving along a Cartesian grid, where one-step movements along the y_1–axis and y_2–axis occur with probabilities λ_1 and λ_2, respectively, while a movement in both directions has probability γ.

Seemingly Unrelated Poisson Regression

The MVP probability model (5.3) together with parameterization

$$\lambda_{ij} = \exp(x'_{ij}\beta_j) - \gamma$$

is often refered to as seemingly unrelated Poisson regression. The model was introduced by King (1989a) who suggested estimation by maximum likelihood. Jung and Winkelmann (1993) give the first and second derivatives of the log-likelihood function.

Applications in econometrics include Jung and Winkelmann (1993) who study the joint determination of the number of voluntary and involuntary job changes over a ten-year period, and Ozuna and Gomez (1994) who study the number of trips to two recreational sites. Applications so far have been limited to the bivariate case, although this is definitely not a binding constraint.

Also, despite the labelling, all previous applications have dealt with data that are multivariate in nature rather than seemingly unrelated proper in the sense of Zellner (1962). This orientation has re-inforced the criticism of the MVP model as being potentially inappropriate, since it imposes non-negative correlation. This a-priori restriction is more of an drawback for multivariate data then it would be for SURE or panel data. In response, attention has shifted to multivariate mixing models such as the Poisson-log-normal model discussed below (Chib and Winkelmann, 2001, Gurmu and Elder, 1998).

Another criticism has been based on the restrictive variance assumption of the MVP model: the conditional expectation and conditional variance are assumed to be equal. One response has been to ignore the issue of over- or underdispersion in estimation but allow for valid inference by computing robust standard errors (Jung and Winkelmann, 1993). Alternatively, Winkelmann (2000a) derives a multivariate negative binomial model along the lines of the MVP model. This model allows for overdispersion. It is presented in the next chapter.

A final point of contention, raised by Gurmu and Elder (1998) is whether it is meaningful to assume that $z_{ij} \sim \text{Poisson}(\exp(x'_{ij}\beta_j) - \gamma)$. This specification does not guarantee that the parameter of the z_{ij}-distribution is positive, causing both conceptual and potentially numerical problems. In an alternative parameterization, $z_{ij} \sim \text{Poisson}(\exp(x'_{ij}\beta_j))$, resulting in a marginal distribution of $y_{ij} \sim \text{Poisson}(\exp(x'_{ij}\beta_j)+\gamma)$. Although the two models differ not only in their constant but also in the underlying assumption for the scedastic (variance) function, the interpretation of the regression parameters is the same in both parameterizations, as in either case $\partial \text{E}(y_{ij}|x_{ij})/\partial x_{ij} = \exp(x'_{ij}\beta_j)\beta_j$.

5.1.2 Multivariate Negative Binomial Model

A multivariate negative binomial (MVNB) model can be derived in close analogy to the MVP. Following Winkelmann (2000a), begin with a convolution structure and let

$$y_{ij} = z_{ij} + u_i$$

where z_{ij} and u_i have independent negative binomial distributions. In order to establish the distribution of the sum of two independent negative binomial distributions, recall the probability generating function of the negative binomial distribution from Chap. 2.3.1:

$$\mathcal{P}(s) = [1 + \theta(1 - s)]^{-\alpha}$$

In this specification, $\mathrm{E}(y) = \alpha\theta$ and $\mathrm{Var}(y) = \mathrm{E}(y)(1 + \theta)$. Thus, the sum of two independent negative binomial distributions is again negative binomial only if the two distributions share the common parameter θ. (This property of the negative binomial distribution was also exploited by Hausman, Hall and Griliches (1984, Appendix A), albeit in a different context). Consider a parameterization where

$$z_{ij} \sim \text{Negbin}(\theta = \sigma, \alpha = \lambda_{ij}/\sigma) \tag{5.7}$$

$$u_i \sim \text{Negbin}(\theta = \sigma, \alpha = \gamma/\sigma) \tag{5.8}$$

It follows that z_{ij} has mean $\lambda_{ij} = \exp(x'_{ij}\beta_j)$ and variance $\lambda_{ij}(1+\sigma)$, whereas u_i has mean γ and variance $\gamma(1 + \sigma)$. Thus, z_{ij} and u_i each are Negbin I distributed.

Applying the basic convolution rules to independent random variables, the distribution of $y_{ij} = z_{ij} + u_i$ can be established as

$$\begin{aligned} \mathcal{P}_y(s) &= \mathcal{P}_z(s)\mathcal{P}_u(s) \\ &= [1 + \sigma(1 - s)]^{-\lambda_{ij}/\sigma}[1 + \sigma(1 - s)]^{-\gamma/\sigma} \\ &= [1 + \sigma(1 - s)]^{-(\lambda_{ij}+\gamma)/\sigma} \end{aligned} \tag{5.9}$$

But (5.9) is the probability generating function of a Negbin I distribution with expectation $\mathrm{E}(y_{ij}) = \lambda_{ij} + \gamma$ and variance $\mathrm{Var}(y_{ij}) = (\lambda_{ij} + \gamma)(1 + \sigma)$. It is easy to verify that among the class of negative binomial distributions, only the Negbin I distribution is closed under convolution. The Negbin II distribution, in particular, is not.

Due to the common factor u_i, this model induces correlation between observations for the same individual but different outcomes: For $i = k$ and $j \neq l$

$$\begin{aligned} \mathrm{Cov}(y_{ij}, y_{kl}) &= \mathrm{Var}(u_i) \\ &= \gamma(1 + \sigma) \end{aligned}$$

Following the notation that was introduced for the MVP model, the covariance matrix of the MVNB model can be written in compact form as

$$\mathrm{Var}(y_i) = [\Lambda_i + \gamma\mathbf{1}\mathbf{1}'](1 + \sigma)$$

Note that this covariance matrix differs from the covariance matrix of the MVP model only by a factor of $(1 + \sigma)$. Thus, the MVNB model allows for overdispersion relative to the MVP model as long as $\sigma > 0$. The restriction implied by the MVP model ($\sigma = 0$) can be subject to test.

The joint probability function of the MVNB model for cluster i is obtained along the lines of (5.3):

$$f(y_{i1},\ldots,y_{iJ}) = \sum_{k=0}^{s_i} f_{\mathrm{NB}}(k) \prod_{j=1}^{J} f_{\mathrm{NB}}(y_{ij}-k) \tag{5.10}$$

where $s_i = \min(y_{i1},\ldots,y_{iJ})$ and f_{NB} is the Negbin I probability function. For instance, for $z_{ij} = y_{ij} - k$

$$f_{\mathrm{NB}}(z_{ij}) = \frac{\Gamma(\lambda_{ij}/\sigma + z_{ij})}{\Gamma(\lambda_{ij}/\sigma)\Gamma(z_{ij}+1)} \left(\frac{1}{1+\sigma}\right)^{\lambda_{ij}/\sigma} \left(\frac{\sigma}{1+\sigma}\right)^{z_{ij}} \tag{5.11}$$

The parameters of the model can be estimated by maximizing the corresponding log-likelihood function.

5.1.3 Multivariate Poisson-Gamma Mixture Model

An alternative approach to induce correlation amongst the counts has been pursued by Hausman, Hall and Griliches (1984) (see also Dey and Chung, 1992). In their model, correlation is generated by an individual specific multiplicative error term. The error term represents individual specific unobserved heterogeneity. The mixture multivariate density of $y_i = (y_{i1} \cdots y_{iJ})'$ is obtained after integration

$$f(y_i|x_i) = \int \left[\prod_{j=1}^{J} \frac{\exp(-\lambda_{ij}u_i)(\lambda_{ij}u_i)^{y_{ij}}}{\Gamma(y_{ij}+1)} \right] g(u_i)\mathrm{d}u_i \tag{5.12}$$

If u_i is gamma distributed with $\mathrm{E}(u_i) = 1$ and $\mathrm{Var}(u_i) = \alpha^{-1}$ it can be shown that the joint distribution function of y_i is of a negative binomial form with distribution function.

$$\begin{aligned} f(y_i|x_i) &= \left(\prod_{j=1}^{J} \frac{(\lambda_{ij})^{y_{ij}}}{\Gamma(y_{ij}+1)} \right) \frac{\alpha^\alpha}{\Gamma(\alpha)} \int e^{-u_i(\lambda_{i.}+\alpha)} u^{y_{i.}+\alpha-1} \mathrm{d}u_i \\ &= \left(\prod_{j=1}^{J} \frac{(\lambda_{ij})^{y_{ij}}}{\Gamma(y_{ij}+1)} \right) \frac{\Gamma(y_{i.}+\alpha)}{\Gamma(\alpha)} \alpha^\alpha (\lambda_{i.}+\alpha)^{-(y_{i.}+\alpha)} \end{aligned} \tag{5.13}$$

where $y_{i.} = \sum_{j=1}^{J} y_{ij}$ and $\lambda_{i.} = \sum_{j=1}^{J} \lambda_{ij}$. Note that this model is very closely related to the univariate Poisson-gamma mixture leading to the univariate negative binomial distribution. The only difference is that mixing is over a common variable u_i rather than over independent gamma variable u_{ij}. The similarity is also seen in the marginals of the multivariate Poisson-gamma model that are univariate negative binomial with $\mathrm{E}(y_{ij}) = \lambda_{ij}$ and $\mathrm{Var}(y_{ij}) = \lambda_{ij} + \gamma\lambda_{ij}^2$ where $\gamma = \alpha^{-1}$ (i.e., of the Negbin II variety).

The covariance between outcomes for a given individual can be derived as follows:

$$\begin{aligned}
\mathrm{Cov}(y_{ij}, y_{il}) &= \mathrm{E}_u \mathrm{Cov}(y_{ij}, y_{il}|u_i) + \mathrm{Cov}_u[\mathrm{E}(y_{ij}|u_i), \mathrm{E}(y_{il}|u_i)|u_i] \\
&= 0 + \mathrm{Cov}_u(\lambda_{ij}u_i, \lambda_{il}u_i) \\
&= \gamma\lambda_{ij}\lambda_{il}, \quad j \neq l
\end{aligned}$$

In compact form, the covariance matrix for individual i is given by

$$\mathrm{Var}(y_i) = \Lambda_i + \Lambda_i \gamma \mathbf{1}\mathbf{1}' \Lambda_i$$

where $\Lambda_i = \mathrm{diag}(\lambda_{ij})$ as before.

Hence, the multivariate Poisson-gamma model allows for overdispersion, and within-individual correlation. As for the MVP and MVNB models, the covariances are non-negative. In contrast to the two previous models, the multivariate Poisson-gamma model does not have an "equi-covariance" property. Rather, within individual covariances are an increasing function of the product of the expected values λ_{ij} and λ_{il}. This could be a useful feature for modeling non-negative random variables. In particular, it eliminates the strict upper bound to the correlation that was observed for the MVP distribution.

A potential disadvantage of this model is that the covariances are not determined independently of the dispersion. Hence, a finding of a significant γ can be as much an indicator of overdispersion in the data as it might be an indicator of correlation (or both). In the MVNB model, by contrast, these two features of the data can be identified, and thus estimated, separately. We also note that all multivariate models discussed so far require covariances to be non-negative. Depending on the application, this can be an undesirable feature, and a more general model is discussed in the next chapter.

Finally, note that for $J = 2$ the multivariate Poisson-gamma mixture model of Hausman, Hall and Griliches (1984) is identical to the bivariate negative binomial model attributed to Marshall and Olkin (1990) (See also Munkin and Trivedi, 1999). Its joint probability distribution function is given by

$$\begin{aligned}
f(y_1, y_2|x_1, x_2) = &\frac{\Gamma(y_1 + y_2 + \alpha)}{y_1! y_2! \Gamma(\alpha)} \left(\frac{\lambda_1}{\lambda_1 + \lambda_2 + 1} \right)^{y_1} \left(\frac{\lambda_2}{\lambda_1 + \lambda_2 + 1} \right)^{y_2} \\
&\times \left(\frac{1}{\lambda_1 + \lambda_2 + 1} \right)^{\alpha}
\end{aligned}$$

which is easily seen to be a special case of (5.13).

5.1.4 Multivariate Poisson-Log-Normal Model

Assume that conditionally on a $(J \times 1)$ vector of individual and outcome specific random effects $\varepsilon_i = (\varepsilon_{i1} \cdots \varepsilon_{iJ})'$ the distribution of y_i is independent Poisson

$$f(y_i|\varepsilon_i) = \prod_{j=1}^{J} f_p(y_{ij}|\lambda_{ij}\exp(\varepsilon_{ij})) \tag{5.14}$$

where f_p is the Poisson distribution function. Since ε_i is unobserved, the model is not complete. In analogy to the univariate Poisson-log-normal model, Aitchison and Ho (1989) suggested a multivariate extention where ε_i is J-variate normal distributed

$$f(\varepsilon_i|\Omega) = \phi_J(-0.5\text{diag}\Omega, \Omega) \tag{5.15}$$

and Ω is the covariance matrix. Aside from the random effects, the model is thus characterized by the parameters $\psi = (\lambda, \Omega)$. The importance of the non-zero mean specification depending on the diagonal elements of Ω is explained below. Aitchison and Ho (1989), as well as Good and Pirog-Good (1989) who considered a bivariate Poisson-log-normal distribution, restricted their attention to the case without regressors but the model can be readily extended to the case with regressors by letting $\lambda_{ij} = \exp(x'_{ij}\beta)$.

To understand this specificationn note that the conditional mean and variance of the outcomes are given by

$$\text{E}(y_{ij}|\varepsilon_{ij}) = \text{Var}(y_{ij}|\varepsilon_{ij}) = \lambda_{ij}\exp(\varepsilon_{ij})\,.$$

This allows one to derive the expectation and variance of the marginal joint distribution of y_i without integration. A simple reparameterization facilitates the analysis. Let $u_{ij} = \exp(\varepsilon_{ij})$ and $u_i = (u_{i1}\cdots u_{iJ})'$. The assumption on ε_i implies that $u_i \sim \text{MVLN}_J(\mathbf{1}, \Sigma)$, a multivariate log-normal distribution with mean vector $\mathbf{1}$ and covariance matrix Σ where $\sigma_{ij} = \exp(\omega_{ij}) - 1$ and thus $\Sigma = \exp(\Omega) - \mathbf{11}'$. Hence,

$$y_{ij}|\lambda_{ij}, u_{ij} \sim \text{Poisson}(\lambda_{ij}u_{ij})$$

and the model is in the form of a Poisson-log-normal distribution.

To derive the marginal moments, let $\lambda_i = (\lambda_{i1}\cdots\lambda_{iJ})'$ and $\Lambda_i = \text{diag}(\lambda_i)$. Then by the law of the iterative expectations one obtains

$$\text{E}(y_i|\lambda_i, \Omega) = \lambda_i \tag{5.16}$$

and

$$\text{Var}(y_i|\Lambda_i, \Omega) = \Lambda_i + \Lambda_i[\exp(\Omega) - \mathbf{11}']\Lambda_i \tag{5.17}$$

Hence, the covariances between the counts are represented by the terms

$$\text{Cov}(y_{ij}, y_{kl}) = \lambda_{ij}(\exp(\omega_{jl}) - 1)\lambda_{kl},\ j \neq l,\ i = k$$

which can be positive or negative depending on the sign of ω_{jl}, the (j,l) element of Ω. The correlation structure of the counts is thus unrestricted. Moreover, the model allows for overdispersion as long as $\omega_{ii} > 0$. Note, however, that the marginal distribution of the counts y_i cannot be obtained by direct computation, requiring as it does the evaluation of a J -variate integral of the Poisson distribution with respect to the distribution of ε_i

$$f(y_i|\lambda_i, \Omega) = \int \prod_{j=1}^{J} f_P(y_{ij}|\lambda_{ij}, \varepsilon_{ij})\phi(\varepsilon_i| - 0.5\text{diag}\Omega, \Omega)\text{d}\varepsilon_i \tag{5.18}$$

where f as above is the Poisson probability function conditioned on $(\lambda_{ij}, \varepsilon_{ij})$ and ϕ is the J-variate normal distribution. This J-dimensional integral cannot be solved in closed form for arbitrary Ω. For $J = 2$, Munkin and Trivedi (1999) discuss estimation by simulated maximum likelihood. However, this is strictly speaking not necessary as one could obtain a one-dimensional integral through a factorization of ϕ into a conditional and a marginal distribution and then apply Gauss-Hermite quadrature. A simulation method based on Markov chain Monte Carlo that works well for high-dimensional problems, is presented in Chap. 5.1.4.

If Ω is a diagonal matrix, the J-variate integral reduces to the product of J single integrals

$$f(y_i|\lambda_i, \Omega) = \prod_{j=1}^{J} \int f_P(y_{ij}|\lambda_{ij}, \varepsilon_{ij})\phi(\varepsilon_{ij}| - 0.5\omega_{jj}, \omega_{jj})\text{d}\varepsilon_{ij} \tag{5.19}$$

a product of J independent univariate Poisson-log normal densities. See Chap. 4.2.1 for a discussion of the univariate Poisson-log-normal model. For $\Omega = 0$ the joint probability simplifies to a product of J independent Poisson densities.

Discussion

There are several ways to generalize the multivariate Poisson log-normal model. The considerations are similar to those for selectivity models. First, the marginal distribution of ε_{ij} may be known but not normal. In this case, one can apply results in Lee (1983) and Weiss (1999) to generate a multivariate distribution in which the random variables are allowed to correlate (see also Chap. 4.5.1). If $F(\varepsilon_{ij})$ is the cumulative marginal distribution function of ε_{ij}, then the transformed random variable

$$v_{ij} = \Phi^{-1}(F(\varepsilon_{ij})),$$

where Φ^{-1} is the inverse cumulative density function of the standard normal distribution, is standard normal distributed. To introduce correlation, assume that the joint distribution $f(v_i) = \phi_J(0, D)$ is multivariate normal with covariance matrix D. Clearly, ε_i is multivariate normal only if F is the normal distribution. In particular, D is usually not the covariance matrix of ε_i. As such, the interpretation of the covariance structure is somewhat difficult. However, this set-up has generated a multivariate distribution for ε_i with known marginal cumulative distribution functions equal to $F(\cdot)$ and unrestricted covariance structure. No application of this method to multivariate count data is known at this stage.

An alternative possibility is to relax the strong distributional assumptions. A method based on squared polynomial series expansions for the unknown density of the correlated errors is suggested in Gurmu and Elder (1998). The method was originally developed by Gurmu, Rilstone and Stern (1998) for the univariate case. Gurmu and Elder extend the method to the bivariate case. There is some doubt whether this method could be succesfully applied to high dimensional multivariate data.

A final alternative is to abandon distributional assumption altogether and specify first and second order moments of the joint distribution of $u_i = \exp(\varepsilon_i)$, and thus y_i, only. This is discussed in Chap. 5.1.6.

5.1.5 Latent Poisson-Normal Model

A latent Poisson-normal model for bivariate correlated counts is presented in van Ophem (1999). In this model, count data are interpreted as realizations of an underlying (latent) normally distributed variable. One problem is that the support of count data distributions is unbounded. To make it a well defined problem, assume that $y = 0, 1, \ldots, K$, where K is an upper bound. This restriction does not matter for estimation, however, as actual data are always finite (van Ophem, 1999).

Consider the following mapping from a standard normal variable u_1 to the count variable y_1:

$$y_1 = k \text{ iff } \eta_{k-1} < u_1 < \eta_k \quad k = 0, 1, \ldots, K$$

where $\eta_{-1} = -\infty$ and $\eta_K = \infty$. Thus

$$P(y_1 = k) = \Phi(\eta_k) - \Phi(\eta_{k-1})$$

and

$$P(y_1 \leq k) = \Phi(\eta_k)$$

or, conversely,

$$\eta_k = \Phi^{-1}[P(y_1 \leq k)]$$

This relationship defines η_k uniquely for any marginal distribution $P(y_1 = k)$. So far, the model has only been re-parameterized without changing its substance. However, now assume that for a second count variable y_2, a similar procedure gives

$$\mu_m = \Phi^{-1}[P(y_2 \leq m)]$$

If u_1 and u_2 are bivariate normal with correlation ρ, then we can write the joint cumulative probability function as

$$P(y_1 \leq k, y_2 \leq m) \int_{-\infty}^{\eta_k} \int_{-\infty}^{\mu_m} \phi(u_1, y_2; \rho) \mathrm{d}u_2 \mathrm{d}u_1$$

where ϕ is the bivariate normal density with means 0, variances 1 and correlation ρ. Moreover, the likelihood contribution can be calculated as

$$\begin{aligned} P(y_1 = k, y_2 = m) = & P(y_1 \leq k, y_2 \leq m) - P(y_1 \leq k-1, y_2 \leq m) \\ & -P(y_1 \leq k, y_2 \leq m-1) \\ & +P(y_1 \leq k-1, y_2 \leq m-1) \end{aligned} \quad (5.20)$$

To summarize, this model has well-specified marginal distributions, in this case Poisson, and a correlation structure that allows both for positive and negative correlations. The correlation results from a latent bivariate normal distribution with correlation ρ. If the only goal of the analysis is to allow for negative correlations it is not clear whether this model offers an advantage over the relatively simpler Poisson-log-normal model. The Poisson-log-normal model has no Poisson marginal distributions, though. Hence the latent Poisson-normal model has an advantage if one strongly believes in Poisson marginals. In practice, however, these restrictive marginals are likely to speak against rather than for this model.

A serious limitation of the latent Poisson-normal model is its focus on the bivariate case. Extentions to higher dimensional multivariate data appear impractical. A major advantage of the model is its versatility: the approach can easily be adopted to any bivariate discrete random variable.

5.1.6 Moment-Based Methods

A parametric model with correlated errors was introduced in Chap. 5.1.4, where it was assumed that

$$y_{ij}|\lambda_{ij}, u_{ij} \sim \text{Poisson}(\lambda_{ij} u_{ij})$$

and

$$u_i = \begin{pmatrix} u_{i1} \\ \vdots \\ u_{iJ} \end{pmatrix} \sim \text{MVLN}(\mathbf{1}, \Sigma)$$

where MVLN denotes the multivariate normal distribution with expected value $\mathbf{1}$ and covariance matrix Σ. This model is appropriate for panel data, where Σ could reflect either serial or contemporaneous correlations, as well as for genuine multivariate data. Maximum likelihood estimation of this model in general requires simulation methods.

Alternatively, Gourieroux, Monfort and Trognon (1984b) discuss semiparametric estimation of this model. Their approach differs slightly from the one employed here, as it is based on a bivariate Poisson distribution with common additive factor (i.e., in their model there are two sources of intra-cluster correlation, one being the common additive factor and the other being mixing over correlated errors). However, this is an inconsequential complication that can be dropped for ease of exposition. For the same reason, we follow Gourieroux, Monfort and Trognon (1984b) and focus on the bivariate case. Hence, the model can be written as

$$\begin{pmatrix} y_{i1}|u_{i1} \\ y_{i2}|u_{i2} \end{pmatrix} \sim \text{ independently Poisson } \begin{pmatrix} \exp(x'_{i1}\beta_1)u_{i1} \\ \exp(x'_{i2}\beta_2)u_{i2} \end{pmatrix}$$

with

$$\mathrm{E}(u_{i1}) = \mathrm{E}(u_{i2}) = 1$$

and

$$\mathrm{Var}\begin{pmatrix} u_{i1} \\ u_{i2} \end{pmatrix} = \Sigma$$

Hence, only the first two moments of the mixing distribution are specified. Let $\lambda_{i1} = \exp(x'_{i1}\beta_1)$ and $\lambda_{i2} = \exp(x'_{i2}\beta_2)$. The correlated random effects introduce within cluster correlation among $y_i = (y_{i1}, y_{i2})'$ as

$$\mathrm{Var}\begin{pmatrix} y_{i1} \\ y_{i2} \end{pmatrix} = \begin{pmatrix} \lambda_{i1} & 0 \\ 0 & \lambda_{i2} \end{pmatrix} + \begin{pmatrix} \lambda_{i1} & 0 \\ 0 & \lambda_{i2} \end{pmatrix} \Sigma \begin{pmatrix} \lambda_{i1} & 0 \\ 0 & \lambda_{i2} \end{pmatrix} \tag{5.21}$$

is not a diagonal matrix. Gourieroux, Monfort and Trognon (1984b) suggest estimating β_1 and β_2 by non-linear least squares minimizing

$$\sum_{i=1}^{n} \left[(y_{i1} - \exp(x'_{i1}\beta_1))^2 + (y_{i2} - \exp(x'_{i2}\beta_2))^2\right]$$

The estimators $\hat{\beta}_1$ and $\hat{\beta}_2$ can be used to define the residuals

$$\hat{w}_{i1} = y_{i1} - \exp(x'_{i1}\hat{\beta}_1)$$

$$\hat{w}_{i2} = y_{i2} - \exp(x'_{i2}\hat{\beta}_2)$$

Moreover, consistent estimators of σ_{11}, σ_{12} and σ_{22}, the elements of Σ, are obtained by applying ordinary least squares to

$$\hat{w}_{i1}^2 - \exp(x'_{i1}\hat{\beta}_1) = \sigma_{11}\exp(2x'_{i1}\hat{\beta}_1) + \text{ disturbance}$$

$$\hat{w}_{i2}^2 - \exp(x'_{i2}\hat{\beta}_2) = \sigma_{22}\exp(2x'_{i2}\hat{\beta}_2) + \text{ disturbance}$$

$$\hat{w}_{i1}\hat{w}_{i2} = \sigma_{12}\exp(x'_{i1}\hat{\beta}_1)\exp(x'_{i2}\hat{\beta}_2) + \text{ disturbance}$$

Having obtained an estimator $\hat{\Sigma}$, Gourieroux, Monfort and Trognon (1984b) recommend the use of quasi-generalized pseudo maximum likelihood (QGPML) in order to reach the lower bound of the asymptotic covariance matrix of pseudo maximum likelihood estimators. QGPML solves the weighted non-linear least squares problem

$$\min \sum_{i=1}^{n} (y_{i1} - \lambda_{i1}, y_{i2} - \lambda_{i2}) \left[\widehat{\mathrm{Var}}\begin{pmatrix} y_{i1} \\ y_{i2} \end{pmatrix}\right]^{-1} \begin{pmatrix} y_{i1} - \lambda_{i1} \\ y_{i2} - \lambda_{i2} \end{pmatrix}$$

where

$$\widehat{\mathrm{Var}}\begin{pmatrix} y_{i1} \\ y_{i2} \end{pmatrix}$$

is obtained from (5.21) using $\hat{\sigma}$, $\hat{\beta}_1$ and $\hat{\beta}_2$.

5.2 Panel Data Models

The most common application of multivariate models is to panel data. Panel data, also refered to as longitudinal data, provide repeated measurements (over time) on a given characteristic for a sample of individuals or households or firms. Examples are pre- and post treatment studies in a controlled experiment, such as the number of epileptic seizures before and after treatment (Diggle, Liang, and Zeger, 1995). In social sciences, panel count data, such as the number of days absent from work in a given year, are observed in repeated linked household surveys, for example the U.S. Panel Study of Income Dynamics or the German Socio Economic Panel. So far, the leading application of panel count data models in the econometrics literature is to firm level data on patent numbers (Hausman, Hall and Griliches, 1984). Recent contributions include Cincera (1997), Crépon and Duguet (1997a, 1997b), Blundell, Griffith and van Reenen (1995, 1999), and Montalvo (1997).

Methods for panel count data differ from standard count data models in at least one of three ways. Firstly, they address the non-standard form of the covariance matrix of the observations that arises since the assumption of independent observations is most likely invalid. Secondly, and relatedly, they provide a richer framework than univariate data for addressing the issue of unobserved heterogeneity (see Chap. 4.2). Thirdly, models for panel count data allow the introduction of dynamic elements, such as a lagged dependent variable, into the regression part of the model.

Panel methods typically also differ from genuine multivariate count data models. First, panel data models are usually somewhat more restrictive in their covariance structure, as they assume that dependence is generated by unobserved heterogeneity that is specific to the individual but constant over time. Secondly, panel data models explicitly consider the possibility that the unobserved individual heterogeneity factor is correlated with one or more explanatory variables. In this situation, conditional models are required.

To give an example for the type of modeling issues encountered for panel data, consider the determinants of patent numbers. It is likely that differences in technological opportunities or operating skills may affect the observed number of patents. And yet, these firm specific factors are typically not captured by the explanatory variables. If the firm specific unobservables are correlated over time, a plausible assumption to start with, they will cause a positive correlation among the repeated observations of a single firm. One special, and most commonly assumed, case is that of a constant firm effect. This can be seen as a limiting case of correlated effects, where the correlation is perfect. In addition, this firm effect may be correlated with explanatory variables. By construction, this must be so if the explanatory variables include a lagged dependent variable. But correlation can arise in other situations as well.

The literature has followed three basic approaches for dealing with individual specific effects in panel count data:

1. Robust methods

2. Fully parametric random effects models
3. Fully parametric fixed effects models

An example for a robust method is the standard Poisson regression model. The Poisson estimator remains consistent if the misspecification is limited to the covariance structure without affecting the mean. This result follows since the Poisson distribution is a linear exponential family and the Poisson maximum likelihood estimator is therefore a pseudo maximum likelihood estimator (see Chap. 4.10.3). However ignoring the correlation will bias the standard errors and cause the estimator to be inefficient. The first problem can be addressed by estimating a robust variance matrix using the sandwich formula. For the second problem, one needs to find a different class of models, such as the fully parametric Poisson-gamma mixture model of Chap. 5.1.3. Such a model is preferred in terms of efficiency. The benefit of the robust standard error approach is, however, that it remains consistent when the fully parametric model may be inconsistent because of misspecification of the distribution of the unobserved heterogeneity.

Both methods break down and lead to biased estimators if the misspecification stems from an unobserved individual specific effect that is correlated with explanatory variables. Consistency of the robust Poisson model requires that the mean function be correctly specified. But this requirement is violated if the individual specific effect and the explanatory variables are correlated. The same argument applies to the random effects estimator. In this case a fixed effects model that conditions on the individual specific effects would be appropriate.

5.2.1 Fixed Effects Poisson Model

Consider a Poisson model with intercept heterogeneity:

$$\begin{aligned}\tilde{\lambda}_{it} &= \exp(x_{it}'\beta + \varepsilon_i) \\ &= \exp(x_{it}'\beta)\exp(\varepsilon_i) \\ &= \lambda_{it}\alpha_i\end{aligned} \tag{5.22}$$

where $i = 1, \ldots, n$ indexes the individual (or household, or firm) and $t = 1, \ldots, T$ indexes time. We now use the subscript t rather than j in order to emphasize the panel structure. For similar reasons, the individual specific multiplicative effect is now denoted as α_i (rather than u_i).

The fixed effects model treats the α_i's as parameters that need to be estimated jointly with β. As hinted at in the preceeding comments, the advantages of the fixed effects model are twofold:

1. the population distribution of α_i does not need to be specified. This avoids inconsistency of a misspecied random effects model;

2. the individual specific error term α_i may be correlated with the explanatory variables x_{it}.

Of course, a random effects model is more efficient if correctly specified as it reaches the efficiency bound of a maximum likelihood estimator. Relative to the fixed effects model, it has n additional degrees of freedom. Moreover, random effects model have the advantage that time invariant regressors can be included. And yet, the assumptions of correct functional form for the distribution of α_i and no correlation with the explanatory variables are likely violated in many practical situations using non-experimental data. The assumptions can be tested using Hausman's (1978) method (see also (3.75).

In order to estimate a fixed effects Poisson model one could simply include n individual specific dummy variables, that is, one intercept for each individual. This may be difficult when n is large and T is small as is the case in almost all applications: $(n+k)$ parameters need to be estimated. For large n this is likely to exceed software restrictions. However, an inspection of the log-likelihood function reveals that this problem does not arise in the fixed effects Poisson model as analytical expressions for α_i can be derived and used to concentrate the likelihood as a function of β. Following Cameron and Trivedi (1998, Chap. 9.3.1.), the joint conditional density for individual i is (see also Blundell, Griffith, and Windmeijer, 2002)

$$
\begin{aligned}
f(y_i|\alpha_i,\beta) &= \prod_{t=1}^{T} \exp(-\alpha_i\lambda_{it})(\alpha_i\lambda_{it})^{y_{it}}/y_{it}! \\
&= \exp\left(-\alpha_i\sum_{t=1}^{T}\lambda_{it}\right)\prod_{t=1}^{T}\alpha_i^{y_{it}}\prod_{t=1}^{T}\lambda_{it}^{y_{it}}/\prod_{t=1}^{T}y_{it}! \qquad (5.23)
\end{aligned}
$$

where $\lambda_{it} = \exp(x_{it}'\beta)$. The log-likelihood contribution of individual i is therefore

$$
\ell(\alpha_i,\beta) = -\alpha_i\sum_{t=1}^{T}\lambda_{it} + \ln\alpha_i\sum_{t=1}^{T}y_{it} + \sum_{t=1}^{T}y_{it}\ln\lambda_{it} - \sum_{t=1}^{T}\ln y_{it}! \qquad (5.24)
$$

Differentiating with respect to α_i and setting equal to zero yields

$$
\begin{aligned}
\hat{\alpha}_i &= \frac{\sum_{t=1}^{T} y_{it}}{\sum_{t=1}^{T}\lambda_{it}} \\
&= \frac{\bar{y}_i}{\bar{\lambda}_i} \qquad (5.25)
\end{aligned}
$$

We can substitute this expression into the first-order condition for β, that can thus be written independently of α_i as

$$
\sum_{i=1}^{N}\sum_{t=1}^{T}\left(y_{it} - \frac{\bar{y}_i}{\bar{\lambda}_i}\lambda_{it}\right)x_{it} \equiv 0 \qquad (5.26)
$$

Alternatively, the analytical expression for α_i can be used to rewrite (5.23) as a concentrated likelihood function for individual i

$$\begin{aligned} L_i(\beta) &= \exp\left(-\sum_{t=1}^{T} y_{it}\right)\prod_{t=1}^{T}\left(\frac{\sum_{t=1}^{T} y_{it}}{\sum_{t=1}^{T}\lambda_{it}}\right)^{y_{it}}\prod_{t=1}^{T}\lambda_{it}^{y_{it}}/\prod_{t=1}^{T} y_{it}! \\ &\propto \prod_{t=1}^{T}\left(\frac{\lambda_{it}}{\sum_{t=1}^{T}\lambda_{it}}\right)^{y_{it}} \end{aligned}$$

Again, α_i has disappeared. The concentrated likelihood of the Poisson model with fixed effects is seen to be of multinomial form with probability function

$$f(y_i) = \frac{(\sum_t y_{it})!}{\prod_t y_{it}!}\prod_{t=1}^{T} p_{it}^{y_{it}} \tag{5.27}$$

where

$$p_{it} = \frac{\lambda_{it}}{\sum_{t=1}^{T}\lambda_{it}}$$

and

$$\sum_{t=1}^{T} p_{it} = 1$$

This probability function differs only slightly from the usual multinomial logit model, as in general all y_{it}'s are allowed to be nonzero and greater than one, instead of only one value being one for the choice that is made and all others being zero.

The concentrated log-likelihood function of the sample can be written as

$$\ell(\beta) = \sum_{i=1}^{N}\sum_{t=1}^{T} y_{it}\ln p_{it} \tag{5.28}$$

with first-order conditions for a maximum

$$\begin{aligned} \frac{\partial\ell(\beta)}{\partial\beta} &= \sum_{i=1}^{N}\sum_{t=1}^{T}\frac{y_{it}}{p_{it}}\frac{\partial p_{it}}{\partial\beta} \\ &= \sum_{i=1}^{N}\sum_{t=1}^{T} y_{it}x_{it} - \sum_{i=1}^{N}\sum_{t=1}^{T} y_{it}\frac{\sum_s x_{is}\exp(x_{is}'\beta)}{\sum_s \exp(x_{is}'\beta)} \\ &= \sum_{i=1}^{N}\sum_{t=1}^{T} y_{it}x_{it} - \sum_{i=1}^{N}\sum_{t=1}^{T} x_{it}\exp(x_{it}'\beta)\frac{\sum_s y_{is}}{\sum_s \exp(x_{is}'\beta)} \\ &= \sum_{i=1}^{N}\sum_{t=1}^{T} x_{it}\left(y_{it} - \lambda_{it}\frac{\bar{y}_i}{\bar{\lambda}_i}\right) \\ &\equiv 0 \end{aligned}$$

This is, of course, the same first-order condition as (5.26).

The derivation by concentrated likelihood has an interesting implication for the asymptotic distribution of β that is pointed out by Cameron and Trivedi (1998) and Blundell, Griffith, and Windmeijer (2002). First, note that the relevant limiting operation for typical panel data is $n \to \infty$ for fixed T. Clearly, the number of parameters α_i increases step for step with increases in n. Thus, they cannot be estimated consistently unless $T \to \infty$ as well which is excluded. The "incidental parameter problem" says that in non-linear models, such as the logit model, the inconsistency of α_i causes inconsistency of β (See Hsiao, 1986). While it was sometimes suggested that the same problem applies for the Poisson model, the result that the likelihood function can be rid of α_i by concentration shows that this is not the case, and $\hat{\beta}$ is consistent for fixed T and $n \to \infty$.

Conditional Maximum Likelihood

Recall the basic problem: how can one obtain an estimator for β without having to estimate each α_i. While concentration of the likelihood function is one possibility, an alternative approach, popularized by Hausman, Hall and Griliches (1984), is to condition the likelihood contribution of individual i on the individual specific sum $\sum_{t=1}^{T} y_{it}$. Since observations are independently Poisson distributed conditional on α_i, the distribution of this sum is easily established as

$$\sum_{t=1}^{T} y_{it} \sim \text{Poisson}\left(\alpha_i \sum_{t=1}^{T} \lambda_{it}\right)$$

Next, consider the joint density for the i-th individual. We can write

$$f(y_{i1}, \ldots, y_{iT}) = f\left(y_{i1}, \ldots, y_{iT}, \sum_{t=1}^{T} y_{it}\right)$$

since the sum of counts for individual i over time is fully determined by its components and thus adds no new information. Thus, for the conditional distribution

$$\begin{aligned} f\left(y_{i1}, \ldots, y_{iT} | \sum_{t=1}^{T} y_{it}\right) &= \frac{f(y_{i1}, \ldots, y_{iT}, \sum_{t=1}^{T} y_{it})}{f(\sum_{t=1}^{T} y_{it})} \qquad (5.29) \\ &= \frac{e^{-\sum_t \lambda_{it}} \prod_t \lambda_{it}^{y_{it}}}{\prod_t y_{it}!} \Big/ \frac{e^{-\sum_t \lambda_{it}} (\sum_t \lambda_{it})^{\sum_t y_{it}}}{(\sum_t y_{it})!} \end{aligned}$$

Terms involving α_i have canceled out. The resulting probability expression turns out to be again of the multinomial form as the conditional probabilities are proportional to

$$f\left(y_{i1}, \ldots, y_{iT} | \sum_{t} y_{it}\right) \propto \prod_{t=1}^{T} \left(\frac{\lambda_{it}}{\sum_{t=1}^{T} \lambda_{it}}\right)^{y_{it}}$$

This is exactly the same probability as the concentrated likelihood. The concentrated and conditional likelihood functions have the same first-order conditions, and produce the same maximum likelihood estimator. Thus, two apparently quite different approaches lead to the same outcome: in each year, events occur independently with probability $p_{it} = \exp(x_{it}'\beta)/\sum_t \exp(x_{it}'\beta)$. This model explains the share of counts in each year as a proportion of the number of total counts in T years.

Robust Estimation of the Fixed Effects Poisson Model

Maximum likelihood estimation of the fixed effects Poisson model requires that the underlying probability distribution is in fact of the Poisson type. For instance, overdispersion cannot be accommodated by this model. Interestingly, the multinomial distribution is a linear exponential family as well. Hence, the pseudo maximum likelihood result by Gourieroux, Monfort and Trognon (1984) can be invoked to establish consistency of the fixed effects Poisson model as long as the conditional expectation function is correctly specified (see also Wooldridge, 1999). This estimator can be interpreted as a moment estimator that solves the moment conditions

$$\sum_{i=1}^{N}\sum_{t=1}^{T}\left(y_{it} - \frac{\bar{y}_i}{\bar{\lambda}_i}\lambda_{it}\right)x_{it} \equiv 0$$

Blundell, Griffith and Windmeijer (2002) call this model a "within groups mean scaling model". In this situation, moment estimator, Poisson maximum likelihood and Poisson conditional likelihood are all the same. Note also that the estimator for α_i implicit in this formula is a moment estimator as well, since from $\mathrm{E}(y_{it}) = \lambda_{it}\alpha_i$ we get that

$$\alpha_i = \mathrm{E}\left(\frac{y_{it}}{\lambda_{it}}\right)$$

which leads to the moment estimator $\hat{\alpha}_i = \bar{y}_i/\bar{\lambda}_i$.

If one chooses to interpret the model within the pseudo likelihood framework, the estimated standard errors need to be adjusted accordingly in order to obtain valid inference. The covariance matrix can be estimated consistently using

$$\mathrm{Var}_{\mathrm{PML}}(\hat{\beta}) = \hat{\mathcal{J}}^{-1}\hat{\mathcal{I}}\hat{\mathcal{J}}^{-1}$$

where

$$\hat{\mathcal{J}} = \sum_{i=1}^{n}\left(\sum_{t=1}^{T} x_{it}x_{it}'\frac{\bar{y}_i}{\bar{\lambda}_i}\lambda_{it} - \frac{1}{T}\sum_{t=1}^{T}\sum_{s=1}^{T} x_{it}x_{is}'\frac{\bar{y}_i}{\bar{\lambda}_i}\lambda_{it}\lambda_{is}\right)$$

and

$$\hat{\mathcal{I}} = \sum_{i=1}^{n}\sum_{t=1}^{T}\sum_{s=1}^{T} x_{it}x_{is}'\left(y_{it} - \frac{\bar{y}_i}{\bar{\lambda}_i}\lambda_{it}\right)\left(y_{is} - \frac{\bar{y}_i}{\bar{\lambda}_i}\lambda_{is}\right)'$$

(See Cameron and Trivedi, 1998).

5.2.2 Fixed Effects Negative Binomial Model

In the presence of overdispersion, a potentially more efficient estimator can be based on the fully parametric fixed effects negative binomial model that was introduced by Hausman, Hall and Griliches (1984). They discuss estimation of the model by conditional maximum likelihood. As for the fixed effects Poisson model, the conditioning is on the individual specific sums $\sum_{t=1}^{T} y_{it}$. In order to derive a closed form expression for the joint conditional probability distribution for individual i, it is necessary that the probability distribution of $\sum_{t=1}^{T} y_{it}$ can be expressed in closed form. As shown in Chap. 5.1.2, a sum of independent negative binomial random variables is again negative binomial distributed if and only if the component distributions are of Negbin I type with probability generating function

$$\mathcal{P}(s) = [1 + \theta(1 - s)]^{-\delta}$$

and common parameter θ. Constrained by this requirement, Hausman, Hall and Griliches (1984) suggest the parameterization $\delta = \lambda_{it}$ and $\theta = \theta_i$, an individual specific fixed effect. In this parameterization, the probability function for observation y_{it} is given by

$$f(y_{it}) = \frac{\Gamma(\lambda_{it} + y_{it})}{\Gamma(\lambda_{it})\Gamma(y_{it} + 1)} \left(\frac{1}{1 + \theta_i}\right)^{\lambda_{it}} \left(\frac{\theta_i}{1 + \theta_i}\right)^{\lambda_{it}} \tag{5.30}$$

To fully appreciate the role of the individual specific effect θ_i in this model, note that

$$\mathrm{E}(y_{it}|\theta_i) = \lambda_{it}\theta_i$$

and

$$\mathrm{Var}(y_{it}|\theta_i) = \lambda_{it}(\theta_i + \theta_i^2) = \mathrm{E}(y_{it}|\theta_i)(1 + \theta_i)$$

Thus, this Negbin I-type model introduces a time invariant variance-to-mean ratio. With $\theta_i = \exp(\alpha_i)$, we could as well write

$$\mathrm{E}(y_{it}|\alpha_i) = \exp(\alpha_i + x_{it}'\beta)$$

and

$$\mathrm{Var}(y_{it}|\alpha_i) = \exp(\alpha_i + x_{it}'\beta)(1 + \exp(\alpha_i))$$

Hence, the α_i's are not just differential intercepts in the mean function – they also appear also as a separate shifter in the variance function. From this, it follows that the α_i's play a different role than x_{it}, and it becomes logically impossible to interpret these terms as a representation of omitted explanatory variables. This aspect limits the usefulness of the model for use in genuine panel count data applications.

In order to preserve the standard structure of a fixed effects panel data model, one might be tempted to let instead $\delta = \exp(\alpha_i + x_{it}'\beta)$. Unfortunately, this parameterization is unsuitable for computational reasons as α_i fails then to drop out of the conditional likelihood function. But this contravenes the purpose of the whole exercise.

Putting aside these caveats regarding the interpretation of the model for a moment, we will now show that the θ_i's indeed disappear from the individual specific likelihood contribution. First, for a given individual i, the y_{it} are independent over time, such that

$$f\left(\sum_{t=1}^{T} y_{it}\right) = \frac{\Gamma(\sum_t \lambda_{it} + \sum_t y_{it})}{\Gamma(\sum_t \lambda_{it})\Gamma(\sum_t y_{it}+1)} \left(\frac{1}{1+\theta_i}\right)^{\sum_t \lambda_{it}} \left(\frac{\theta_i}{1+\theta_i}\right)^{\sum_t y_{it}}$$

while

$$f(y_{i1},\ldots,y_{iT}) = \prod_{t=1}^{T} \frac{\Gamma(\lambda_{it}+y_{it})}{\Gamma(\lambda_{it})\Gamma(y_{it}+1)} \left(\frac{1}{1+\theta_i}\right)^{\lambda_{it}} \left(\frac{\theta_i}{1+\theta_i}\right)^{y_{it}}$$

Therefore,

$$f\left(y_{i1},\ldots,y_{iT} | \sum_{t=1}^{T} y_{it}\right) = \frac{\Gamma(\sum_t \lambda_{it})\Gamma(\sum_t y_{it}+1)}{\Gamma(\sum_t \lambda_{it}+\sum_t y_{it})} \times \prod_{t=1}^{T} \frac{\Gamma(\lambda_{it}+y_{it})}{\Gamma(\lambda_{it})\Gamma(y_{it}+1)} \tag{5.31}$$

All terms involving θ_i have cancelled out of the conditional model, and the remaining parameters can be easily estimated. This model is available in econometric software packages such as Stata, Release 6 (StataCorp., 1999) and Limdep, Release 7 (Greene, 1995).

5.2.3 Random Effects Count Data Models

Alternatively, a random effects model for count panel data can be formulated. Hausman, Hall and Griliches (1984) derive a Poisson model with gamma distributed individual specific effect $u_i = \exp(\varepsilon_i)$. The derivation of this model is very similar to the derivation of the univariate negative binomial model in Chap. 4.2.2. The difference is that unobserved heterogeneity is now individual specific, that is, modeled as u_i rather than u_{it}. As shown in Chap. 5.1.3,if u_i is independently gamma distributed with parameters (γ,γ) (i.e., with mean 1 and variance $1/\gamma$) the joint marginal distribution of $y_i = (y_{i1}\cdots y_{iT})'$ is of negative binomial form with

$$f(y_i) = \frac{\Gamma(\sum_t y_{it}+\gamma)}{\Gamma(\gamma)} \left(\frac{\gamma}{\gamma+\sum_t \lambda_{it}}\right)^{\gamma} \frac{1}{(\gamma+\sum_t \lambda_{it})^{\sum_t y_{it}}} \prod_{t=1}^{T} \left(\frac{\lambda_{it}^{y_{it}}}{y_{it}!}\right)$$

In addition, Hausman, Hall and Griliches (1984) have introduced a random effects negative binomial model. As for the fixed effects negative binomial model, the starting point is a Negbin I model as in (5.30). Now, assume that $1/(1+\theta_i)$ is distributed as beta(a,b). With this assumption, θ_i can be integrated out and, after some algebra, the resulting joint probability function for individual i can be written as

$$f(y_i) = \frac{\Gamma(a+b)\Gamma(a+\sum_t \lambda_{it}))\Gamma(b+\sum_t y_{it})}{\Gamma(a)\Gamma(b)\Gamma(a+b+\sum_t \lambda_{it})+\sum_t y_{it})} \times \prod_t \frac{\Gamma(\lambda_{it}+y_{it})}{\Gamma(\lambda_{it})y_{it}!} \tag{5.32}$$

5.2.4 Moment-Based Methods

Brännäs and Johannson (1996) discuss moment based estimation of a panel data model with both serial *and* contemporaneous correlation. Let

$$\mathrm{E}(y_{it}|\lambda_{it}, u_{it}) = \mathrm{Var}(y_{it}|\lambda_{it}, u_{it}) = \lambda_{it}u_{it}$$

Denote by Σ_{ii} the variance matrix of $u_i = (u_{i1} \cdots u_{iT})'$ where i indexes individuals and t indexes time. Hence, Σ_{ii} contains the within-individual serial correlations. Some restrictions, such as covariance stationarity of a $AR(1)$ process, can be imposed. Similarly, Σ_{ij} is the covariance matrix between u_i and u_j, $i \neq j$. The diagonal elements of Σ_{ij} are contemporaneous correlations. Thus, the classical SURE model is obtained if both Σ_{ii} and Σ_{ij} are diagonal matrices. For $\Sigma_{ij} = 0$, this model is a multivariate extension of Zeger's (1988) time-series model (see Chap. 5.3).

Under the assumptions of the general model,

$$\mathrm{Var}(y_i) = \Lambda_i + \Lambda_i \Sigma_{ii} \Lambda_i$$

where $\Lambda_i = \mathrm{diag}(\lambda_{it})$ as before. However, in addition,

$$\mathrm{Cov}(y_i, y_j) = \Lambda_i \Sigma_{ij} \Lambda_j \qquad i \neq j$$

Brännäs and Johannson (1996) estimate the parameters of the model by GMM.

5.2.5 Dynamic Panel Count Data Models

There has been substantial recent interest in methods for panel count data with correlated individual specific effects and weakly exogenous regressors. The literature includes Montalvo (1997), Crépon and Duguet (1997a), Blundell, Griffith and van Reenen (1995) and Blundell, Griffith and Windmeijer (2002). With correlated individual specific effects, estimation requires the use of fixed effects. It was shown in Chap. 5.2.1 that the fixed effects Poisson estimator solves the first-order conditions

$$\sum_{i=1}^{N}\sum_{t=1}^{T} x_{it}\left(y_{it} - \frac{\lambda_{it}}{\bar{\lambda}_i}\bar{y}_i\right) = 0$$

Moreover, if we define $u_{it} = y_{it} - \lambda_{it}$, and thus $\bar{u}_i = \bar{y}_i - \bar{\lambda}_i$, then the term in parentheses is a "mean scaled" residual

$$u_{it}^* = u_{it} - \frac{\lambda_{it}}{\bar{\lambda}_i}\bar{u}_i$$

This situation is very similar to the within transformation of the residual in the linear model. Consistency of the estimator requires that

$$\mathrm{E}(x_{it}u_{it}^*) = 0$$

This condition, however, is violated under weak exogeneity, such as is the case in the presence of a lagged endogenous variable that is predetermined. Formally, a regressor is said to be predetermined when it is not correlated with current and future residuals but correlated with past residuals:

$$\mathrm{E}(x_{it}u_{it+j}) = 0 \quad j \geq 0$$

$$\mathrm{E}(x_{it}u_{it-s}) \neq 0 \quad s \geq 1$$

Conversely, we can say that current outcomes on y_{it} (or u_{it}) depend on past and current $x's$ only. In other words, under the assumption of weak exogeneity, we have that

$$\mathrm{E}(y_{it}|x_{i1}, \ldots, x_{it}, \alpha_i) = \exp(x_{it}'\beta)\alpha_i$$

The conditioning refers only to past and current values of x. This condition is weaker than strict exogeneity, conditional on α_i, where

$$\mathrm{E}(y_{it}|x_{i1}, \ldots, x_{iT}, \alpha_i) = \exp(x_{it}'\beta)\alpha_i$$

Only under strict exogeneity does the corresponding Poisson log-likelihood function factor into the product of independent individual likelihood contributions for given α_i. With weak exogeneity, the fixed effects Poisson, or mean-scaled, estimator is inconsistent as x_{it} is correlated with u_{it}^* through $\bar{u}_i$. While panel data can handle correlation between $\bar{u}_i$ and $\bar{x}_i$, they still require that x_{it} be uncorrelated with the time varying error u_{is}, i.e., $\mathrm{E}(x_{it}u_{is}) = 0$ for all t and s.

Hence, alternative methods are required. The problem is to find a transformation that eliminates the multiplicative fixed effect and at the same time generates useable moment conditions. Following Chamberlain (1992), Blundell, Griffith and van Reenen (1995) and Montalvo (1997) apply the following reasoning. A residual u_{it} can be implicitly defined through

$$y_{it} = \exp(x_{it}'\beta)\alpha_i + u_{it} \tag{5.33}$$

Under weak exogeneity, u_{it} has conditional expectation

$$\mathrm{E}(u_{it}|x_{i1}, \ldots, x_{it}, \alpha_i) = 0$$

Re-writing (5.33) for $t + 1$, and solving for α_i

$$\alpha_i = \frac{y_{it+1} - u_{it+1}}{\exp(x_{it+1}'\beta)}$$

Substituting back into (5.33) we obtain

$$\begin{aligned} y_{it} &= \exp(x'_{it}\beta)\left(\frac{y_{it+1} - u_{it+1}}{\exp(x'_{it+1}\beta)}\right) + u_{it} \\ &= y_{it+1}\exp[(x_{it} - x_{it+1})'\beta] + v_{it} \end{aligned} \tag{5.34}$$

where

$$\begin{aligned} v_{it} &= u_{it} - u_{it+1}\exp[(x_{it} - x_{it+1})'\beta] \\ &= y_{it} - y_{it+1}\exp[(x_{it} - x_{it+1})'\beta] \end{aligned} \tag{5.35}$$

The transformed variable v_{it} is uncorrelated with past and current values of x:

$$\begin{aligned} \mathrm{E}[v_{it}|x_{i1},\ldots,x_{it}] &= \mathrm{E}[u_{it} - u_{it+1}\exp[(x_{it} - x_{it+1})'\beta]|x_{i1},\ldots,x_{it}] \\ &= -\mathrm{E}[\mathrm{E}[u_{it+1}|x_{i1},\ldots,x_{it+1}]\exp[(x_{it} - x_{it+1})'\beta]|x_{i1},\ldots,x_{it}] \\ &= 0 \end{aligned}$$

Thus, if we define v_i to be the vector

$$v_i = \begin{bmatrix} y_{i1} - y_{i2}\exp[(x_{i1} - x_{i2})'\beta] \\ y_{i2} - y_{i3}\exp[(x_{i2} - x_{i3})'\beta] \\ \\ y_{iT-1} - y_{iT}\exp[(x_{iT-1} - x_{iT})'\beta] \end{bmatrix}$$

and a matrix of instruments Z_i as

$$Z_i = \begin{bmatrix} z_{i1} & 0 & \cdots & 0 \\ 0 & z_{i2} & \cdots & 0 \\ \vdots & \vdots & \ddots & 0 \\ 0 & 0 & 0 & z_{iT-1} \end{bmatrix}$$

where $z_{it} = (1, x_{i1}, \ldots, x_{it})$, the GMM estimator of β is obtained by minimizing

$$\left[\sum_{i=1}^{n} v'_i Z_i\right] \widehat{W}_n^{-1} \left[\sum_{i=1}^{n} Z'_i v_i\right]$$

where the weight matrix is given by

$$\widehat{W}_n = \frac{1}{n}\sum_{i=1}^{n} Z'_i \hat{v}_i \hat{v}'_i Z_i$$

Alternatively, Wooldridge (1997a) proposes to eliminate the fixed effects using the transformation

$$u^{\dagger}_{it} = \frac{y_{it}}{\lambda_{it}} - \frac{y_{it+1}}{\lambda_{it+1}} \tag{5.36}$$

which equals v_{it} divided by λ_{it}. Windmeijer (2000) shows that this transformation is applicable for endogenous as well as weakly exogenous regressors.

He also suggests that the failure of the Wooldridge transformation for non-negative regressors (in which case the associated β would go to infinity) can be overcome by transforming the x's first around their grand mean.

An alternative approach to deal with weakly endogenous regressors in panel count data is pursued by Blundell, Griffith and Windmeijer (2002) who use pre-sample information to form instruments for GMM estimation based on the mean-scaling model.

5.3 Time-Series Count Data Models

Pure time series count data can be seen as a special case of panel count data where $n = 1$ and T is large. Examples from the previous count data literature include the number of strikes per month (Buck, 1984), the number of bank failures per year (Davutyan, 1989) and the founding rate of national labor unions (Barron, 1992). In practice, the absence of a cross-sectional dimension makes a substantial difference, and developments of specialist time series models have been pursued independently of, and in most cases preceeding, those of panel models. The main concern of this literature has been a parsimonious and yet flexible correlation structure.

Dependence across time periods can be modeled in one of two ways. The first way is the introduction of an explicit lag structure in the endogenous count variable. This approach is also referred to as an "observation-driven" model (Firth 1991). The alternative is a "parameter-driven" model where time-series characteristics are introduced by correlated unobserved heterogeneity, following Zeger (1988), who augments the Poisson model by a multiplicative error term that follows an autoregressive process. This approach introduces both overdispersion and autocorrelation into y_t. Zeger proposes estimation of the model parameters by quasi-likelihood in the tradition of generalized linear models.

The observation driven approach is pursued by Al-Osh and Alzaid (1987, 1988) who define a fully parametric framework for modelling integer valued process with serial correlation. Al-Osh and Alzaid (1987) considers the case of integer valued autoregression, whereas Al-Osh and Alzaid (1988) deals with integer valued moving averages. In either case is the transition model characterized through a stopped-sum distribution (i.e., "binomial mixing" or "binomial thinning"). A synthesis of the two approaches that combines the INAR(1) structure with additional dependence from correlated errors is proposed by Brännäs and Hellström (2002).

An extensive survey of these methods, including an analysis of the performance of the estimators in simulation studies, is provided by Jung (1999). Ronning and Jung (1992), Brännäs (1994), and Böckenholt (1999) give applications of integer valued modeling in econometrics. See also Jung and Liesenfeld (2001).

Time Series with Correlated Multiplicative Errors

A time series model with correlated multiplicative error was proposed by Zeger (1988). This model can be seen as a special case of the multivariate Poisson model with correlated errors that was discussed in Chap. 5.1.4 and in Chap. 5.1.6. Recall that in the multivariate Poisson-log-normal model

$$\mathrm{E}(y_{it}|u_{it}) = \lambda_{it}u_{it}$$

and

$$\mathrm{Var}(y_{it}|u_{it}) = \lambda_{it}u_{it}$$

where

$$u_i = (u_{i1} \cdots u_{iT})' \sim \mathrm{MVLN}(\mathbf{1}, \Sigma)$$

For a time series, $n = 1$. Without further assumptions, the parameters of this model cannot be identified from a pure time-series. For instance, Σ, a symmetric $(T \times T)$ matrix, has $T(T+1)/2$ different elements. While an unrestricted covariance matrix can be estimated with multivariate data, restrictions are needed for time series data. For instance, Zeger (1988) considers a covariance stationary process where $\mathrm{Cov}(u_t, u_{t+\tau}) = \sigma_u(\tau)$. Under this assumption,

$$\Sigma = \begin{pmatrix} \sigma^2 & \sigma(1) & \ldots \sigma(t-1) \\ \sigma(1) & \sigma^2 & \ldots \sigma(t-2) \\ \vdots & \vdots & \ddots \quad \vdots \\ \sigma(t-1) & \sigma(t-2) & \ldots \quad \sigma^2 \end{pmatrix} \tag{5.37}$$

Zeger (1988) adopts a quasi-, rather than maximum, likelihood framework. Thus, the assumption of log-normality is dropped, and only the first two moments of the distribution of y_{it} are taken into consideration. Zeger's approach is based on the score function

$$D'V^{-1}(y - \lambda) = 0 \tag{5.38}$$

where $D = \mathrm{d}\lambda/\mathrm{d}\beta$ is of dimension $(T \times k)$, y and λ are of dimension $(T \times 1)$, and

$$V = \mathrm{Var}(y) = \Lambda + \Lambda\Sigma\Lambda$$

where $\Lambda = \mathrm{diag}(\lambda_t)$. For independent observations, V is a diagonal matrix with diagonal element λ_t and the score equations reduce to the sum of the individual scores. In a time series context, however, Σ has non-zero off-diagonal elements as specified above.

The estimator that solves (5.38) has the well defined asymptotic distribution of a quasi-likelihood estimator under arbitrary forms of the covariance matrix (See, for instance McCullagh and Nelder, 1989, Chap. 9, Zeger, 1988). In practice, V is unknown and thus requires estimation. Zeger (1988) suggests the moment estimators

$$\hat{\sigma}^2 = \frac{\sum_{t=1}^{T} \tilde{y}_t^2 - \hat{\lambda}_t}{\sum_{t=1}^{T} \tilde{y}_t^2}$$

and

$$\hat{\sigma}(\tau) = \frac{\sum_{t=\tau+1}^{T} \tilde{y}_t \tilde{y}_{t-\tau}}{\hat{\sigma}^2 \sum_{t=\tau+1}^{T} \hat{\lambda}_t \hat{\lambda}_{t-\tau}}$$

where $\tilde{y}_t = y_t - \hat{\lambda}_t$. An iterative process can be used for estimation.

Alternatively, one may want to approximate the general covariance matrix Σ, and thus V, by a more parsimonious parameterization that follows for instance from a stationary autoregressive process. In this way, one can also avoid the repeated computation of the inverse of V, a matrix of dimension $(T \times T)$. First, note that V can be rewritten as

$$\begin{aligned} V &= (\Lambda + \sigma^2 \Lambda^2)^{1/2} R (\Lambda + \sigma^2 \Lambda^2)^{1/2} \\ &= D^{1/2} R D^{1/2} \end{aligned}$$

where R is the autocorrelation matrix of u and $D = \Lambda + \sigma^2 \Lambda^2$. Take, for instance, the case where u_t is assumed to follow a first-order autoregression (This case is also discussed in Wun, 1991). Then,

$$R^{-1} = \frac{1}{1-\rho} \begin{pmatrix} 1 & -\rho & 0 & \dots & 0 \\ -\rho & 1+\rho^2 & -\rho \dots & 0 & \\ 0 & -\rho & 1+\rho^2 & -\rho \dots & 0 \\ \vdots & \vdots & \vdots & \ddots & \vdots \\ 0 & 0 & \dots & -\rho & 1 \end{pmatrix}$$

(or, alternatively, $R^{-1} = L'L$ where L is the matrix that applies the autoregressive filter, i.e., $Ly = y_t + \alpha y_{t-1}$, $t > 1$). Thus, the inverse of V can be computed as

$$V^{-1} = D^{-1/2} R^{-1} D^{-1/2}$$

which is a considerable simplification.

Brännäs and Johannson (1994) consider estimation of the model by pseudo maximum likelihood. As long as the mean function is correctly specified, the Poisson model remains consistent but the asymptotic covariance matrix of the estimator needs to be adjusted.

Integer Valued Autoregression

Another model for time series count data is the integer valued autoregressive model (INAR), due to Al-Osh and Alzaid (1987) and McKenzie (1988) (see also Ronning and Jung, 1992). A random variable y follows a first order INAR process with Poisson marginals (written $y \sim$INAR(1)) if

$$y_t \overset{\mathrm{d}}{=} \alpha \circ y_{t-1} + \varepsilon_t \tag{5.39}$$

where

$$y_{t-1} \sim \text{Poisson}(\lambda)$$

$$\varepsilon_t \sim \text{Poisson}((1-\alpha)\lambda)$$

$$\varepsilon_t, y_{t-1} \text{ independent}$$

$$\alpha \circ y_{t-1} = \sum_{i=1}^{y_{t-1}} d_i$$

$$\alpha \in [0,1]$$

and

$$\{d_i\} \text{ i.i.d. with } P(d_i = 1) = 1 - P(d_i = 0) = \alpha\,.$$

The symbol "$\stackrel{d}{=}$" stands for "is equally distributed as". Equation (5.39) defines a somewhat unusual relationship as y_t is a random variable even as α, y_{t-1}, and ε_t are known. In the remainder of this part, equality signs will have the same interpretation, although the explicit notation using "d" is dropped for simplicity.

In (5.39), $\alpha \circ y_{t-1}$ is a mixture of a binomial distribution and a Poisson distribution. For independent d_i and y_{t-1}, the mixture operation $'\circ'$ is called *binomial thinning* (McKenzie, 1988). It replaces the scalar multiplication in the continuous AR(1) model. $\alpha \circ y_{t-1}$ denotes the number of elements out of $t-1$ that survive to period t. The probability of survival is given by α. By the rules for convolutions (See Appendix A) $\alpha \circ y_{t-1} \sim \text{Poisson}(\alpha\lambda)$.

This model has the following interpretation: the innovation process $\{\varepsilon_t\}$ gives the number of new elements entering the process. The total number of elements in t is the sum of surviving and newly entering elements with marginal distribution $y_t \sim \text{Poisson}(\lambda)$. (The INAR(1) process has the following properties:

i) $0 \circ y = 0,\ 1 \circ y = y$
ii) $\text{E}(\alpha \circ y) = \alpha \text{E}(y)$
iii) $\underbrace{\alpha \circ \cdots \circ \alpha}_{k\text{-times}} \circ y = \alpha^k \circ y$

From (5.39) and ii), it follows that

$$\text{E}(y_t|y_{t-1}) = \alpha y_{t-1} + (1-\alpha)\lambda\,. \tag{5.40}$$

Like for the first order autoregressive process with normally distributed innovations, the conditional expectation of y_t is linear in y_{t-1}. However, the regression is not linear in the parameters. Also, there is an additional source of randomness: given ε_t and y_{t-1}, y_t is still a (displaced binomial distributed) random variable.

Using iii) and recursive substitution, (5.39) can be rewritten as:

$$\begin{aligned} y_t &= \alpha \circ y_{t-1} + \varepsilon_t \\ &= \alpha \circ (\alpha \circ y_{t-2} + \varepsilon_{t-1}) + \varepsilon_t \\ &= \alpha \circ (\alpha \circ (\alpha \circ y_{t-3} + \varepsilon_{t-2}) + \varepsilon_{t-1}) + \varepsilon_t \\ &\;\;\vdots \end{aligned}$$

i.e.,

$$y_t = \alpha^\tau \circ y_{t-\tau} + \sum_{j=0}^{\tau-1} \alpha^j \circ \varepsilon_{t-j} \,. \tag{5.41}$$

The marginal distribution of the INAR(1) process is then given by

$$y_t = \sum_{j=0}^{\infty} \alpha^j \circ \varepsilon_{t-j} \,. \tag{5.42}$$

The effect of $\{\varepsilon_t\}$ on y_t is reduced exponentially with increasing lag length. (5.41) implies for the auto-covariance structure:

$$\begin{aligned} \gamma(\tau) &= \mathrm{Cov}(y_{t-\tau}, y_t) \\ &= \mathrm{Cov}\left(y_{t-\tau}, \alpha^\tau \circ y_{t-\tau}\right) + \mathrm{Cov}\left(y_{t-k}, \sum_{j=0}^{\tau-1} \alpha^j \circ \varepsilon_{t-j}\right) \\ &\overset{(ii)}{=} \alpha^\tau \mathrm{Var}(y_{t-\tau}) + \sum_{j=0}^{\tau-1} \alpha^j \mathrm{Cov}(y_{t-\tau}, \varepsilon_{t-j}) \\ &= \alpha^\tau \gamma(0) \end{aligned}$$

The auto-correlations $\rho(\tau) = \gamma(\tau)/\gamma(0)$ are, in contrast to those of the Gaussian process, restricted to the positive interval (0,1). The INAR(1)-Poisson process is stationary for $\alpha \in (0,1)$. For $y_0 \sim \text{Poisson}(\lambda)$ it holds $\forall t$ that

$$\begin{aligned} &\mathrm{E}(y_t) = \lambda \\ &\mathrm{Cov}(y_t, y_{t-\tau}) = \alpha^\tau \lambda \;, \tau = 0, 1, \ldots \end{aligned}$$

In particular, for $\tau = 0$, the typical Poisson property of equidispersion follows. Estimation can proceed by maximum likelihood. The INAR(1) model has Markovian property

$$f(y_t | y_{t-1}, y_{t-2}, \ldots) = f(y_t | y_{t-1})$$

and thus the joint distribution of the sample can be factored as

$$f(y_t, y_{t-1}, y_{t-2}, \ldots) = f(y_t | y_{t-1}) f(y_{t-1} | y_{t-2}) \ldots f(y_0)$$

The conditional distribution of y_t given y_{t-1} is a binomial-Poisson mixture, the probabilities of which are given by

$$f(y_t|y_{t-1}) = \exp[-(1-\alpha)\lambda](1-\alpha)^{y_{t-1}+y_t}\lambda^{y_t} \\ \times \sum_{k=0}^{\min(y_t,y_{t-1})} \frac{\alpha^k y_{t-1}!}{(1-\alpha)^{2k}\lambda^k k!(y_t-k)!(y_{t-1}-k)!}$$

Denoting the factor in the second line by B_t, the joint distribution of the process can be written as

$$f(y_0, y_1, \ldots, y_T) = \frac{((1-\alpha)\lambda)^{y_0+y_1}B_1}{\exp((2-\alpha)\lambda)} \prod_{t=2}^{T} \frac{y_{t-1}!(1-\alpha)^{y_{t-1}+y_t}\lambda^{y_t}B_t}{\exp((1-\alpha)\lambda)} \quad (5.43)$$

The parameters α, λ, and y_0 can be estimated by maximizing the corresponding likelihood. The starting value problem, which is the more severe the shorter the time series, is discussed in detail in Ronning and Jung (1992). Brännäs (1995a) shows, how the INAR(1) model can be extended in order to include explanatory variables.

Example

Kennan (1985) analyses the frequency of, and duration between, contract strikes in the United Kingdom. The observations are from January 1968 to December 1976. The empirical mean of the series is 5.5, the empirical variance 13.4. The empirical overdispersion indicates that an INAR(1) process with Poisson marginals cannot be appropriate since this would require equidispersion. The actual time series is plotted in Fig. 5.1.

To illustrate the method, Fig. 5.2 displays a simulated INAR(1) process for $\alpha = 0.5$. The starting value corresponds to the observed value $y_0 = 5$ and the expected value of the marginal distribution is equal to the empirical mean. The simulation of the Poisson and binomial variables used algorithms from Kennedy and Gentle (1980). The mean and variance of the simulated time series are 5.6 and 4.9, respectively. The series thus corresponds much closer to the postulated equality of mean and variance. Also, it is clearly a stationary process with mean reversion.

INAR(1) Process With Unobserved Heterogeneity

The model (5.39) can be extended to allow for negative binomial marginals, and thus unoabserved heterogeneity and overdispersion McKenzie, 1986). Consider again the basic relationship

$$y_t = \alpha \circ y_{t-1} + \varepsilon_t$$

Fig. 5.1. Kennan's Strike Data

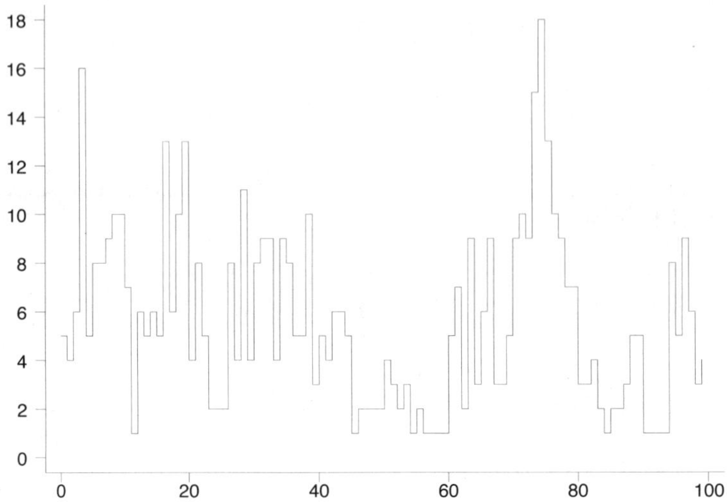

Fig. 5.2. Simulated INAR(1) Time Series for $\alpha = 0.5$

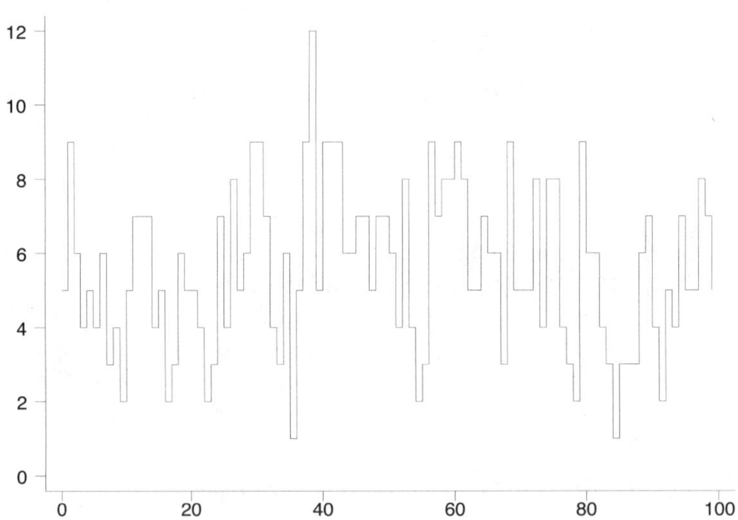

Assume that y_{t-1} has a negative binomial distribution with generic parameterization Negbin(δ, θ) (i.e., this is not a mean parameterization; rather, $\mathrm{E}(y_{t-1}) = \delta\theta$). In analogy to the Poisson case, one would be interested in a thinning operation $\circ$ that preserves the negative binomial distribution for $z_{t-1} = \alpha \circ y_{t-1}$. In the Poisson case we assumed that conditional on y_{t-1} and α, $z_{t-1} \sim$ binomial(y_{t-1}, α). McKenzie (1986) suggests that randomizing α through an indendent beta-distribution Be$(\gamma, \delta - \gamma)$ has a similar effect. In particular, $z_{t-1}|y_{t-1}, \gamma, \delta$ has a beta-binomial distribution and beta-binomial thinning results. It can be shown that the unconditional distribution of z_{t-1} is negative binomial Negbin(γ, θ). If, moreover, $\varepsilon_t \sim$ Negbin$(\delta - \gamma, \theta)$, an independent distribution, then it follows that the marginal distribution of y_t is Negbin(δ, θ).

Böckenholt (1999) discusses estimation of an INAR(1) process where unobserved heterogeneity is represented by a finite mixture, and where, conditional on the latent class, the process has all the standard properties of (5.39), including the Poisson marginals.

6

Bayesian Analysis of Count Variables

The existing econometrics literature on count data models has largely ignored the Bayesian paradigm of inference. Likewise, in Zellner's (1971) influential book on Bayesian inference in econometrics, the Poisson regression model is not mentioned. The probable reasons for this neglect are computational complexities that in the past made the Bayesian analysis of count data models appear unattractive. However, increased computer power now allows for fast evaluation of posterior distributions by simulation methods. The basic approaches to Bayesian inference by simulation are discussed in this chapter.

In Bayesian econometrics the interest centers around the posterior distribution $\pi(\theta|y)$ which is a product of the likelihood function $f(y|\theta)/f(y)$ and a prior distribution $g(\theta)$

$$\pi(\theta|y) = \frac{f(y|\theta)g(\theta)}{f(y)} \tag{6.1}$$

where

$$f(y) = \int_{\Theta} f(y|\theta)g(\theta)\mathrm{d}\theta \tag{6.2}$$

does not depend on θ and is a normalizing constant of integration, also called *marginal likelihood.* This constant is often difficult to evaluate, and so is, as a consequence, the posterior distribution. The standard approach is to omit the normalizing constant and write

$$\pi(\theta|y) \propto f(y|\theta)g(\theta) \tag{6.3}$$

If the right hand side is the kernel of a known distribution, the normalizing constant can be infered from there. Alternatively, recent simulation based methods do not require an evaluation of the normalizing constant at all and thus are much more versatile.

In contrast to classical inference, Bayesian methods condition on the data and model the parameter as a random variable. While much of the debate on the relative merits of the Bayesian over the frequentist approaches has been cast in philosophical terms, some of the recent literature has shifted the focus of the debate towards practical aspects: using recent simulation methods, the Bayesian approach can provide relatively simple solutions in models where frequentists methods fail, or at best, are difficult to implement. More on this below.

6.1 Bayesian Analysis of the Poisson Model

A standard result of a closed form posterior distribution exists for the Poisson model without covariates. Suppose $\{y_i\}, i = 1, \ldots, n$ is a random sample from a Poisson distribution with mean λ, and that the prior distribution of λ is a gamma distribution with parameters $\alpha \geq 0$ and $\beta \geq 0$. The gamma distribution is the conjugate prior for the Poisson parameter, and

$$\begin{aligned} g(\lambda|y) &\propto \left(\prod_{i=1}^{n} e^{-\lambda} \lambda^{y_i} \right) \frac{\alpha^{\beta}}{\Gamma(\alpha)} \lambda^{\alpha-1} e^{-\lambda\beta} \\ &\propto e^{-\lambda(\beta-n)} \lambda^{\alpha+n\bar{y}-1} \end{aligned} \tag{6.4}$$

Hence, the posterior distribution of λ is a gamma distribution with parameters $\tilde{\alpha} = \alpha + n\bar{y}$ and $\tilde{\beta} = (\beta + n)$. Recall that the mean of the prior gamma distribution is given by $\mathrm{E}_0(\lambda) = \alpha/\beta$. Therefore, the posterior mean $\tilde{\alpha}/\tilde{\beta}$ can be written as

$$\begin{aligned} \mathrm{E}_{\pi}(\lambda|y, \alpha, \beta) &= \frac{\alpha + n\bar{y}}{\beta + n} \\ &= \frac{\beta}{\beta + n} \mathrm{E}_0(\lambda) + \frac{n}{\beta + n} \bar{y} \end{aligned}$$

The Poisson-gamma model is an example for a common result in Bayesian statistic, namely that the posterior mean is a weighted average of prior mean and sample mean. The weight given to the sample mean is an increasing function of the number of observations.

No conjugate prior exists for the $(k \times 1)$ parameter vector β in the Poisson regression model where the likelihood is proportional to

$$L(\beta|y, x) \propto \prod_{i=1}^{n} \exp[-\exp(x_i'\beta)] \, [\exp(x_i'\beta)]^{y_i} \tag{6.5}$$

Even with a noninformative prior, this expression is not the kernel of any known parametric distribution for β. There are two solutions. One is the use of approximation methods as, for instance, in Albert and Pepple (1989). The other is the evaluation of the exact posterior distribution using simulation methods. Consider approximation first. Let $\hat{\beta}$ be the mode of the posterior density, i.e., the maximum likelihood estimator. If the logarithm of this density is expanded in a second-order Taylor's series expansion around $\hat{\beta}$, we obtain

$$\ln \pi(\beta, y, x) \approx \ln L(\hat{\beta}) - \frac{1}{2}(\beta - \hat{\beta})' H(\beta - \hat{\beta}) \tag{6.6}$$

where H is minus the expected Hessian matrix evaluated at $\hat{\beta}$. Thus, the posterior of β is approximately multivariate normal with mean $\hat{\beta}$ and covariance matrix H^{-1}.

Next, assume a normal prior for β

$$g(\beta) = \phi(\beta | \beta_0, B_0^{-1}) \tag{6.7}$$

where β_0 denotes the prior mean, and B_0^{-1} the prior precision matrix, the inverse of the covariance matrix. That is,

$$g(\beta) \propto \exp[-1/2(\beta - \beta_0)' B_0^{-1} (\beta - \beta_0)]$$

In order to impose a reasonably vague prior, it is common to let $\beta_0 = 0$ and $B_0^{-1} = 10^{-2} I_k$, where I_k is the k-dimensional identity matrix. The posterior density $\pi(\beta|y)$ is then proportional to

$$\prod_{i=1}^{n} \exp[-\exp(x_i'\beta)]\, [\exp(x_i'\beta)]^{y_i} \exp[-1/2(\beta - \beta_0)' B_0^{-1} (\beta - \beta_0)] \tag{6.8}$$

and not in standard form.

Again, an approximation method could be used. Instead, in order to obtain exact results, we now consider simulation of the posterior density by the Metropolis-Hastings (MH) algorithm (See Chib and Greenberg, 1995). This is a special case of Markov chain Monte Carlo simulation. For a given target density $f(\psi)$, the MH algorithm is defined by

(1) a proposal density $q(\psi^\dagger | \psi)$ that is used to supply a proposal value $\psi^\dagger$ given the current value ψ, and
(2) a probability of move that is defined as

$$\alpha(\psi, \psi^\dagger) = \min \left\{ \frac{f(\psi^\dagger) q(\psi | \psi^\dagger)}{f(\psi) q(\psi^\dagger | \psi)}, 1 \right\}. \tag{6.9}$$

Hence, if $f(\psi^\dagger) q(\psi|\psi^\dagger) > f(\psi) q(\psi^\dagger|\psi)$ the chain moves to $\psi^\dagger$. Otherwise, it moves with probability $0 < \alpha(\psi, \psi^\dagger) < 1$. If rejected, the next sampled value is taken to be ψ.

For the MH algorithm to work efficiently, the choice of proposal density q is critical. Following Chib, Greenberg, and Winkelmann (1998), the proposal distribution for the Poisson regression model can be based on the mode $\hat{\beta}$ and curvature $V_\beta = [-H_\beta]^{-1}$ of $\ln \pi(\beta|y)$ where these quantities are found using a few Newton-Raphson steps with gradient vector

$$g_\beta = B_0^{-1}(\beta - \beta_0) + \sum_{i=1}^{n} [y_i - \exp(x_i'\beta)] x_i$$

and Hessian matrix

$$H_\beta = -B_0^{-1} - \sum_{i=1}^{n} \exp(x_i'\beta)x_i x_i'$$

The proposal can be obtained by reflecting the current value around the modal value $\hat{\beta}$ and then adding a Gaussian increment with variance τV_β (τ is a scalar that is adjusted in trial runs in order to obtain acceptance rates between 40 and 60 percent). The resulting proposal density is

$$q(\beta, \beta^\dagger|y) = \phi(\beta^\dagger|\hat{\beta} - (\beta - \hat{\beta}), \tau V_\beta)$$

To draw from the proposal density, we simply compute

$$\beta^\dagger = \hat{\beta} - (\beta - \hat{\beta}) + \tau \text{chol}(V_\beta)'\text{rndn}(k, 1)$$

where $P = \text{chol}(V_\beta)$ gives the Cholesky (upper-triangular) decomposition of V_β such that $V_\beta = P'P$, and $\text{rndn}(k, 1)$ is a $(k \times 1)$ vector of standard normal pseudo-random numbers.

Finally, the probability of move is given in terms of the ratio of density ordinates

$$\alpha(\beta, \beta^\dagger|y) = \min\left\{\frac{\pi(\beta^\dagger|y)}{\pi(\beta|y)}, 1\right\}, \tag{6.10}$$

since the proposal density is symmetric in $(\beta, \beta^\dagger)$ and hence cancels. In practice, the algorithms goes through a large number of iterations: 2,000 or 10,000 are some common values. In addition, it is recommended to preceed the proper sampling from the posterior by a burn-in phase of a given number of iterations (500, say), in order to reduce the influence of arbitrary starting values and let the algorithm move to its main area of support. Finally, the posterior sample can be analysed in order to report any distributional characteristics of choice, such as (posterior) mean, standard deviation, median, percentiles, or credibility intervals.

Discussion

The example of posterior simulation in the Poisson regression model illustrates well the potential of the method. Part of the simplicity arises since an evaluation of the normalizing constant is not required. This method has some direct additional benefits. For instance, inequality constraints on parameters can be imposed without problem: if sampled values fall within the inadmissable area, they are simply dropped. Likewise, it is also very simple to simulate the posterior distribution of a (possibly complicated) function of the parameters. In contrast to maximum likelihood, where the invariance property applies to the modal estimates, but standard errors need to be derived using asymptotic properties and the delta rule, the simulation approach immediately provides the full posterior distribution of the function, including correct standard errors, percentiles, etc. The approach is easily extended to other prior distributions. Finally, modified Poisson distributions (such as truncated, censored) can be introduced by simply adjusting the likelihood function in (6.8).

6.2 A Poisson Model with Underreporting

In the Poisson regression model, the basic simulation tool was the Metropolis-Hastings step to draw from the posterior distribution of β. The power and versatility of Markov chain Monte Carlo can be substantially increased by combining MH-simulation with Gibbs sampling and data augmentation. For more detailed references on Markov chain Monte Carlo, see Chib and Greenberg (1996) and Gamerman (1997).

The following application to a Poisson model with underreporting illustrates these possibilites. Count data models with underreporting have been discussed in Chap. 4.8. The presentation here follows Winkelmann (1996b) who re-analysed the model in a Bayesian framework.

Let y_i^* denote the total number of events during a fixed time period T for individual i, and assume that the likelihood function $f(y_i^*|\beta)$ is of standard Poisson form, i.e.,

$$f(y_i^*|\beta) = \frac{\exp(-\exp(x_i'\beta))\exp(x_i'\beta)^{y_i^*}}{y_i^*!}$$

If y_i^* was observed, the algorithm of the previous section could be used to obtain the posterior distribution of β.

With underreporting, y_i^* is unobsered. We observe the number of reported events y_i which, conditional on y_i^*, is binomial distributed

$$f(y_i|y_i^*, p_i) = \frac{y_i^*!}{(y_i^* - y_i)!y_i!} p_i^{y_i}(1-p_i)^{y_i^*-y_i} \tag{6.11}$$

where p_i gives the individual probability of reporting an event. The structure of the model becomes more apparent once we write down the joint posterior distribution of β, p and y^*, where y^* has been included among the parameters, a case of data augmentation:

$$\pi(y^*, p, \beta|y, x) \propto f(y|y^*, p, \beta)f(y^*|\beta)g(\beta)g(p) \tag{6.12}$$

The following prior distributions g can be used:

$$g(\beta) \sim \phi(\beta_0, B_0^{-1}) . \tag{6.13}$$

and

$$g(p) \sim \mathcal{U}(0,1) \tag{6.14}$$

where $\mathcal{U}(0,1)$ is the standard uniform distribution. The resulting joint posterior distribution of y_i^*, p_i, and β is then proportional to

$$\pi(y^*, p, \beta | y, x) \propto \prod_{i=1}^{n} \exp\left\{y_i^* x_i' \beta - \exp(x_i'\beta)\right\} \frac{p_i^{y_i}(1-p_i)^{y_i^* - y_i}}{(y_i^* - y_i)! y_i!}$$
$$\times \exp(-1/2(\beta - \beta_0)' B_0 (\beta - \beta_0)) \qquad (6.15)$$

While it is intractable to derive analytically the marginal posterior distributions for the parameters of interest from (6.15), the MCMC approach allows to simulate the joint posterior density. One could contemplate a direct "brute-force" simulation of (6.15) using the MH method described in the previous section. However, it will be problematic to obtain a suitable proposal density and the approach is likely to be costly and inefficient.

A superior algorithm is to sample the joint posterior by successively sampling through its full conditional distributions. This is also frequently refereed to as Gibbs sampling. The gains from Gibbs sampling are most evident when some of the full conditional distributions can be simulated from standard distributions, as is the case in this example. The three required full conditional distributions in this case are

$$[y^* | p, \beta, y, x], \; [p | y^*, \beta, y, x], \; [\beta | y^*, p, y, x]$$

Inspection of the joint posterior (6.15) reveals that the full conditional posterior of y^* is given by

$$\pi(y^* | \beta, p, y, x) \propto \prod_{i=1}^{n} \frac{[\exp(x_i'\beta)(1-p_i)]^{y_i^*}}{(y_i^* - y_i)!}$$

This is the kernel of a Poisson distribution shifted by y_i, with parameter $\lambda_i^* = \exp(x_i'\beta)(1-p_i)$. Pseudo random numbers from this distribution can be readily generated using the Knuth (1969) algorithm.

The full conditional distribution of p is of beta form, while the full conditional distribution of β (that depends on y^* only) is the same as the posterior distribution of the Poisson regression model and can thus be simulated using the MH algorithm discussed in the previous section. This last step exactly reflects the effect of data augmentation: data augmentation replaces unobserved values by simulated values and thus restores the model in standard form.

The sampling process is initiated with values in the support of the posterior density. The algorithm runs in cycles through the three full conditional densities, where the conditioning values for the parameters are updated as soon as a more recent value becomes available. As for the MH algorithm, the simulated values after an initial burn-in phase are retained as a sample from the target joint posterior distribution. It also should be noted that the MH-step required in the simulation of β involves only one draw in each cycle of the Gibbs sampler.

6.3 Estimation of the Multivariate Poisson-Log-Normal Model by MCMC

Recall the multivariate Poisson-log-normal model presented in Chap. 5.1.4. Conditionally on a $(J \times 1)$ vector of individual and outcome specific random effects $\varepsilon_i = (\varepsilon_{i1} \cdots \varepsilon_{iJ})'$ the distribution of y_i is independent Poisson

$$f(y_i|\varepsilon_i) = \prod_{j=1}^{J} f_P(y_{ij}|\lambda_{ij} \exp(\varepsilon_{ij})) \tag{6.16}$$

Moreover

$$f(\varepsilon_i|\Omega) = \phi_J(-0.5\text{diag}\Omega, \Omega) \tag{6.17}$$

where ϕ_J is the J-variate normal density with covariance matrix Ω. The details of this model structure were discussed before. For the purposes of the present chapter, the important aspect is that the likelihood function requires the evaluation of a J-variate integral of the Poisson distribution with respect to the distribution of ε_i

$$f(y_i|\lambda_i, \Omega) = \int \prod_{j=1}^{J} f_P(y_{ij}|\lambda_{ij}, \varepsilon_{ij})\phi(\varepsilon_i| - 0.5\text{diag}\Omega, \Omega)\text{d}\varepsilon_i$$

and hence estimation by maximum likelihood may not be possible for large J. Chib and Winkelmann (2001) show how the joint posterior distribution of $\psi = (\beta, \Omega)$ can be obtained using MCMC for arbitrary J (An application of the Gibbs sampler to a univariate Poisson-log-normal model, i.e., for $J = 1$, is provided by Plassmann and Tideman, 2001). Suppose that the parameters (β, Ω) independently follow the prior distributions

$$\beta \sim N(\beta_0, B_0^{-1}), \quad \Omega^{-1} \sim \text{Wish}(\nu_0, R_0),$$

where (β_0, B_0, v_0, R_0) are known hyperparameters and $\text{Wish}(\cdot,\cdot)$ is the Wishart distribution with ν_0 degrees of freedom and scale matrix R_0.

Using data augmentation for the random effects ε, the parameters can be blocked as ε, β, and Ω after which the joint posterior is simulated by recursively sampling the full conditional distributions

$$[\varepsilon|y, \beta, \Omega]; \;\; [\beta|y, \varepsilon]; \;\; [\Omega^{-1}|\varepsilon], \tag{6.18}$$

using the most recent values of the conditioning variables at each step.

In order to sample ε from the target density

$$\pi(\varepsilon|y, \beta, \Omega) = \prod_{i=1}^{n} \pi(\varepsilon_i|y_i, \beta, \Omega)$$

consider the i-th target density

$$\pi(\varepsilon_i|y_i, \beta, \Omega) = c_i\phi(\varepsilon_i|\Omega) \prod_{j=1}^{J} \exp(-\tilde{\lambda}_{ij})\tilde{\lambda}_{ij}^{y_{ij}} \tag{6.19}$$

where $\tilde{\lambda}_{ij} = \exp(x'_{ij}\beta_j + \varepsilon_{ij})$. Draws from this conditional density can be obtained using the MH-algorithm. The proposal density is taken to be multivariate-t with parameters that are tailored to those of the target $\pi(\varepsilon_i|y_i, \beta, \Omega)$. Let $\hat{\varepsilon}_i = \arg\max \ln \pi(\varepsilon_i|y_i, \beta, \Omega)$ and $V_{\varepsilon_i} = (-H_{\varepsilon_i})^{-1}$ be the inverse of the Hessian of $\ln \pi(\varepsilon_i|y_i, \beta, \Omega)$ at the mode $\hat{\varepsilon}_i$. Then, the proposal density is given by $q(\varepsilon_i|y_i, \beta, D) = f_T(\varepsilon_i|\hat{\varepsilon}_i, V_{\varepsilon_i}, \nu)$, a multivariate-$t$ density with ν degrees of freedom (where ν is a tuning parameter).

The sampling of β, conditional on ε, follows the standard Poisson procedure discussed in Chap. 6.1. Finally, Ω^{-1} is sampled by MH from a target density proportional to

$$\pi = f_W(\Omega^{-1}|v_0, R_0^{-1}) \prod_{i=1}^{n} \phi(\varepsilon_i| - 0.5\text{diag}\Omega, \Omega) \tag{6.20}$$

where f_W denotes again the Wishart density. See Chib and Winkelmann (2001) for further details and applications.

6.4 Estimation of a Random Coefficients Model by MCMC

With panel data, parameter heterogeneity becomes a possible issue. While most of the literature has dealt with intercept heterogeneity, this is not an intrinsic constraint. In fact, among linear model, the class of random coefficient models explicitly introduce heterogeneity for all regression coefficients of the model, including the slopes (see, e.g., Greene, 1993). Unfortunately, few of the known results for the linear model carry over to non-linear models such as the Poisson model. In fact, such a class of count data models may be intractable using maximum likelihood based methods. However, Chib, Greenberg, and Winkelmann (1998) show how a random coefficients model can be analysed in a Bayesian framework using MCMC. As in the previous chapter, data augmentation and Gibbs sampling are the key elements of the algorithm.

The main elements of the model are as follows: The distribution of $y_i = (y_{i1} \cdots y_{iT})'$ conditional on (b_i, β) is assumed to be independent Poisson

$$f(y_i|\beta, b_i) = \prod_{t=1}^{T} \frac{\exp(-\tilde{\lambda}_{it})\, \tilde{\lambda}_{it}^{y_{it}}}{y_{it}!} .$$

where

$$\tilde{\lambda}_{it} = \exp(x'_{it}\beta + w'_{it}b_i)$$

β are fixed coefficients and b_i are random coefficients. It should be stressed that w is not a subset of x. In a pure random coefficients model, all explanatory variables are part of w and x does not appear in the conditional mean function.

The random coefficients have a multivariate normal distribution

$$b_i \sim N_q(\eta, \Omega)$$

The (unknown) vector η represents the corresponding fixed effects. The joint density of $y = (y_1, ..., y_n)$ conditional on β, η and Ω (but unconditional on b_i) is given by

$$\begin{aligned} f(y|\beta, \eta, \Omega) &= \prod_{i=1}^{n} f(y_i|\beta, \eta, \Omega) \\ &= \prod_{i=1}^{n} \int \prod_{t=1}^{T} f(y_{it}|\beta, b_i)\, \phi(b_i|\eta, \Omega)\, d\, b_i \end{aligned} \tag{6.21}$$

If one were to proceed with maximum likelihood estimation (6.21) could be maximized with respect to β, η and D as long as there are few random coefficients and hence a low-dimensional integral structure.

Here, the interest is in the joint posterior. It is given by

$$f(\beta, \eta, b, \Omega|y) \propto f(y|\beta, b) f(\beta) f(b|\eta, \Omega) f(\eta) f(\Omega)$$

with priors

$$\beta \sim N_k(\beta_0, B_0^{-1})\ ,\ \eta \sim N_k(\eta_0, M_0^{-1}),\ \ \Omega^{-1} \sim \text{Wish}(v_0, R_0)\ ,$$

where Wish is the Wishart distribution. The joint posterior can then be rewritten as (up to a proportionality constant)

$$\phi(\beta|\beta_0, B_0^{-1}) \phi(\eta|\eta_0, M_0^{-1})\, f_W(\Omega^{-1}|v_0, R_0) \prod_{i=1}^{n} f(y_i|\beta, b_i)\, \phi(b_i|\eta, \Omega)$$

To simulate the posterior distribution by Markov Chain Monte Carlo, the following simple conditional structure is used.

- $[\beta|y, b, \Omega]$ is proportional to $[\beta] \times [y|\beta, b]$ because the conditional density of the observable data is independent of Ω. Its density is proportional to

  $$\exp\left(-\frac{1}{2}(\beta - \beta_0)' B_0 (\beta - \beta_0)\right) \prod_{i=1}^{n} \prod_{t=1}^{T} \exp(-\tilde{\lambda}_{it}) \tilde{\lambda}_{it}^{y_{it}}$$

- The conditional distribution $[\eta|y, \beta, b, \Omega]$ is proportional to $[\eta] \times [b|\eta, \Omega]$, independent of y and β, and is given by

  $$N_q(\hat{\eta}, M_1^{-1})$$

  where $\hat{\eta} = M_1^{-1}(M_0 \eta_0 + \sum_{i=1}^{n} \Omega^{-1} b_i)$ and $M_1 = (M_0 + n\Omega^{-1})$.
- The conditional distribution $[b|y, \beta, \Omega]$ factors into a product of conditionally independent distributions

  $$[b|y, \beta, \eta, \Omega] = \prod_{i=1}^{n} [b_i|y_i, \beta, \eta, \Omega]$$

  where $[b_i|y_i, \beta, \eta, \Omega]$ is proportional to $[b_i|\eta, \Omega]\ [y_i|\beta, b_i]$ with density given by

$$\exp\left(-\frac{1}{2}(b_i - \eta)'\Omega^{-1}(b_i - \eta)\right)\prod_{t=1}^{T}\exp(-\tilde{\lambda}_{it})\tilde{\lambda}_{it}^{y_{it}}$$

- The conditional distribution $[\Omega^{-1}|y, \beta, \eta, b]$ is proportional to $[\Omega^{-1}]\,[b|\eta, \Omega]$ wich can be shown to be of Wishart form

$$\text{Wish}\left(n + v_0, \left[R_0^{-1} + \sum_{i=1}^{n}(b_i - \eta)(b_i - \eta)'\right]^{-1}\right)$$

Except for the conditional distributions of η and Ω^{-1} these are unknown distribution families that are simulated using the Metropolis-Hastings algorithm. Successively sampling the distributions

$$[\beta|y, \eta, b, \Omega] \rightarrow [\eta|y, \beta, b, \Omega] \rightarrow [b|y, \beta, \Omega] \rightarrow [\Omega^{-1}|y, \beta, b]$$

generates then a Markov Chain sample that converges to the target density.

7

Applications

7.1 Accidents

The study of accident rates is one of the earliest applications of count data analysis: Bortkiewicz (1898) established that the number of deaths from mule kicks in the Prussian army during the period 1875-1894 could be well described by a Poisson distribution with mean of 0.7 corps per year. Although later studies have shifted their interest to different causes of death, the same principles apply. Accidents, by their very definition, come possibly closest to the notion of randomness required for processes such as the Poisson process.

A common feature of accident studies is that the choice of the right risk group is usually a non-trivial matter. For instance, Evans and Graham (1990) investigate the effect of child restraint use legislation on child mortality in car accidents using state level data. Their risk set is defined as total vehicle miles traveled (per year and state, in thousands). Michener and Tighe (1992) in a similar analysis investigate the effects of speed limits, minimum drinking age, and mandatory seatbelt laws on the number of fatal accident. In addition to vehicle miles traveled, they consider the number of registered drivers and number of registered vehicles as indicators of the size of the risk set. Rose (1990) analyzes the determinants of air-traffic incidents per number of scheduled departures per year (in thousands).

As discussed in Chap. 3.1.5 these scale variables can be either included in the regression as logarithmic offset with unit coefficient (as in Rose, 1990) or the coefficient can be given free for estimation (as in Evans and Graham, 1990, and Michener and Tighe, 1992). In the former case, the interpretation of the regression function $\exp(x_i'\beta)$ is in terms of an accident rate.

The focus of the above studies, and numerous others of similar kind, is on evaluating public policy: how successful was past (traffic) safety legislation in reducing the number of accidents. Frequently, such studies use aggregate data. Another type of analysis has pursued the insurance aspect of accidents. For instance, based on individual level data, it would be interesting to estimate the actuarial risk associated with insuring a certain individual with given characteristics and record. Although the data are different, the employed methods are essentially the samc. For instance, Dionne and Vanasse (1992) use driving records and Negbin II regressions in order to predict individual claim frequencies, taking into account individuals characteristics and driving histories. Dionne et al. (1994, 1995) extend this analysis by controlling for various aspects of the driver's medical condition.

Applications of count data models to accident numbers are by no means limited to those involving transport. Feinstein (1989) uses pseudo-maximum likelihood Poisson regressions in order to model the number of "incidents" at U.S. nuclear power plants. Bauer et al. (1998) are interested in the determinants of the number of workplace accidents using firm level data for Germany. Ruser (1991, 1993) studies the number of workdays lost due to occupational injury.

7.2 Crime

Grogger (1990a) analyzes the short-term deterring effect of capital punishment. To be specific, he investigates whether the daily number of homicides in California in the early 1960s were lower on days that followed an execution (or on days leading up to one, as they were usually reported in the newspapers with a lead), than on other days. Using Poisson and negative binomial regression models, the null-hypothesis of no effect cannot be rejected.

Kelly (2000) analyzes the effects of inequality on the number of property and violent crimes. Data are at the U.S. county level for the year 1990. The employed technique is Poisson quasi-likelihood estimation. Kelly finds that for violent crime, there is a large impact of inequality even after controlling for the effects of poverty, race and family composition. There is a negative effect of policy activity on crime that persists once the possible endogeneity of police activity is accounted for in an instrumental variable GMM procedure.

Finally, Plassmann and Tideman (2001) estimate generalized Poisson models to examine the dynamic effects of right to carry laws on reported homicides, rapes, and robberies, again using data for the U.S. with variation over time and across U.S. states. They conclude that such laws appear to have statistically significant deterrent effects on the numbers of reported murders, rapes, and robberies.

7.3 Trip Frequency

A natural application of count data modeling arises when one is interested in finding out what determines the number of trips taken by a person (or household) over a specific period of time. Such data occur, for example, in empirical studies in the fields of environmental economics and regional economics. In the former, trip frequency can be used to estimate the value that a recreational site, such as a hiking trail, a swimming beach, or a river for fishing provides to its users. Such an estimate can enter into a cost-benefit analysis when elimination of such a recreational site is considered. In regional economics, and urban planning in particular, one is interested in the number of trips to a particular shopping site, and how this number is affected by the distance, the characteristics of the shopping site, and the location and attributes of alternative sites in the region.

The demand for recreational trips was analyzed, among others, by Creel and Loomis (1990), Ozuna and Gomez (1994, 1995), Shonkwiler and Shaw (1996), Gurmu and Trivedi (1996), Haab and McConnell (1996), and Hellström (2002, Chap. 4). The prevailing approach is the so-called "travel cost method". The goal of these models is to estimate a conventional downward sloping demand function. The 'quantity demanded' is the number of trips taken to a site during a given period of time, and the 'price' is the travel cost of reaching the site. Price variation derives from the fact that individuals live at different distances from the site. Those living nearby have lower cost and would be expected to undertake more trips. Formally, let y_i, the number of trips to a single site by individual (or household) i, have a count data distribution with

$$\mathrm{E}(y_i) = \exp(\beta_1 tc_i + \beta_2 d_i)$$

where tc_i are the cost per trip and d_i are various socio-economic characteristics, including income. The parameters β_1 and β_2 can be estimated in the usual way, once the model has been specified and a random sample of potential users has been surveyed. If the sample is taken at the site, corresponding adjustments have to be made to account for the endogenous sampling (see Chap. 4.4.2).

Presumably, β_1 is negative. The consumer surplus for individual i can then be calculated as the integral under the demand function from tc_i to tc_i^*, where tc_i^* is the cost that would choke of demand, i.e., reduce the demand to (approximately) zero. In the log-linear form used here, where demand approaches zero only in the limit, we have

$$\begin{aligned} cs_i &= \int_{tc_i}^{\infty} \exp(\beta_1 tc_i + \beta_2 d_i) d\,tc_i \\ &= \frac{1}{\beta_1} \exp(\beta_1 tc_i + \beta_2 d_i)\Big|_{tc_i}^{\infty} \\ &= -\frac{\mathrm{E}(y_i)}{\beta_1} \end{aligned}$$

This would be the access value for individual i that would be lost if the site was closed. Aggregation over all affected individuals would give the overall consumer surplus lost that should be counted among the cost of a site-closure.

Extensions of this model recognize that the "use value" of one site may be affected by the availability and location of other similar sites, and their respective attributes. This is addressed in so-called multiple site models. Also, one has to recognize that the value of a site may not be limited to its use-value. Even non-users might attach a values to the availability of a site, and different methods are required to determine the "non-use value".

The second area where trip frequencies have been subject to count data modeling in the past is for the number of trips to a number of alternative retail locations (Flowerdew and Aitkin, 1982, Okoruwa, Terza, and Nourse, 1988, Barmby and Doornik, 1989). Again, distance from the site is an important explanatory variable. In addition, the effect of the size of the retail location (or shopping mall) is of interest, following the notion of a gravity model that the attractive force of a location is directly proportional to its size and inversely proportional to the distance or travel time to that location. Estimation of such a model using Poisson or negative binomial regression models is straightforward. In these models, one can also control for further socio-economic individual specific variables that may affect shopping behavior. A further application of such gravity models using count data is the modeling of spatial flows (for instance of people) as in Smith (1987).

7.4 Health Economics

Count data models have a wide applicability in health economics and, more generally, in health sciences. A number of separate application areas can be distinguished.

Firstly, count data models are used to model the frequency or intensity of a health problem. For instance, using individual level patient records from a controlled experiment, Diggle, Liang and Zeger (1995) and Chib, Greenberg and Winkelmann (1998) estimate the effect of a drug treatment on the number of epileptic seizures over a given period. Mullahy and Portney (1990) use survey data from the 1979 National Health Interview Survey to estimate the effect of smoking on the number of days of respiratory illness during a two-week recall period. Jorgensen et al. (1999) study the number of emergency room visits for respiratory diseases, again using count data regression models. Böhning et al. (1999) provide an epidemiological application to the number of teeth affected by caries.

A second, related area of research concerns the occurrence and incidence of patterns of behavior that are perceived or known to be potentially harmful and "unhealthy". One such issue, related to public health, is the freqency of sexual intercourse of teenagers (for instance, Moffatt and Peters, 2000). Other examples are the determinants of cigarette or alcohol consumption (for instance, Yen, 1999, and Kenkel and Terza, 2001).

A third area of application comprises the utilization of health services. Often, household survey data provide information on variables such as the number of doctor consultations over the previous three or twelve months. Such data are relatively inexpensive to obtain and, although lacking information on the cost of the service or the associated diagnosis, they constitute a good first indicator of health care utilization. In fact, it has been argued that the number of doctor consultations can be used as a proxy for health status per se (Møller Danø, 1998). Early papers include Cameron et al. (1988) and Pohlmeier and Ulrich (1995). This type of analysis has seen a veritable explosion of activity in recent years. See, for example, Schellhorn (2002), Jimenéz-Martin, Labeaga and Martinez-Granado (2002), Doorslaer, Koolman and Jones (2002), and Winkelmann (2003).

The number of doctor consultations is not the only indicator of this sort. Depending on data availability, a distinction can be made between visits to a general practitioner, to a specialist, to a dentist (Melkersson and Olsson, 1999, Guo and Melkersson, 2002), or to a non-doctor health professional (Gurmu and Elder, 1998). Alternatively, other aspects of health utilization can been studied, such as the number of hospital outpatient department visits (Freund et al., 1996), the number of emergency room visits (Freund et al., 1996), the number of hospital inpatient days (Freund et al., 1996, Geil et al., 1997), or the amount of home care received by the elderly (Gameren and Woittiez, 2002).

In related types of research using count data methods, Grootendorst (1995) studies the usage of prescription drugs and Jensen, 1987, studies the discovery of new drugs. Finally, one can count among the health related studies those that address worker absenteeism due to illness, or worker absenteeism in general (Vistnes, 1997, Winkelmann, 1999, Barmby, Nolan and Winkelmann, 1999).

Studies of health outcomes often attempt to estimate the effect of a treatment, controlling for the general effects of socio-economic characteristics (e.g., age, sex, ethnicity, labor force status). The paper by Melkersson and Olsson (1999) is an example of a treatment study as it estimates the effect of preventive dental care during childhood and adolescence on dental health (measured by the number of visits to a dentist) as an adult. Møller Danø (1999) estimates the effects of unemployment on health, contributing to the wider literature on the "social cost of unemployment".

Studies of health utilization frequently draw their motivation from microeconomic theory and the analysis of demand and supply in the market for health. Three leading areas of interest are the sign and size of the income effect, the role of health insurance (as health insurance implicitly determines the relative price of health services and thereby the substitution effect), and the effect of supply (physician density) on demand, if any. Examples in this literature include Grootendorst (1995), Gerdtham (1997), and Winkelmann. Grootendorst finds, using Canadian data, that the removal of co-payments for prescription medicines increased the utilization of prescription drugs. Winkelmann (2003) reports that a 1997 German health care reform that increased the co-payments by up to 200 percent, reduced the demand for doctor visits by around 10 percent.

Geil et al. (1997) establish that private insurance has no effect on the number of admissions to a hospital in Germany, while Cameron et al. (1988) find for Australia that the usage of health services increased with coverage of the insurance policies. Freund et al. (1996) use differences-in-differences to estimate the impact of Medicaid changes where fee-for-service coverage is replaced by managed care. They report significant reductions in the hospital outpatient usage under managed care. Finally, Pohlmeier and Ulrich (1995) find some evidence for supplier-induced demand: a higher density of physicians leads to increased usage, as measured by the number of doctor consultations.

Count data based empirical research in the health area typically encounters the full array of methodological problems discussed in this book. Most research is based on single outcome measures but sometimes several outcomes are modelled jointly. For instance, Riphahn, Wambach and Million (2003) use the multivariate Poisson-log-normal model to jointly estimate the determinants of the number of visits to a doctor and the number of visits to a hospital. Their model allows in addition for individual specific random effects, as their application is based on panel data. See also Gurmu and Elder (1998).

Of course, as in other applications, overdispersion is common in health data. In many cases, the excess of zeros is so great that it cannot be accommodated by the negative binomial distribution or similar single-index models. As a consequence, two-part models, and hurdle models in particular, have become the method of choice in empirical applications (Pohlmeier and Ulrich, 1995). One interesting aspect of those models is that they sometimes may be given a structural interpretation. For instance, as far as the number of doctor visits during a given period of time is concerned, the hurdle part can explain the decision to contact a general practitioner (GP), i.e., the onset of a sickness spell. Once a GP has been contacted, further referrals follow a different process that is to a considerable degree determined by the decisions of the GP rather than the individual. Recently, Santos Silva and Windmeijer (1999) have pointed out that hurdle models are unnecessarily restrictive as they allow at most one sickness spell during the given period of time. Instead, one can use the more general framework of compound count data distributions to jointly model the number of sickness spells and the number of referrals per spell.

In an application of zero inflated models, Gameren and Woittiez (2002) estimate the determinants of the demand for home care by the elderly. The dependent variable is the number of shifts of home care received per week. The zero inflated model has an interesting interpretation in this context, because in the case considered by the authors, the Netherlands in 1996, demand was rationed as was evidenced by substantial waiting lists. Hence, there are two types of non-users, those without demand and those with demand but rationed by waiting lists. The model in principle allows to disentangle these two effects.

Another problem in empirical health economics is the potential endogeneity of explanatory variables. This problem has moved to the forefront of recent research. Solutions have been proposed among others by Freund et al. (1996), Mullahy and Portney (1990), and Windmeijer and Santos Silva (1997) with particular applications in health economics in mind. With endogeneity, inconsistency will arise and the desired interpretation of the estimates as a causal relation becomes inadmissible. An example for the problem of endogeneity is given by the effect of insurance coverage. If individuals can choose their coverage then economic theory predicts a process of "adverse selection". With imperfect and asymmetric information individuals whose high health risk is known to themselves but not to others, including the insurer, will choose the high coverage insurance policy. A related problem is that of "moral hazard": high coverage may lead to negligent behavior and reduced preventive care on the part of the individual.

In either case, the observed insurance effect will not necessarily measure the causal behavioral response of insurance on health utilization. In order to address this problem, instrumental variable and switching regression estimators have been proposed. Freund et al. (1996) use state variation in changes to Medicaid laws as an instrument. Schellhorn (2002), using data for Switzerland, estimates the effect of choosing a higher deductible on the number of doctor visits. Clearly, those with low expected use will benefit from selecting a high deductible. The results indicate that the effect of choosing a higher deductible is overestimated when this self selection is not controlled for. Depending on the method, all of the observed difference in utilization can be explained by self-selection.

7.5 Demography

The main application of count data models in demography is the analysis of individual fertility, as measured by the number of children ever born or the number of children living in a household. Examples for recent applications are Nguyen-Dinh (1997) and Al-Qudsi (1998a, 1998b). The *Journal of Population Economics* devoted a symposium to fertility studies using count data models (Winkelmann and Zimmermann, 2000).

Modeling fertility produces a number of interesting methodological issues. These include, in no particular order, the frequent presence of underdispersion (Winkelmann and Zimmermann, 1994, Winkelmann, 1995), the influence of infertility and social norms (as opposed to individual choice based on economic factors), and the question of how to account for the fact that women may not have yet completed their childbearing age (Caudill and Mixon, 1995, McIntosh, 1999).

A number of approaches have been taken in order to deal with incomplete fertility. The most radical one is to consider older women only, for instance those aged 45 or older, in order to bypass the problem. Examples are Winkelmann (1995) and Mayer and Riphahn (2000). This "method" has a couple of drawbacks, however. First, the omission of data on the current child-bearing generation generates a substantial lag in the collection of evidence on fertility patterns. This becomes more of a problem if fertility behavior is rapidly changing over cohorts. Second, the method cannot be used if the number of children is based on household composition data (such as in Famoye and Wang, 1997, and Kalwij, 2000), as children typically leave the household once they reach adulthood.

Instead, one can include a variable such as age, or age-at-marriage, or the number of fertile years, as logarithmic offset in the regression (see Chap. 3.1.5). Alternatively, one can consider models where the number of children observed for women with incomplete fertility is interpreted as a lower bound of completed fertility. A corresponding censored probability model is relatively simple to establish. Such models due to Caudill and Mixon (1995) and McIntosh (1999) were discussed in Chap. 4.4.1. The two contributions differ in the way "completion status" is determined. In Caudill and Mixon it is based on age whereas in McIntosh it is based on an additional survey question on desired fertility.

When modeling the determinants of fertility there are strong reasons to believe that the standard assumption of a homogeneous exponential mean function is violated. These include the possibility of infertility (i.e., the outcome of zero children that results from processes other than choice), and the potential influence of social norms. For instance, in many societies, to have an only child is considered to be socially undesirable whereas to have two children is considered desirable. That "zeros" are different can also be seen in aggregate data. For instance, Santos Silva and Covas (2000) point out that in developed countries the average number of children per couple has fallen while the percentage of childless couples has remained relatively stable.

The offshoot of these considerations is that the homogeneity assumption underlying the exponential mean function of the count data model may be wrong in which case the standard Poisson-based estimator is inconsistent. Thus, the literature has considered alternative data generating processes. Two recent contributions to this area of research include Santos Silva and Covas (2000) and Melkersson and Roth (2000). Both papers model completed fertility ((de-facto) married women aged 40 or older in Portugal and (de-facto) married women aged 45 or older in Sweden, respectively). The two papers make, however, different assumptions on the processes that give extra weight to the outcomes zero, one or two.

Santos Silva and Covas combine a hurdle-at-zero model with inflation (or deflation) at one for the positive count data part.

$$f(y_i|x_i) = \begin{cases} g_1(0|x_i) & \text{for } y_i = 0 \\ (1-g_1(0|x_i))(\omega + (1-\omega)g_2(1|x_i)) & \text{for } y_i = 1 \\ (1-g_1(0|x_i))(1-\omega)g_2(y_i|x_i) & \text{for } y_i = 2,3,4,\ldots \end{cases}$$

Specifically, they assume that $g_1(y|\beta_1)$ is a generalized Poisson distribution (see Chap. 4.10.2), $g_2(y|\beta_2)$ is a truncated-at-zero generalized Poisson distribution, and

$$\omega = \frac{(\theta-1)g_2(y|\beta_2)}{1-(1-\theta)g_2(y|\beta_2)}$$

In this way, the generalized Poisson distribution without hurdle and in-/deflation is obtained for $\beta_1 = \beta_2$ and $\theta = 1$.

Melkersson and Roth (2000) devise a model that inflates both the "zero" and the "two" outcomes. The zero-and-two inflated model has the following probability distribution function

$$f(y_i|x_i) = \begin{cases} \omega_0 + (1-\omega_0-\omega_2)g(0|x_i) & \text{for } y_i = 0 \\ \omega_2 + (1-\omega_0-\omega_2)g(2|x_i) & \text{for } y_i = 2 \\ (1-\omega_0-\omega_2)g(y_i|x_i) & \text{for } y_i = 1,3,4,\ldots \end{cases}$$

where $g(y_i|x_i)$ is a proper count data distribution. In principle, the ω's can be negative, representing a shortfall of zero's or two's relative to the base model, as long as some inequality restrictions are observed (e.g., $\omega_0 > g(0|x_i)(\omega_2 - 1)/(1-g(0|x_i)))$. Of course, if the ω's are modeled as a logit-function of covariates as in Melkersson and Roth (2000), then zero- or two-deflation is precluded.

The findings of Melkersson and Roth and Santos Silva and Covas cast doubt on the assumption of a homogeneous count process in these situations, and suggest a re-interpretation of the phenomenon of underdispersion that is so characteristic of completed fertility data. Here, underdispersion stems from differences between the various components of the model, rather than from a mere departure from the Poisson variance function. Hence, the earlier practice of modeling fertility using count data models with generalized variance function (for instance, Winkelmann and Zimmermann, 1994, and Wang and Famoye, 1997) might be misguided, as a violation of the mean function leads to inconsistent parameter estimates.

A research area of substantive interest is the dynamic interaction between child-bearing and employment status over the life-cycle. While count data are certainly less than ideal to address such simultaneity, the contribution by Kalwij (2000) offers substantial progress in that direction. He makes the identifying assumption that a woman's employment status remains unchanged after birth of the first child. This assumption is supported by some simple descriptive evidence for Dutch women.

Under this assumption, the simultaneous choice of having at least one child and employment can be modeled using cross-section data only in a bivariate probit or multinomial logit framework, whereas the number of children for those who have at least one child is modeled conditional on employment. An important finding, using data from a Dutch household survey, is that the effects of educational attainment on the observed fertility pattern runs via the effects of educational attainment on female employment status, which in turn significantly affects the fertility behavior of households. The direct effect of educational attainment on the presence and number of children is found to be relatively small.

Mayer and Riphahn (2000) and Atella and Rosati (2000) use standard count data models to address novel questions related to the determinants of fertility. Meyer and Riphahn analyze the fertility adjustment of Guestworkers in Germany. In particular, they are interested in the effect of the variable "fertile years in Germany", which is, by assumption, the number of years between the age of 15 and the age of 40 that an immigrant woman has spent in Germany. Using individual level data on completed fertility from the German Socio-Economic Panel, the evidence favors an "assimilation" hypothesis (a gradual decline to the lower fertility levels of German-born women) over a "disruption" hypothesis (an initial drop in fertility below native levels with subsequent catch-up). It is interesting to note that contrary to the well known identification problem that arises in the study of earnings assimilation, fertility assimilation as defined by the authors in fact does allow to disentangle cohort and assimilation effects even from pure cross-section data. This is so because a given arrival cohort can differ at any point in calendar time in the number of fertile years spent in Germany (by virtue of differences in age at arrival in Germany).

Finally, Atella and Rosati (2000) build a model of fertility decisions in the context of a developing country where children are a means of intergenerational asset transfer. In such a model fertility does not only depend on the expected survival rate of children but also on the uncertainty associated with this survival rate. The empirical analysis using data from India shows that increased uncertainty leads to lower fertility levels.

7.6 Marketing and Management

Count data regressions become increasingly common in marketing and management as well. A prime example is the analysis of consumer behavior in studies that attempt to explain and predict purchase frequencies or amounts (for instance, Wedel et al. 1993, Robin, 1993, Ramaswamy, Anderson and DeSarbo, 1994, Brockett, Golden and Panjer, 1996). A related problem is that of modeling consumer brand choice (for instance, Gupta, 1988, Dillon and Gupta, 1996). In such models, the number of purchases of a certain brand is modeled conditional on the total number of purchases of a given item (over a year, say). The resulting model, similar to the conditional likelihood approach of the fixed effects Poisson model, is of a multinomial logit form.

Shonkwiler and Harris (1996) estimate a trivariate Poisson-gamma mixture model for the 1988 number of retail stores in three different sectors (Building materials and garden supply; Clothing; and Furniture) in each of 242 rural U.S. communities having populations between 100 and 5,000. The explanatory variables are the population size, the square root of the population size, per-capita income and the population density.

Finally, there are a number of applications related to the financial sector. Davutyan (1989) performs a time series analysis of the number of failed banks per year in the U.S. for 1947 - 1981, relating the bankruptcy risk to factors such as a measure of the absolute profitability of the economy, the relative profitability of the banking sector, as well as aggregate borrowing from the

Federal Reserve. Greene (1998) estimates a count data model using individual level data on the number of major derogatory reports in a sample of credits card holders. And Jaggia and Thosar (1993) study the determinants of the number of bids received by 126 U.S. firms that were targets of tender offers during the period 1978-1985 and whose management resisted takeover.

7.7 Labor Mobility

Labor mobility is a pervasive feature of market economies. Individuals typically hold several jobs during their working career. Topel and Ward (1992) report an average of 9 job changes during lifetime for male workers in the U.S. Own calculation for the German labor market, based on the *German Socio–Economic Panel*, indicate a distinctly lower average mobility of 3 male lifetime job changes. The sources of international differences in labor mobility are a research topic of substantial interest. A related question is why labor mobility differs so much between individuals within a country. And what can these differences tell us about the operation of the labor markets?

Explaining the variation in individual labor mobility has been a topic of interest in applied labor economics for more than 40 years. Early studies are Long and Bowyer (1953) and Silcock (1954). Recent studies include Börsch-Supan (1990), Topel and Ward (1992), Jung and Winkelmann (1993), and Winkelmann and Zimmermann (1993a, 1993b, 1994, 1998). The existing literature reports the following stable empirical findings.

1. Individual variation in mobility, as measured for instance by the variance of the number of job changes during a given period, is great. Hall (1982, p. 716) paraphrases this observation for the U.S.: "Though the U.S. labor market is justly notorious for high turnover (...) it also provides stable, near-lifetime employment to an important fraction of the labor force."
2. Most of the job changes occur at early stages of the career. In the US, an average of two out of three lifetime job changes occur during the first ten years after entering the job market (Topel and Ward, 1992). In Germany, about one out of two job changes falls within the first ten career years (own calculations using the German Socio-Economic Panel).
3. Labor mobility reduces with increasing tenure. Or, using Silcock's (1954, p. 430) words, "the amount of wastage decreases as the lenght of service increases".

Employment can be characterized in many ways: by occupation, employer, location, and position on the job ladder, to name but a few. Mobility in a broad sense is a change in any of these categories. Some types of mobility affect several categories at a time. For instance, a change of employer may require both a move to another city or region, and a change in occupation. On the other hand, geographic dislocation and moves on the job ladder may occur within a single firm. Regional mobility is studied in Börsch-Supan (1990).

Lazear (1990) addresses the issue of intra-firm job mobility. Here, as in Topel and Ward (1991) and Jung and Winkelmann (1993), labor mobility is defined as a change of employer. This event is referred to as a "job change".

7.7.1 Economics Models of Labor Mobility

Most analyzes of the determinants of individual labor mobility are in one way or another based on the *human capital theory* (See Becker, 1962, and Mincer 1962). The human capital theory states that workers invest in productivity enhancing skills as long as the cost is less then the present value of the expected future benefits. The return to human capital depends on the wage which, in competitive labor markets, equals marginal productivity (for a given type of human capital).

The human capital approach has been mainly used to model the dynamics of individual earnings over the life cycle. In particular, the theory implies a wage growth over the life cycle since initial earnings disadvantages during the time of human capital investments (which optimizing behavior places at early stages of the life cycle) are joined by higher wages during later periods. Further, if individuals are observed in a cross section, observed earnings differentials can be explained by variations in the stock of human capital, as measured by variations in the years of schooling, labor market experience, and tenure.

For the analysis of labor mobility, it is important to distinguish between two types of human capital. The first is *general* human capital, which is acquired through the education system. The second is *firm specific* human capital. It is acquired with current tenure and, as opposed to general human capital, it may not be transferable across employers. Thus, firm specific human capital creates a wedge between actual wages and potential outside wages. An increase in the wedge through larger firm specific investments reduces mobility.

The implications of human capital for labor mobility have been further explored by Jovanovic (1979a, 1979b) and Mincer and Jovanovic (1981). These authors emphasize the importance of imperfect information and heterogeneity. In particular, it is assumed that each worker has a nondegenerate productivity distribution across different firms or jobs. Human capital effects enter the model by determining the location (and possibly dispersion) of this distribution, and its shift over time. Further, the models are based on the following decision rule: a job change occurs if the expected present value of an alternative job is higher than the expected present value of the incumbent job (or if the difference exceeds transaction costs in case they exist).

A job change requires new information that changes the expectations of either the incumbent job or the outside offers. Two model types have been developed. In the first version, job changes occur as a result of new information about the current match (Jovanovic, 1979a). In this view, jobs are considered to be *experience* goods. The value of the match is unknown a priori but reveals itself by experiencing the match. Thus, the experience provides new information which is for instance processed using a Bayesian updating rule. A job separation occurs if, compared to the initial evaluation, the present match is revealed to have a lower expected present value.

In the second type of models, job changes occur as a result of new information about the outside offers (Jovanovic, 1979b). In this view, jobs are considered to be *inspection* goods, i.e., the value of a match is known prior to the match. Here, job changes occur as new information about better paying outside jobs arises. The arrival rate of new information increases with the search effort which in turn increases search costs. Thus, an optimal search strategy can be established.

These models predict the following effects of standard human capital variables like education and labor market experience on labor mobility. The effect of education, as measured by the years of schooling, is ambiguous. First, as far as education creates general human capital, it should increase both inside and outside opportunities, i.e. (potential) wages, proportionally und thus leave mobility unaffected. Second, better general education creates skills that allow for a faster accumulation of firm specific human capital. Thus, for given tenure, individuals with higher education have a lower mobility. Third, in markets with imperfect information, better educated individuals should be better able to collect and process information. They tend to have a higher search efficiency and therefore lower transaction costs and higher mobility.

The models unambiguously predict an inverse relation between tenure and mobility. The negative correlation arises due to a wedge created by the accumulation of firm specific human capital over time or, if jobs are seen as experience goods, due to the operation of a sorting process.

Separating tenure and experience effects may be impossible. As Mincer and Jovanovic (1981) point out, a distinction has to be made between true experience effects and indirect effects via job tenure. Let the propensity to change job m be a function of both tenure ten and experience ex. Then

$$\frac{\mathrm{d}m}{\mathrm{d}ex} = \frac{\partial m}{\partial ten} \cdot \frac{\mathrm{d}ten}{\mathrm{d}ex} + \frac{\partial m}{\partial ex} \,. \tag{7.1}$$

Only $\partial m/\partial ex$ is a genuine experience effect. It is complemented by an indirect tenure effect since tenure grows with experience. Clearly, $0 < \mathrm{d}ten/\mathrm{d}ex < 1$, and mobility declines with experience also if there is no true experience effect but only a tenure effect, unless one controls for tenure. With count data this is generally not possible as job tenure (at the time of the job change) is not observed. Hence, the two effects are not separately identified. The reduced form effect of labor market experience picks up the combined effects of experience and tenure.

7.7.2 Previous Literature

Börsch-Supan (1990) studies the influence of education on labor and regional mobility using data from the PSID. Observations on 736 male individuals are available for the period 1962-1982. Estimating Poisson regression models, Börsch-Supan finds that an increase in education reduces labor mobility, while it increases regional mobility. The reduction in labor mobility with increasing education is of considerable magnitude: The lowest education level has a predicted number of job changes that is about 50 percent higher than the predicted number of job changes for the highest education level. The conditional effect (i.e. after controlling for other characteristics in a multiple regression and evaluating the remaining variables at their sample means) is greater than the marginal effect obtained in a cross tabulation. Further, Börsch-Supan finds that experience has a negative effect on both types of mobility. As mentioned earlier, this finding might reflect tenure effects that cannot be controled for.

Merkle and Zimmermann (1992) use a German sample of labor force participants drawn from the unemployment register in 1977. The 1610 selected individuals answered questions on the number of employers and the previous number of unemployment spells during a five year period preceding the interview. The data are censored from above at five. Using Poisson and negative binomial regression models for censored data, Merkle and Zimmermann (1992) find that both the number of job changes and the number of unemployment spells increase with the education level, whereas these variables are affected in a concave way by previous labor market experience. Thus, their evidence is in conflict with the findings of Börsch-Supan. This apparent contradiction can be resolved when considering the differences in the sampling schemes. Sampling from the stock of unemployed as opposed to sampling from the labor force already tends to select less skilled individuals with a higher propensity to unstable labor relations. Within this group, better educated people might have higher re-employment chances, reducing their overall time spent in unemployment and increasing their turnover.

A further study of interest is Ebmer (1990) who looks at the determinants of offer arrival frequencies. The process of job mobility may be decomposed into two steps. In a first step, offers are made to the individual at a certain rate. In a second step, the individual decides whether or not to accept the offer. Usually, data on offer arrivals are not available. In Germany and Austria, however, job offers both for unemployed and employed individuals are mainly administered through a public placement service. Ebmer (1990) uses data on offers provided by the Austrian placement service, and, using Poisson and Negbin models, finds that the offer arrival rate falls with elapsed unemployment duration, which he interprets as discriminating behavior of labor exchange officials. Furthermore, his dataset allows to test for the assumption of Poisson arrival rates. This assumption is common in the search literature. The hypothesis is rejected although one cannot exclude that rejection is due to unobserved heterogeneity.

7.7.3 Data and Descriptive Statistics

The following sections illustrate the use of count data models for studying labor mobility in an empirical application using data from the *German Socio-Economic Panel* (GSOEP). Wagner, Burkhauser and Behringer (1993) provide a short introduction to the data set. The annual panel was first collected in 1984. The basic sampling units are households. The sample included 5921 households in 1984. Within each household, every person aged 16 or older is interviewed, resulting in 12,245 person records for 1984. The selection of households is stratified by nationality: One subsample consists of a random sample of the population living in Germany which is not of Turkish, Yugoslave, Greek, Italian or Spanish nationality. The proportion of non-Germans in this subsample of 9076 individuals is 1.5%. The second subsample of size 3169 includes 33% Turks, 18% Yugoslaves, 15% Greeks, 20% Italians and 13% Spaniards (Deutsches Institut für Wirtschaftsforschung 1990). All in all, the GSOEP oversamples the foreign population whose overall proportion was 7.5% in 1984 (Statistisches Bundesamt 1985).

The dependent variable is the number of employers and the number of unemployment spells during the ten year period 1974-84. This information is collected retrospectively in the first wave of the panel. In order to ensure that the analysis is based on persons with a reasonably strong labor force attachment, the sample is restricted to persons in employment in 1984 whose work career started before 1974. Women are excluded in order to minimize complications due to non-participation spells. Non-participation is known to be empirically relevant for women, and yet unobservable in the type of data studied here. Finally, self-employed persons and civil servants are excluded. The resulting sample has 1962 observations.

Using the information on the number of employers and the number of unemployment spells, two measures of labor mobility can be derived. First, assume that

i) people do not return to the same job (or employer) after a spell of unemployment, and
ii) individuals have been employed at the beginning of the period.

Then the number of employers minus the number of unemployment spells minus 1 measures the number of direct job-to-job transitions (without an intervening unemployment spell). Under the same assumptions, the number of indirect job changes (job-to-unemployment-to-new job transition) is simply equal to the total number of unemployment spells. A cross tabulation of **direct job changes** and **unemployment spells** is given in Table 7.1

There is a slight positive correlation between the two types of mobility ($\rho = 0.06$). For instance, the proportion of individuals having experienced at least one **direct job change** is greater for the group of individuals that did experience one **unemployment spell** than for the group that did experience no unemployment. The same holds true for **unemployment spells** vs.

Table 7.1. Frequency of Direct Changes and Unemployment

		Direct Job Changes												
		0	1	2	3	4	5	6	7	8	9	10	12	Total
U	0	1102	301	105	25	20	5	1	2	1		2		1564
n	1	146	79	21	10	1	3	2	1		1			264
e	2	34	16	6	6	2	2	1	1				1	69
m	3	20	4	1		2								27
p	4	7	2		2									11
l	5	6	2											8
o	6	2									1			3
y	7	2												2
m	8	3												3
e	9	3												3
n	10	7												7
t	15	1												1
	Total	1333	404	133	43	25	10	4	4	1	2	2	1	1962

direct job changes. For both **direct job changes** and **unemployment spells** the mode is at zero. The means are 0.54 and 0.37, respectively (See Table 7.2). The variance–mean relation is 2.16 for **direct job changes** and 3.32 for **unemployment spells**, indicating a tendency for overdispersion at the marginal level. This appears to provide a first check of the (non-)validity of the Poisson regression model, since conditional overdispersion violates the Poisson assumption. However, overdispersion at the marginal level is (theoretically) compatible with mean-variance equality conditional on covariates.

Sec. 7.7.1 defined the primary empirical question: What can individual characteristics tell us about individual propensities towards mobility, measured by the frequency of future **direct job changes** and **unemployment spells** ? The theoretical arguments developed in Chap. 7.7.1 suggested the main variables of interest: **Education** as measured by the years of schooling and previous professional **experience**. Further variables which have been used in the literature to control for individual heterogeneity in wages and mobility are occupational status, nationality, family status and union membership. The corresponding dummy variables are (Yes=1; Sample means in parentheses) **Qualified White Collar** (0.137), **Ordinary White Collar** (0.059), **Qualified Blue Collar** (0.501), **Ordinary Blue Collar** (0.304), **German** (0.668), **Single** (0.077), and **Union** (0.429). Exact definitions and measurement issues are given in the notes to Table 7.2

Table 7.2 displays the 'gross' effect of these variables on the two types of labor mobility. The mean values in the 1st and 3rd column give the average number of **direct job changes** (**unemployment spells**) during the ten year period 1974–84 for the various classifications.

The most visible effect is certainly the strong reduction of mobility with increased labor market **experience**. Individuals at the beginning of their career (less than 5 years of experience) have on average 3 times more **direct job changes**, and almost 2 times more **unemployment spells** over the next ten years, than individuals with more than 25 years of professional experience. Furthermore, the amount by which the mobility is reduced decreases with experience, i.e., there exists a convex pattern between experience and mobility.

Table 7.2. Mobility Rates by Exogenous Variables

	Direct Changes[1]		**Unemployment**[2]		
	Mean	Std.Dev.	Mean	Std.Dev.	Obs.[3]
by Occupational Status[4]					
Qualified White Collar	0.498	1.032	0.212	0.638	269
Ordinary White Collar	0.566	0.999	0.257	0.777	113
Qualified Blue Collar	0.540	1.120	0.431	1.285	983
Ordinary Blue Collar	0.553	1.069	0.377	1.036	597
by Nationality[5]					
German	0.466	0.974	0.367	1.194	1311
Foreign	0.688	1.270	0.390	0.928	651
by Family Status[6]					
Single	0.651	1.246	0.697	1.671	152
Married	0.530	1.071	0.348	1.049	1810
by Union Status[7]					
Union	0.440	0.964	0.273	0.898	841
Nounion	0.615	1.163	0.450	1.243	1121
by Professional Experience[8]					
-5 Years	0.954	1.478	0.578	1.361	372
6-15 Years	0.543	0.988	0.384	1.059	672
16-25 Years	0.407	0.965	0.259	0.935	659
26+ Years	0.274	0.735	0.338	1.195	259
by Educational Attainment[9]					
-10 Years	0.585	1.132	0.405	0.997	511
11-12 Years	0.514	1.027	0.447	1.351	876
13-18 Years	0.567	1.304	0.247	0.799	478
19+ Years	0.402	0.640	0.187	0.507	97
Total	0.539	1.080	0.372	1.112	1962

Source: *German Socio-economic Panel*, own calculations.

Notes:

1. **Direct Changes** give the number of direct job changes an individual has experienced during the period 1974-1984. A direct job change is defined by the number of employers minus the number of unemployment spells minus one. The information is obtained through a retrospective question.
2. **Unemployment** gives the number of unemployment spells an individual has experienced during the period 1974-1984. As *Direct Changes*, the information is obtained through a retrospective question.

3. Number of observations in the sample. The total sample size is 1962. The selection was conditional on being male, being part of the labor force during the period 1974-1984, and on being neither self-employed nor civil servant.
4. **Occupational Status** is measured upon entry into the labor market, i.e., it is the status in the first job.
5. The distribution of the **Nationality** reflects that the *German Socio-economic Panel* is a stratified panel: Foreigners are oversampled as compared to their share of the labor force in Germany. However, the sampling is exogenous and not choice based.
6. An individual is classified as **Single** if he is and always was a single, i.e. widowers and divorced are classified as married.
7. **Union** membership in 1985. Included are members of unions and comparable professional organizations.
8. Professional **experience** uses information on the year of entrance into the labor market, subtracting the latter from 1974, the start of the ten year period.
9. To obtain a continuous measure of the **Educational Attainment** the years of schooling are calculated using information on the various degrees obtained by an individual, and attributing to every degree a "typical" time it requires. For instance, a university degree takes on average 18 years of schooling. The years of schooling measure also includes the time spent in professional education, as long as it is a part of special training programs ("Lehre").

The effect of **education** on mobility is less uniform. Comparing individuals with less than 10 years of schooling and individuals with 13 to 18 years of schooling, there is almost no change in the average number of **direct job changes**. The number of **unemployment spells**, by contrast, is reduced by 40% for the more educated individuals.

Germans, union members, married individuals and qualified white collar workers have on average less **direct job changes** than foreigners, non-union members, singles and ordinary white collar or blue collar workers, respectively. The number of **unemployment spells** is higher for blue collar workers, singles, and non-union members than for white collar workers, married individuals, and union members, respectively. Nationality seems to have no effect on the frequency of unemployment.

Although the descriptive statistics provide some valuable information on the interaction between the variables, an interpretation in the light of the aforementioned theories is problematic. While the theoretical models establish *specific* effects, or effects that hold ceteris paribus, the descriptive statistics display the *gross* effects which mix specific contributions and contributions due to correlations with other explanatory variables. Thus, a multiple count data regression analysis is required in order to estimate the specific effect of a unit change in one explanatory variable on the expected number of job changes, holding everything else constant. Moreover, it allows to predict the mobility behavior for any given individual. Most importantly, though, it provides information on the underlying data generating process, i.e. the stochastic process governing mobility.

7.7.4 Regression Results

This chapter reports the results of various estimated models for the labor mobility data. For simplicity, we restrict our attention here to one of the two mobility measures, the number of direct job changes, from now on or simplicity referred to as the **number of job changes**. The models differ in the assumption on the underlying probability processes. To ensure comparability, the set of explanatory variables is kept identical in all cases. The explanatory variables include **education, experience, squared experience, union, single, German, qualified white collar, ordinary white collar,** and **qualified blue collar** worker.

The following models were estimated with **number of job changes** as dependent variable:

- Poisson
- Poisson-log-normal
- Negbin I, Negbin II and Geck$_k$
- robust Poisson
- Poisson-logistic
- hurdle Poisson
- probit-Poisson-log-normal
- finite mixture Poisson and finite mixture Negbin
- zero-inflated Poisson and zero-inflated Negbin

The full set of estimation results for the various models are listed in Tables D.1 – D.7 in Appendix D.

Poisson Results

The Poisson model is specified with a log-linear conditional expectation function. This means that the coefficients can be interpreted as semi-elasticities. Take the point estimate of -0.138, pertaining to the education effect, for illustrative purposes. Since the education variable is scaled (division by 10), we find that the estimated effect of 10 additional years of schooling is a reduction of the number of job changes by approximately 13.8 percent. The exact effect would be $[\exp(-0.138)-1]\times 100 = -12.9$, a 12.9 percent reduction. Similarly, based on the point estimate, each single additional year of education would reduce the number of job changes by 1.4 percent.

Sometimes, it is meaningful to compute absolute rather than relative marginal effects. This is in particular the case if one wants to compare marginal effects across models, where some of the models (such as the hurdle Poisson model or the zero-inflated Poisson model) may not have a log-linear conditional expectation function. We know from Chap. 3.1.4 that

$$\frac{d\,\mathrm{E}(y_i|x_i)}{d\,x_{ij}} = \exp(x_i'\beta)\beta_j$$

Thus, the marginal effect depends on the point in the covariate space where it is to be computed. It is common to take the sample mean, i.e., replace x_i by $\bar{x}$. We find for example that the marginal effect of education at the mean of the regressors is -0.067.

Of course, the point estimate of -0.138 and its associated relative or absolute marginal effects are subject to sampling variability. Indeed, one finds that **education** has no significant effect on **direct job changes**, since the t-ratio for the null hypothesis of no effect is about one, based on the Poisson standard errors estimated from inverting the Hessin matrix of the log-likelihood function (the computation of the standard error of the semi-elasticity and the marginal effect would need to be based on the delta rule).

Substantively, the result of 'no-effect' is compatible with the human capital view that education increases *general* human capital which in turn promotes outside and inside job opportunities alike. In other words, the level of education does not affect the probability of finding (being offered) a new job that is preferable to the current one. The finding is in contrast to Börsch-Supan (1990), who reports a negative and significant effect of the level of education on labor mobility. One possible explanation for the discrepancy is that he includes all job changes, also those with intervening spell of unemployment, whereas the results here are for the number of direct job-to-job transitions only.

The convex experience-mobility profile implied by the point estimates of the second order-experience polynomial is very plausible. It conforms to the stylized fact that job changes are much more likely to occur early in ones career. One year after entering the workforce, the expected job change rate has decreased 7.5 percent relative to the initial rate. The predicted job change rate further decreases with each additional year of experience, but at a decreasing rate. After 32 years of experience, the effect of experience on mobility reaches zero. If, instead of computing relative or percentage effects, one was interested in absolute changes, one would need to compute the marginal experience effect as follows: Let ex denote the variable 'experience', and $exsq$ denote the variable 'experience squared'. Then

$$\frac{\partial \mathrm{E}(y_i|x_i)}{\partial ex_i} = \exp(x_i'\beta)[\beta_{ex} + 2\beta_{exsq}ex_i] \tag{7.2}$$

This partial derivative depends on x_i. Evaluating (7.2) at the sample means, the marginal effect is given by -0.022. For an average individual, an additional year of experience decreases the expected number of job changes by 0.022.

Union membership reduces the expected number of job changes by 29 percent or, evaluating the effect as above, by 0.131 job changes during the ten year period, relative to non-unionized workers. This specific effect is smaller than the gross effect of 0.175 displayed in Table 7.2, reflecting the interactions between the variables. Finally, German nationality reduces the expected number of direct job changes, while the remaining variables have no significant effect on mobility.

While the previous remarks referred to the results of the Poisson regression, the findings display a remarkable robustness accross the various specifications. Table 7.3 compares the results for ten of the estimated models. The signs and the significance levels of the coefficients are mostly identical.

Table 7.3. Direct Job Changes: Comparison of Results

	(1)	(2)	(3)	(4)	(5)	(6)	(7)	(8)	(9)	(10)
Education	o	o	o	o	−−	o	−−	o	o	o
Experience	−−	−−	−−	−−	−−	−−	−−	o	−−	−−
Experience^2	++	++	++	++	+	++	o	o	o	o
Union	−−	−−	−−	−−	−	−−	−−		−−	o
Single	o	o	o	o	o	o		o	o	o
German	−−	−−	−−	−−	−	−−	−−		−−	−−
Qual. Wh.C.	o	o	o	o	o	o	o		o	o
Ord. Wh.C.	o	o	o	o	o	o	o		o	o
Qual. Bl.C.	+	o	o	o	o	o	o		o	+

Notes:

Dependent Variable: Direct Job Changes.
++ Positive sign and significant at the 5%-level.
+ Positive sign and significant at the 10%-level.
−− Negative sign and significant at the 10%-level.
− Negative sign and significant at the 5%-level.
o Insignificant.
(1) Poisson Regression.
(2) Robust Poisson Regression.
(3) Generalized Event Count Model (GEC_k)
(4) Hurdle Poisson (0/1).
(5) Hurdle Poisson (1+).
(6) Poisson lognormal.
(7) Poisson-Logistic Regression: Job Offers (Overlapping).
(8) Poisson-Logistic Regression: Acceptance.
(9) Two-components Poisson (Group 1)
(10) Two-components Poisson (Group 2)

What conclusions should we draw from this evidence? One might be tempted to argue that given the uniformity of the results and the different complexities of the models, the recommendation would be to choose the simplest model, in this case the standard Poisson regression. This approach would even find theoretical support by the consistency property of the Poisson regression as a PMLE. Therefore, the similarity of the findings in this application does not come too much as a surprise. However, this is only a part of the story. It neglects two important purposes of the econometric analysis: The use of the model for prediction and as a tool for learning about the underlying data generating process. We therefore proceed by investigating three further issues.

First, we assess which of the models has the best fit, a purely statistical exercise. Second, we investigate one possible reason for the superior performance of some models, applying the concept of marginal probability effects to show that a certain minimum flexibility in the distributional assumption is needed in order to account for the effect of the explanatory variables on the outcome distribution. And third, we return to a main theme of this book, namely that generalized models can be informative on interesting aspects of an underlying count mechanism, and that therefore structural inferences can be made. We will see to what extent this actually applies in the present context.

7.7.5 Model Performance

The Poisson regression model assumes that events (here: job changes) occur randomly over time, with a constant process intensity that is a deterministic function of individual covariates. The Poisson-log-normal, Negbin I, Negbin II and GEC_k models allow for unobserved heterogeneity. The remaining models relax the single-index structures in favor of a dual-index, or two-part structure. There are several ways of doing this. The hurdle models state that the intensity of the process switches conditional on the first occurrence. The Poisson-logistic model assumes a two-step process: In a first step, offers reach the individual according to a Poisson process. In a second step, the individual decides whether or not to accept the offer. The two-component models assume that the population consists of two latent groups, each one with its own regression function. The zero-inflated models use a logistic model to augment the probability of a zero relative to the base count data model.

The above models were estimated using an identical set of regressors and the following table lists the log likelihood, evaluated at the maximum likelihood parameter estimates.

Table 7.4. Number of Job Changes: Log Likelihood and SIC

	ℓ	K^1	SIC^2
Poisson	-2044.47	10	4164.76
Poisson-log-normal	-1866.80	11	3817.00
Negbin I	-1873.28	11	3829.96
Negbin II	-1878.63	11	3840.66
GEC_k	-1873.17	12	3837.32
Poisson-logistic (overlapping)[3]	-2039.35	13	4177.26
Poisson-logistic (non-overlapping)[3]	-2043.88	10	4163.58
Hurdle Poisson	-1928.00	20	4007.63
Probit Poisson-log-normal[3]	-1856.70	22	3880.20
two-components Poisson	-1868.16	21	3895.54
two-components Negbin II	-1856.05	23	3886.48
zero-inflated Poisson[3]	-1926.28	20	4004.19
zero-inflated Negbin II[3]	-1866.73	21	3892.68

Notes:

[1] K denotes the number of parameters in the model.
[2] Schwarz information criterion: $SIC = -2\ell + K \ln N$
[3] These models do not nest the Poisson model.

The log likelihood values can be used to formally test models against each other insofar as they are nested. For example, the Poisson model is nested in all unobserved heterogeneity-type models, as well as in the hurdle and the two-components Poisson models. From Tab. 7.4, the Poisson model is rejected by the various tests against any of the more general alternatives. This is clearly due to overdispersion. For instance, the GEC_k estimates a $\hat{sigma}^2$ of 0.892 with a standard error of 0.170. In the absence of over- or underdispersion, $\sigma^2 = 0$, but σ^2 is significantly greater than zero at any conventional significance level. The estimated k is not different from 0 either. However, it is significantly smaller than 1, providing evidence for the presence of a linear variance function as opposed to a quadratic one. Interestingly, the Poisson-lognormal model has a higher log-likelihood than either Negbin I or GEC_k. This suggests that the mixing distribution used to model unobserved heterogeneity is better described through a log-normal distribution than through a gamma distribution. Note, however, that the improved fit comes at the expense of increased computational complexity, since the integration requires numerical quadrature.

There are other nested model pairs in Tab. 7.4. For example, using likelihood ratio tests, the two-components Poisson model is rejected against the two-components Negbin II model (test statistic 24.22, p-value $= 0.000$); the non-overlapping Poisson-logistic model is rejected against the overlapping Poisson-logistic model (test statistic 9.06, p-value $= 0.0285$); and the zero-inflated Poisson model is rejected against the zero-inflated Negbin II model (test statistic 119.1, p-value $= 0.000$).

In other cases, Vuong's test for non-nested hypothesis can be used. For example, the Negbin I model and the Negbin II model are not nested. Since they both nest the standard Poisson model, they are overlapping rather than strictly non-nested, following the terminology of Vuong. Hence, a pre-test is required in order to establish that the two models are not equivalent. In this case, it is sufficient to show that the respective dispersion parameters are significantly different from zero. From Tab. D.2, we see that this is the case indeed. The null-hypothesis $H_0 : \sigma^2 = 0$ can be rejected in each model, using the asymptotic z test for instance. Next, the Vuong statistic proper can be computed. The formula was given in (3.74). The test statistic in this case is 0.999. It has a standard normal distribution, with the critical values being the usual $\alpha/2$ and $1 - \alpha/2$ quantiles. Hence, there is no evidence that the Negbin I model is significantly better than the Negbin II model. Note that this result differs from the conclusion based on the hyper model (GEC_k), where the Negbin II restriction could be rejected but the Negbin I restriction could not. The Vuong test has low power in finite samples. We also find that a test of the Poisson-log-normal model against the Negbin I model is inconclusive. In this case, the Vuong test statistic is 1.255.

Finally, we can compare the models in Tab. 7.4 based on model selection criteria. The models have a different number of parameters, varying between a minimum of 10 and a maximum of 23. Using the Schwarz information criterion as a penalty function, it turns out the Poisson-log-normal model is the best model. The double index models lead to a large increase in the log-likelihood but this increase is more than offset by the larger number of additional parameters that needs to be estimated. In this application, it seems most important to use a model that allows for unobserved heterogeneity and overdispersion. Further generalizations are not dictated by the data. Nevertheless, as we will see next, these generalizations do offer some interesting insights into distributional effects of covariates and the underlying data generating process.

7.7.6 Marginal Probability Effects

It was mentioned earlier that ten additional years of education are predicted to reduce the number of job changes by 0.067, based on the Poisson estimates (Tab. D.1) and the sample means of the explanatory variables. Fig. 7.1 shows how this mean effect arises. Ten additional years of education increase the probability of no job change by about 4 percentage points. The probability of one job change is reduced by 2 percentage points. The probability that a person reports two, three, four and so forth job changes is reduced at diminishing rates.

Fig. 7.1. Poisson Model: Marginal Probability Effect of a Unit Increase in Education

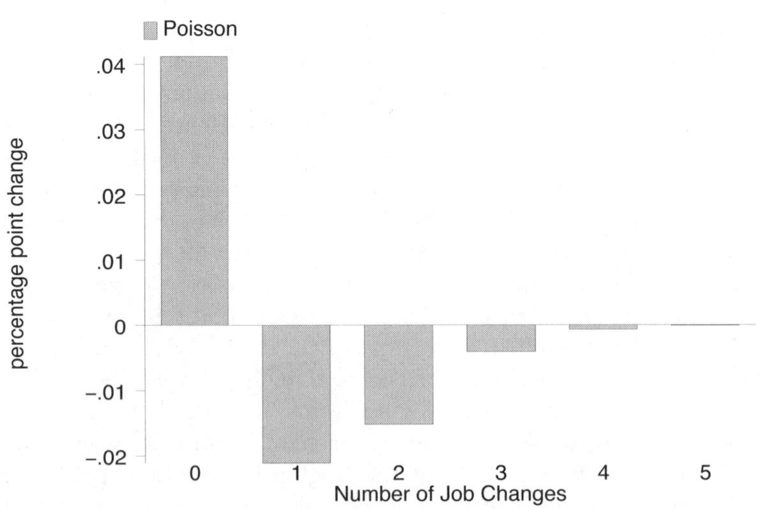

The Poisson marginal probability effects are very restrictive. For example, the sign of the effect can only change once from positive to negative, or vice versa. In order to see what would happen in a more flexible model, consider the hurdle Poisson model instead. The formula for computing the marginal mean effects in this double index model was given in (4.74). The formula for computing the marginal probability effects was given in (4.75). First, Fig. 7.2 shows the predicted probabilities of the Poisson and the hurdle Poisson model. As to be expected, the main difference is a larger probability of a zero in the hurdle model, corresponding to the phenomenon of unobserved heterogeneity/overdispersion/excess zeros in the data.

Fig. 7.2. Predicted Poisson and Hurdle Poisson Probabilities

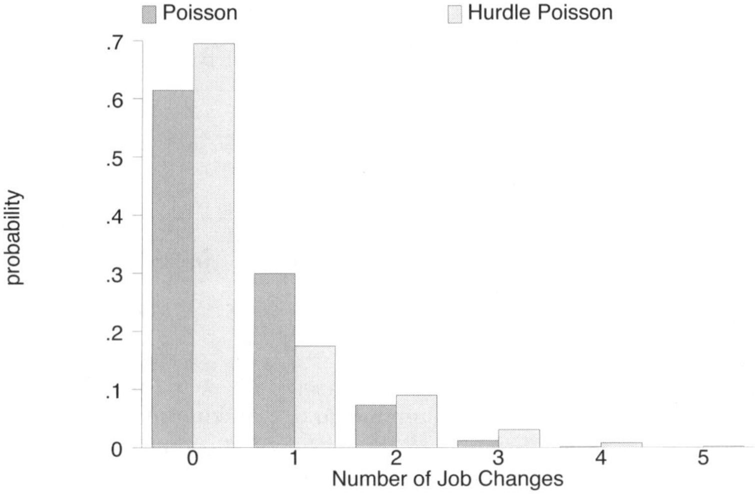

But what can we say about the marginal effects? Consider the education effect as an example. Using (4.74), the estimated coefficients from Tab D.5, and evaluating the covariates at their sample means we obtain an effect of -0.071. This estimated mean effect is quite close to the effect in the Poisson model, -0.067. However, the same overall effect has very different distributional implications, as illustrated in Fig. 7.3. Based on the hurdle model, ten additional years of education reduce the probability of no job change by about 3 percentage points whereas the Poisson model predicts an increase. Similarly, in the hurdle Poisson model, we find that more education increases the probability of one job change, whereas the simple Poisson model predicts a decrease. Using the hurdle model, we come thus to conclusions with regard to marginal probability effects that are diametrically opposite to those obtaines from the Poisson model. This is an illustration of the idea that an explanatory variable may have different marginal probability effects in different parts of the distribution (relative to a single index base model). If one ascribes substantive interest to these single outcomes, employing a sufficiently flexible model (that does not have the single crossing property) becomes imperative.

Fig. 7.3. Marginal Probability Effect of Education: Poisson and Hurdle Poisson

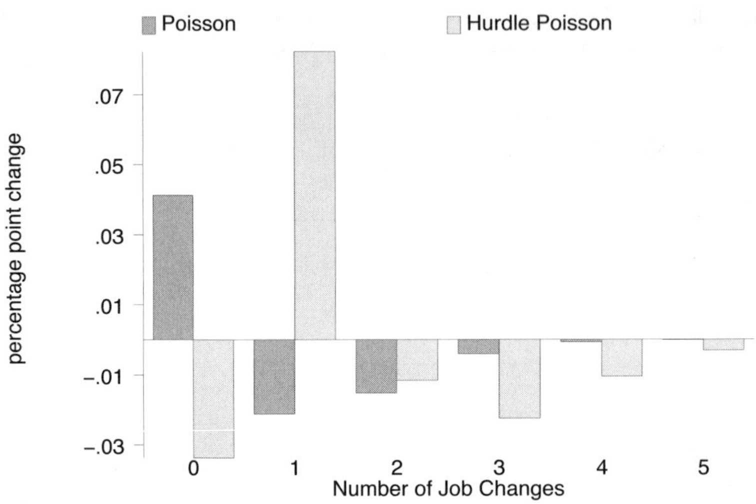

The hurdle model is only one among several possible generalizations. In the class of double index models, zero-inflated and two-components models would have similar advantages. From the perspective of allowing for maximal flexibility in the conditional probability distributions, there is no good reason to stop with double index models. The most general conceivable model would be a regression model based on a multinomial distribution. In this case, each outcome probability is parameterized as a separate function of the explanatory variables, subject to an adding-up constraint. For example, in the multinomial logit model,

$$p_{i1} = \frac{1}{1 + \sum_{k=2}^{J} \exp(x_i'\beta_k)}$$

$$p_{ij} = \frac{\exp(x_i'\beta_j)}{1 + \sum_{k=2}^{J} \exp(x_i'\beta_k)} \qquad j = 2, \ldots, J$$

where $j = 1, 2, \ldots, J$ are the J distinct counts observed in the sample. We immediately see two limitations of this model. First, it will only work if J is relatively moderate, since otherwise parameters will proliferate unduly. This problem could be mitigated in an ad-hoc way by grouping outcomes into classes. Second, the model does not allow the prediction of probabilities (or marginal probability effects) for outcomes that are not observed in the data. Relatedly, the model stands in no correspondance to an underlying count process. In short the multinomial logit model is not a count data model proper. Putting these reservations aside for a moment, one can use the model as a descriptive tool, obtaining the following result for the job change example.

Fig. 7.4. Marginal Probability Effect of Education: Hurdle Poisson and Multinomial Logit

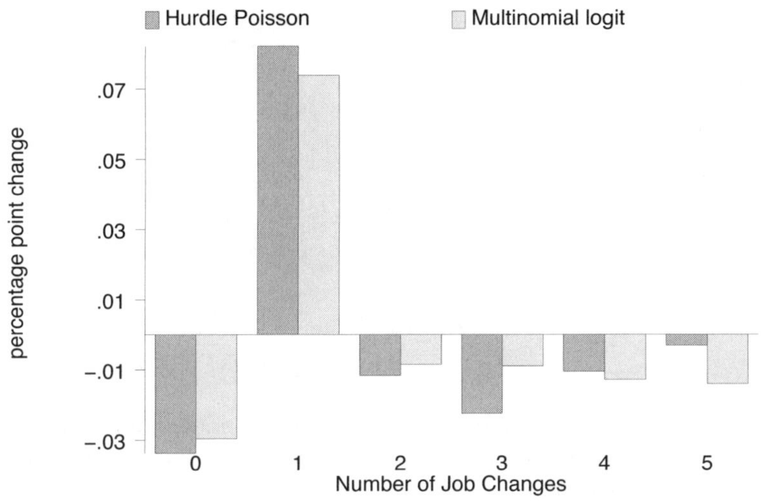

Fig. 7.4 compares the marginal probability effects of education in the hurdle model and in the multinomial logit model, everything as before evaluated at the sample means of the explanatory variables. Surprisingly, the discrepancy between the two response functions are minor. One can conclude that in this example, the double index hurdle model offers sufficient flexibility to capture how education changes the outcome distribution. More general models are not necessary.

An alternative way of capturing "non-standard" responses of whole distributions to changes in an explanatory variable is through quantile regression (Chap. 4.11). Consider the relative mean effect of education, a 13.8 percent decrease in the average number of joh changes per 10 additional years of education. If this effect was spread evenly over the distribution, and since the distribution is non-negative, one would expect to find a negative effect of additional education at all quantiles. Tab. D.8 shows the regression results for the 50-percent, the 75-percent and the 90-percent quantiles. We see that the effect of education is not so uniform. Quite to the contrary, the 50-percent quantile and the 75-percent quantile are positive functions of education. This is also seen from Fig. 7.5.

Fig. 7.5. 50/75/90 Percent Quantiles by Years of Education

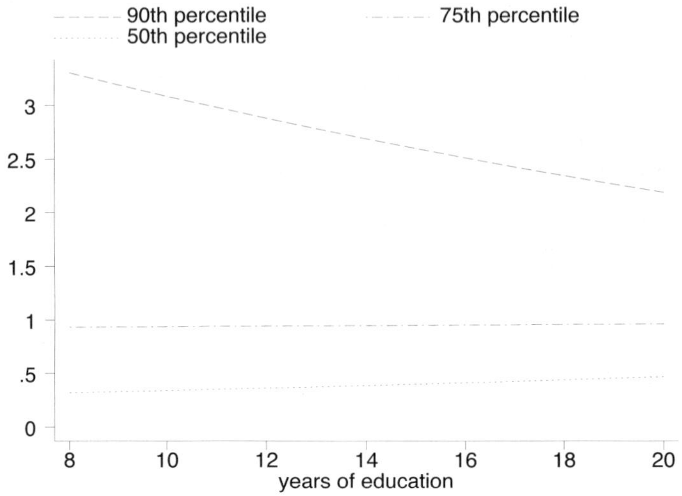

This finding is indeed compatible with the hurdle and multinomial results above. More educated people are on average less mobile but they also have a larger probability of having changed job once, relative to less educated people.

7.7.7 Structural Inferences

Any empirical analysis can have one of two goals. A first goal is to describe the data as well as possible, estimate the effect of explanatory variables on certain key features, such as conditional mean or probability function, and to predict outcomes. A second goal is to learn something about the data generating process. This second, 'structural', approach requires stronger assumptions. Given the validity of these assumptions, one can then draw stronger inferences. In the following, we will discuss the idea behind such structural models in the context of the analysis of labor mobility, using three examples: Poisson-logistic regression, zero-inflated Poisson models, and two-components Poisson models.

Poisson-Logistic Regression

As mentioned earlier, this model combines a latent process for counting event occurrences with an observation mechanism. A combination of the two then leads to the observed counts. In the context of labor mobility, one can reinterpret this set-up as one, where the latent model describes job offers, the reporting mechanism reflects whether a job offer is accepted or not, and the reported counts are equal to the number of accepted offers, i.e., the number of job changes, in a given period of time. Based on Tab. D.4, one may calculate the expected number of job offers for the sample. This number has to be greater than the number of accepted offers. For the overlapping specification, the predicted number of offers is 0.84 as opposed to 0.54 predicted job changes. Thus, a typical individual accepts roughly two out of three job offers.

Zero-Inflated Poisson Regression

A frequent substantive interpretation of the zero-inflated Poisson model is one where 'strategic' zeros are combined with 'incidental' ones. In the context of job search, one could think of this distinction as follows. Some individuals do not look for outside jobs at all, maybe because they are civil servants, or for some other reason. As a consequence, they are unlikely to get any outside offers. These are 'strategic' non-changers, because they have decided to keep their current job. In contrast, other individuals may be 'in the market'. These persons look for outside jobs. Some of them do not find an alternative job in a given period. These are then 'incidental' non-changers.

Within the structure of the zero inflated Poisson model (see Tab. D.7) one can compute the relative frequencies of the two types of workers. The model predicts model predicts 68.4 percent zeros, which is close to the 67.9 percent observed in the sample (this and the following predictions are computed first for each individual, given their covariates. Then, arithmetic means are calculated). The predicted proportion of strategic zeros is 50.9 percent. The probability of an incidental zero can be computed as $(1-0.509)\times 0.356 = 17.5$ percent. Hence, 74 percent of all zeros are strategic and 26 percent are incidental.

Two-Components Poisson Regression

The two-components Poisson model allows inferences to be drawn with respect to two subpopulations. From the results in Tab. D.6, we know that an estimated 93 percent of the population belong to Group 1, whereas 7 percent of the population belong to Group 2. One can compute the mean job change rate for the two groups, based on sample means of the explanatory variables. Group 1 has a mean of 0.31 changes, whereas Group 2 has a mean of 2.8 changes. Thus, most individuals belong to the low-mobility group. One can furthermore study, how the response to explanatory variable differs between the two groups. Take the effect of education as an example. In the low mobility group ten additional years of education increase the number of job changes by a predicted 7.8 percent. In the high mobility group ten additional years of education reduce the number of job changes by a predicted 36.8 percent.

These differential effects are compatible with the results in the previous chapter, for example based on quantile regression, where a large negative effect of education was found at the 90th percentile, and a small positive effect was found at the median.

A

Probability Generating Functions

This appendix is based on Feller (1968, Chap. XI and Chap. XII). Let X be a random variable taking values $j \in I\!N_0$ with $P(X = j) = p_j$. Upper case letters X, Y, and Z denote a random variable, while lower case letters j and k denote a realization. $p_{j \in I\!N_0}$ is called the *probability function*, while $F_{i \in I\!N_0} = P(X \leq i)$ is called the *distribution function.*

Definition 4. .
Let X be a random variable defined over the non-negative integers. The probability generating function (PGF) is given by the polynomial

$$\mathcal{P}^{(X)}(s) = p_0 + p_1 s + p_2 s^2 + \ldots = \sum_{j=0}^{\infty} p_j s^j = \mathrm{E}(s^X) \tag{A.1}$$

The function $\mathcal{P}(s)$ is defined by the $p_j's$ and, in turn, defines the $p_j's$ since a polynomial expansion is unique.

Example: Let X have a binomial distribution function with parameters n and p, $p_j = 0$ for $j > n$ (writing $X \sim B(n,p)$). The probability generating function is given by

$$\mathcal{P}(s) = \sum_{j=0}^{n} \binom{n}{j} (ps)^j q^{n-j} = (q + ps)^n \tag{A.2}$$

If it is not clear out of the context which random variable is meant, we write $\mathcal{P}^{(X)}$ where X is the random variable. An important property of a PGF is that it converges for $|s| \leq 1$ since $\mathcal{P}(1) = \sum_{j=0}^{\infty} p_j = 1$. The PGF can be used to directly derive the probability function of the random variable, as well as its moments. Single probabilities can be calculated as

$$P(X = j) = p_j = (j!)^{-1} \left. \frac{\mathrm{d}^j \mathcal{P}}{\mathrm{d}s^j} \right|_{s=0} \tag{A.3}$$

Example: A binomial distributed random variable has PGF $\mathcal{P}(s) = (q+ps)^n$. Thus,

$$\begin{aligned}
P(X = 0) &= \mathcal{P}(0) = q^n \\
P(X = 1) &= \mathcal{P}'(0) = nq^{n-1}p^1 \\
P(X = 2) &= (2!)^{-1}\mathcal{P}''(0) = (2!)^{-1}n(n-1)q^{n-2}p^2 \\
&\vdots \quad \vdots
\end{aligned}$$

The expectation $\mathrm{E}(X)$ satisfies the relation

$$\mathrm{E}(X) = \sum_{j=0}^{\infty} jp_j = \mathcal{P}'(1) \tag{A.4}$$

Example: A binomial distributed random variable has mean

$$\begin{aligned}
\mathcal{P}'(1) &= np(q+p)^{n-1} \\
&= np
\end{aligned}$$

Calculating first

$$\mathrm{E}[X(X-1)] = \sum_{j=1}^{\infty} j(j-1)p_j = \mathcal{P}''(1) \tag{A.5}$$

the variance is obtained as

$$\begin{aligned}
\mathrm{Var}(X) &= \mathrm{E}[X(X-1)] + \mathrm{E}(X) - [\mathrm{E}(X)]^2 \\
&= \mathcal{P}''(1) + \mathcal{P}'(1) - [\mathcal{P}'(1)]^2
\end{aligned} \tag{A.6}$$

Example: A binomial distributed random variable has variance

$$\begin{aligned}
\mathrm{Var}(X) &= n(n-1)p^2 + np - (np)^2 \\
&= np(1-p)
\end{aligned}$$

Proposition 8. *Let X be a random variable defined over the non-negative integers with probability distribution $P(X = j) = p_j$, $j = 0, 1, \ldots$. Let X_T be a positive random variable with truncated-at-zero probability distribution $P(X_T = j) = p_j/(1 - p_0)$, $j = 1, 2, \ldots$. The probability generating function of the truncated-at-zero distribution of X_T is given by*

$$\mathcal{P}_T(s) = \frac{\mathcal{P}(s) - \mathcal{P}(0)}{1 - \mathcal{P}(0)} \tag{A.7}$$

Proof: (A.7) follows directly from the definition of the probability generating function:

$$\mathcal{P}_T(s) = \mathrm{E}(s^{X_T}) = \sum_{j=1}^{\infty} \frac{p_j}{1-p_0} s^j$$

where $p_0 = \mathcal{P}(0)$.

There exists a close relationship between the probability generating function and the moment generating function $\mathcal{M}(t)$:

$$\mathcal{M}(t) = \mathrm{E}(e^{tX}) = \mathcal{P}(e^t) \tag{A.8}$$

While the moment generating function is a concept that can be used for any distribution with existing moments, the probability generating function is defined for non-negative integers. Since $s = e^t = 1$ if and only if $t = 0$, we obtain $\mathrm{E}(X) = \mathcal{P}'(1) = \mathcal{M}'(0)$.

In the same way as in (A.1) one can define a *bivariate probability generating function.*

Definition 5. .
Let X, Y be a pair of integer-valued random variables with joint distribution $P(X = j, Y = k) = p_{jk}\,,\, j,k \in I\!N_0$. The bivariate probability generating function is given by:

$$\mathcal{P}(s_1, s_2) = \sum_{j=0}^{\infty}\sum_{k=0}^{\infty} p_{jk} s_1^j s_2^k = \mathrm{E}(s_1^X s_2^Y) \tag{A.9}$$

Proposition 9. *The probability generating functions of the marginal distributions $P(X = j)$ and $P(Y = k)$ are $\mathcal{P}(s,1) = E(s^X)$ and $\mathcal{P}(1,s) = E(s^Y)$, respectively.*

Proposition 10. *The probability generating function of $X + Y$ is given by $\mathcal{P}(s,s) = E(s^{X+Y})$.*

Proposition 11. *The variables X and Y are independent if and only if $\mathcal{P}(s_1, s_2) = \mathcal{P}(s_1, 1)\mathcal{P}(1, s_2)$ for all s_1, s_2.*

Probability generating functions can be used to establish the distribution of a sum of independent variables. This is also called a *convolution.* Using **Proposition 10** and **Proposition 11**, the probability generating function of $Z = X + Y$ is given by:

$$\mathcal{P}^{(Z)}(s) = \mathrm{E}(s^Z) = \mathrm{E}(s^{X+Y}) = \mathrm{E}(s^X s^Y) \overset{(\star)}{=} \mathrm{E}(s^X)\mathrm{E}(s^Y) \tag{A.10}$$

where $(\star)$ follows from the independence assumption.

Example: Let X have a binomial distribution function with $B(1,p)$. Consider the convolution $Z = \underbrace{X + \ldots + X}_{n-\text{times}}$. Then:

$$\mathcal{P}^{(Z)}(s) = (q + ps)^n \tag{A.11}$$

Z has a binomial distribution function $B(n,p)$. Conversely, the binomial distribution is obtained by a convolution of identically and independently distributed Bernoulli variables.

B

Gauss-Hermite Quadrature

This appendix describes the basic steps required for a numerical evaluation of the likelihood function of count data models with unobserved heterogeneity of the log-normal type. The method is illustrated for the Poisson-log-normal model, although a similar algorithm can be used to estimate the models with endogenous selectivity presented in Chap. 4.5. Butler and Moffitt (1982) discuss Gauss-Hermite quadrature in the context of a panel probit models. Million (1998) points out that the Poisson-log-normal integral can be approximated using Gauss-Laguerre and Gauss-Legendre polynomials as well, and he evaluates the relative performance of the three methods. Crouch and Spiegelman (1990) discuss numerical integration in the related logistic-normal model.

Starting point for Gauss-Hermite quadrature is the integral

$$\int_{-\infty}^{\infty} f(y|x,\beta,\varepsilon)g(\varepsilon|\sigma^2)\mathrm{d}\varepsilon \tag{B.1}$$

that cannot be solved by analytical methods. However, assume that by appropriate change of variable, B.1 can be brought into the form

$$\int_{-\infty}^{\infty} h(\nu;y,x,\beta,\sigma^2)\exp(-\nu^2)\mathrm{d}\nu \tag{B.2}$$

In this case, Gauss-Hermite quadrature can be applied to numerically evaluate the integral (B.1), and thus the marginal likelihood $L(y|x)$. Once the evaluation has been done, the logarithm $\ln L(y|x)$ can be passed on to a maximizer that uses numerical derivatives in order to find the maximum likelihood estimators $\hat{\beta}$ and $\hat{\sigma}^2$.

The Poisson-log-normal model has the following components (see also Chap. 4.2.1):

$$f(y|\varepsilon) = \frac{\exp(-\exp(x'\beta+\varepsilon))\exp(x'\beta+\varepsilon)^y}{y!}$$

where $\varepsilon \sim N(0,\sigma^2)$, i.e.,

$$f(\varepsilon) = \frac{1}{\sqrt{2\pi}\sigma} e^{-\frac{1}{2}\left(\frac{\varepsilon}{\sigma}\right)^2}$$

Change of variable from ε to ν where

$$\nu = \frac{\varepsilon}{\sqrt{2}\sigma}$$

has inverse $\varepsilon = \nu\sqrt{2}\sigma$ and Jacobian $\mathrm{d}f(\nu)/\mathrm{d}\nu = \sqrt{2}\sigma$. Therefore

$$g(\nu) = \frac{1}{\sqrt{\pi}} e^{-\nu^2}$$

and

$$f(y|\nu)g(\nu) = \frac{\exp(-\exp(x'\beta + \nu\sqrt{2}\sigma))\exp(x'\beta + \nu\sqrt{2}\sigma)^y}{\sqrt{\pi}y!} e^{-\nu^2}$$

Let

$$h_i(\nu) = \frac{\exp(-\exp(x_i'\beta + \nu\sqrt{2}\sigma))\exp(x_i'\beta + \nu\sqrt{2}\sigma)^{y_i}}{\sqrt{\pi}y_i!}$$

where the subscript i reminds us that this function depends on observations y_i and x_i. Then the Gauss-Hermite approximation to the integral $B.1$ is obtained as

$$\begin{aligned} L_i^{gh} &= \int_{-\infty}^{\infty} h_i(\nu)\exp(-\nu^2)\mathrm{d}\nu \\ &\approx \sum_{j=1}^{n} w_j h_i(\nu_j) \end{aligned}$$

where w_j are weights and ν_j are the evaluation points. The likelihood function for n independent observations is given by

$$L^{gh} = \prod_{i=1}^{n}\sum_{j=1}^{n} w_j h_i(\nu_j)$$

Weight factors and abscissas for 20-point quadrature are given in Tab. B.1 (Source: Abramowitz and Stegun, 1964, p. 924).

Table B.1. Abcissas and Weight Factors for 20-point Gauss-Hermite Integration

u_i	w_i
-5.3874809	2.2939000e-13
-4.6036824	4.3993400e-10
-3.9447640	1.0860000e-07
-3.3478546	7.8025500e-06
-2.7888061	0.00022833863
-2.2549740	0.0033243773
-1.7385377	0.024810521
-1.2340762	0.10901721
-0.73747373	0.28667551
-0.24534071	0.46224367
0.24534071	0.46224367
0.73747373	0.28667551
1.2340762	0.10901721
1.7385377	0.024810521
2.2549740	0.0033243773
2.7888061	0.00022833863
3.3478546	7.8025500e-06
3.9447640	1.0860000e-07
4.6036824	4.3993400e-10
5.3874809	2.2939000e-13

Source: Abramowitz and Stegun, 1964, p. 924

C

Software

Most statistical and econometric software distributions contain built-in procedures for standard count data models, such as the Poisson and the negative binomial regression models. Development in the software sector is fast, and specific recommendations risk to become outdated very quickly. Nevertheless, there are a few general points that should be of help to anyone interested in working with count data and estimating the models presented in this book.

Within the econometrics research community, GAUSS traditionally has been the major development tool. GAUSS is mostly a programming environment, but specialised procedures are available both as part of the general distribution, and through web sites and mailing lists. For example, the "count" module allows the estimation of seemingly unrelated regression models, of various types of negative binomial models as well as hurdle Poisson models. Yet, the development of this module has stalled for some time, and the latest models are not available.

Two alternative programs with a much more ambitious offering in this area are STATA and LIMDEP. This appendix is not intended as a comprehensive review of available software for count data, and there may be other software with similar or even broader scope. Yet, the possibilities that these two packages offers should be closely scrutinized by anyone seriously interested in count data applications who wants to apply up-to-date methods without doing the programming for herself. In fact, most of the models discussed in this book are easily estimated with STATA or LIMDEP, providing little support for those who resort to the most basic models in want of available software for the more appropriate ones.

The following short summary refers to STATA release 7.0. This release includes built-in procedures, apart from the standard Poisson and Negbin models (in its various parameterizations, as Negbin I, Negbin II or with more flexible variance function), for zero-inflated Poisson and zero-inflated negative binomial models, and for fixed and random effects panel count data models. Random effects models include the negative binomial panel model (with fixed or random effects) but also the panel Poisson-log-normal model. This procedure can also be used in cross sections to estimate the standard Poisson-log-normal model that frequently has a better fit than the Negbin model. Hurdle Poisson or negative binomial models are not included in the standard distribution. However, they can be estimated using routines on truncated-at-zero models authored by Joseph Hilbe and described in the *Stata Technical Bulletin* Nr. 47. Most procedures include options for the computations of robust standard errors (to perform pseudo maximum likelihood estimation) as well as account for clustered sampling.

The latest version of LIMDEP is release 8.0. Apart from the standard count data models, its capabilities include the estimation of sample selection models by maximum likelihood, parametric models for underreporting where the observed counts represent only the reported fraction of the total events which have occurred, and maximum likelihood estimation of various types of hurdle models and zero-inflated models. LIMDEP and STATA are both quite versatile in the area of count data modelling.

D

Tables

Table D.1. Number of Job Changes: Poisson and Poisson-Log-Normal

	Poisson	Poisson-log-normal	Mean
Constant	0.501**	0.072	1
	(0.158)	(0.227)	
Education$*10^{-1}$	-0.138	-0.120	1.216
	(0.137)	(0.187)	
Experience$*10^{-1}$	-0.770**	-0.846**	1.460
	(0.111)	(0.155)	
Experience$^2 * 10^{-2}$	0.119**	0.127*	2.943
	(0.037)	(0.050)	
Union	-0.292**	-0.324**	0.429
	(0.065)	(0.088)	
Single	-0.050	-0.093	0.077
	(0.108)	(0.153)	
German	-0.368**	-0.390**	0.668
	(0.076)	(0.104)	
Qualified White Collar	0.067	-0.002	0.137
	(0.131)	(0.179)	
Ordinary White Collar	0.185	0.190	0.058
	(0.147)	(0.207)	
Qualified Blue Collar	0.147	0.112	0.501
	(0.082)	(0.114)	
σ^2		1.048**	
		(0.048)	
Log likelihood	-2044.47	-1866.80	
Log likelihood ($\beta_1, \ldots, \beta_9 = 0$)	-2155.40	-1934.53	
Number of Observations	1962		

Source: *German Socio-Economic Panel*, Wave A/1984; own calculations.
Note: Asymptotic standard errors in parentheses.

Table D.2. Number of Job Changes: Negative Binomial Models

	Negbin I	Negbin II	GEC_k
Constant	0.341	0.616**	0.380
	(0.191)	(0.224)	(0.212)
Education$*10^{-1}$	0.008	-0.179	-0.011
	(0.162)	(0.187)	(0.180)
Experience$*10^{-1}$	-0.762**	-0.786**	-0.775**
	(0.139)	(0.152)	(0.144)
Experience$^2 * 10^{-2}$	0.113*	0.118*	0.115*
	(0.046)	(0.048)	(0.047)
Union	-0.274**	-0.308**	-0.283**
	(0.080)	(0.087)	(0.084)
Single	-0.114	-0.054	-0.108
	(0.139)	(0.152)	(0.141)
German	-0.316**	-0.404**	-0.331**
	(0.097)	(0.102)	(0.103)
Qualified White Collar	-0.022	0.043	-0.013
	(0.163)	(0.174)	(0.173)
Ordinary White Collar	0.213	0.188	0.214
	(0.176)	(0.201)	(0.181)
Qualified Blue Collar	0.086	0.132	0.094
	(0.103)	(0.111)	(0.107)
σ^2	0.823**	1.378**	0.892**
	(0.088)	(0.137)	(0.080)
k			0.139
			(0.281)
Log likelihood	-1873.28	-1878.63	-1873.17
Number of Observations	1962		

Source: *German Socio-Economic Panel*, Wave A/1984; own calculations.
Notes: Asymptotic standard errors in parentheses. For $\sigma^2 > 0$ and $k = 0$, the GEC_k model coincides with the Negbin I model. For $\sigma^2 > 0$ and $k = 1$, the GEC_k model coincides with the Negbin II model.

Table D.3. Number of Job Changes: Robust Poisson Regression

	Coefficient	t_{Poisson}	**Robust t-Values** t_{WHITE}	t_{LVF}	t_{QVF}
Constant	0.501	3.167	2.617	2.229	2.304
Education$*10^{-1}$	-0.138	-1.006	-0.823	-0.707	-0.749
Experience$*10^{-1}$	-0.770	-6.929	-4.830	-4.877	-5.055
Experience$^2 * 10^{-2}$	0.119	3.269	2.385	2.301	2.486
Union	-0.292	-4.499	-3.115	-3.167	-3.385
Single	-0.050	-0.460	-0.309	-0.323	-0.326
German	-0.368	-4.843	-2.892	-3.409	-3.503
Qualified White Collar	0.067	0.514	0.343	0.361	0.384
Ordinary White Collar	0.185	1.255	0.964	0.883	0.917
Qualified Blue Collar	0.147	1.794	1.261	1.263	1.308
Log likelihood	-2044.47				
Number of Observations	1962				

Notes:

Three alternative methods to calculate robust standard errors (and thus robust t-values) were given in Chap. 4.10.3. t_{LVF} and t_{QVF} are based on the assumption of a quadratic and linear variance function, respectively, while the White method makes no explicit assumption.

Table D.4. Number of Job Changes: Poisson-Logistic Regression

	a) Overlapping		**b) Non Overlapping**	
Variable	Offers	Acceptance	Offers	Acceptance
Constant	0.812		1.151	
	(3.746)		(9.740)	
Education$*10^{-1}$	-0.322	3.732		-0.260
	-2.073)	(1.582)		(-1.633)
Experience$*10^{-1}$	-0.668	-6.044		-1.068
	(-4.804)	(-1.221)		(-7.678)
Experience$^2 * 10^{-2}$	0.071	3.321		0.175
	(1.382)	(1.132)		(3.920)
Union	-0.291		-0.290	
	(-4.477)		(-4.470)	
Single		0.379		-0.068
		(0.153)		(-0.460)
German	-0.397		-0.355	
	(-5.112)		(-4.708)	
Qualified White Collar	0.069		0.088	
	(0.452)		(0.684)	
Ordinary White Collar	0.178		0.195	
	(1.125)		(1.328)	
Qualified Blue Collar	0.132		0.156	
	(1.389)		(1.919)	
Log likelihood	-2039.35		-2043.88	
Observations	1962			

Notes:

Asymptotic t-values in parentheses.

Table D.5. Number of Job Changes: Hurdle Count Data Models

	Hurdle Poisson		Probit-Poisson-log-normal	
Variable	1+/0	1+	1+/0	1+
Constant	-0.069	1.163	0.269	0.799
	(0.202)	(0.245)	(0.157)	(0.666)
Education$*10^{-1}$	0.133	-0.600**	0.094	-0.764**
	(0.170)	(0.218)	(0.128)	(0.324)
Experience$*10^{-1}$	-0.758**	-0.403**	-0.629**	-0.544
	(0.148)	(0.156)	(0.111)	(0.405)
Experience$^2 * 10^{-2}$	0.107**	0.085	0.098**	0.103
	(0.048)	(0.050)	(0.034)	(0.088)
Union	-0.268**	-0.167*	-0.205**	-0.230
	(0.084)	(0.097)	(0.061)	(0.189)
Single	-0.194	0.192	-0.149	0.195
	(0.149)	(0.147)	(0.114)	(0.249)
German	-0.330**	-0.206**	-0.254**	-0.223
	(0.101)	(0.108)	(0.076)	(0.208)
Qualified White Collar	-0.071	0.271	-0.076	0.283
	(0.170)	(0.196)	(0.125)	(0.285)
Ordinary White Collar	0.239	-0.039	0.200	-0.057
	(0.185)	(0.236)	(0.143)	(0.336)
Qualified Blue Collar	0.069	0.184	0.042	0.167
	(0.109)	(0.117)	(0.081)	(0.172)
σ^2			0.932**	
			(0.156)	
ρ			0.212	
			(0.893)	
Log-likelihood	-1928.00		-1856.70	
Observations	1962			

Notes:

Asymptotic standard errors in parentheses.
Hurdle negbin results are nor displayed because of convergence problems.

Table D.6. Number of Job Changes: Finite Mixture Models

	2-components Poisson		2-components Negbin II	
Variable	group 1	group 2	group 1	group 2
Constant	-0.000	2.229**	1.047*	0.154
	(0.226)	(0.433)	(0.630)	(0.458)
Education$*10^{-1}$	0.078	-0.368	-0.648	0.243
	(0.184)	(0.374)	(0.425)	(0.307)
Experience$*10^{-1}$	-0.857**	-0.541**	-0.371	-1.140**
	(0.172)	(0.231)	(0.346)	(0.296)
Experience$^2 * 10^{-2}$	0.104*	0.081	0.050	0.138
	(0.060)	(0.074)	(0.099)	(0.105)
Union	-0.309**	-0.207	-0.259	-0.328**
	(0.095)	(0.141)	(0.181)	(0.152)
Single	-0.156	0.057	0.200	-0.274
	(0.158)	(0.229)	(0.325)	(0.267)
German	-0.351**	-0.478**	-0.609**	-0.101
	(0.114)	(0.158)	(0.236)	(0.229)
Qualified White Collar	-0.037	0.168	0.320	-0.263
	(0.192)	(0.288)	(0.386)	(0.327)
Ordinary White Collar	0.253	0.103	-1.037	0.609*
	(0.202)	(0.358)	(0.945)	(0.353)
Qualified Blue Collar	0.082	0.317*	0.393	-0.130
	(0.124)	(0.173)	(0.263)	(0.224)
σ^2			2.096**	0.146
			(0.949)	(0.281)
π_1	0.930**		0.395**	
	(0.013)		(0.158)	
Log-likelihood	-1868.16		-1856.05	
Observations	1962			

Notes:

Asymptotic standard errors in parentheses.
Hurdle Negbin I results are nor displayed because of convergence problems.

Table D.7. Number of Job Changes: Zero Inflated Count Data Models

	zero-inflated Poisson		zero-inflated Negbin II	
Variable	logit	Poisson	logit	Negbin II
Constant	1.132	-0.303	0.483*	-7.390**
	(0.245)	(0.529)	(0.255)	(2.777)
Education$*10^{-1}$	-0.583**	1.016**	-0.262	-0.746
	(0.203)	(0.455)	(0.216)	(1.152)
Experience$*10^{-1}$	-0.373**	1.035**	-0.613**	4.535**
	(0.153)	(0.312)	(0.192)	(1.715)
Experience$^2 * 10^{-2}$	0.072	-0.157	0.129**	-.759**
	(0.049)	(0.091)	(0.062)	(0.346)
Union	-0.158	0.293	-0.253**	0.351
	(0.097)	(0.179)	(0.102)	(0.465)
Single	0.151	0.461	0.066	1.368
	(0.154)	(0.297)	(0.169)	(0.962)
German	-0.173	0.435**	-0.236**	1.306
	(0.106)	(0.211)	(0.118)	(0.788)
Qualified White Collar	0.272	0.519	0.178	1.228
	(0.189)	(0.354)	(0.206)	(0.843)
Ordinary White Collar	-0.123	-0.870	0.025	11.967
	(0.243)	(0.711)	(0.203)	(349.869)
Qualified Blue Collar	0.166	0.094	0.151	0.193
	(0.115)	(0.219)	(0.129)	(0.614)
σ^2			1.103	
			(0.146)	
Log-likelihood	-1926.28		-1866.73	
Observations	1962			

Notes:

Asymptotic standard errors in parentheses.

Table D.8. Number of Job Changes: Quantile Regressions

	$Q_z(0.5, x)$	$Q_z(0.75, x)$	$Q_z(0.9, x)$
Constant	-0.181	1.138	1.768
	(0.468)	(0.475)	(0.272)
Education$*10^{-1}$	0.319	0.026	-0.343
	(0.373)	(0.343)	(0.207)
Experience$*10^{-1}$	-1.346	-1.413	-0.721
	(0.249)	(0.258)	(0.196)
Experience$^2 * 10^{-2}$	0.288	0.220	0.066
	(0.069)	(0.083)	(0.054)
Union	-0.388	-0.395	-0.336
	(0.193)	(0.187)	(0.117)
Single	-0.469	-0.191	-0.128
	(0.324)	(0.248)	(0.220)
German	-0.479	-0.522	-0.209
	(0.246)	(0.213)	(0.162)
Qualified White Collar	-0.144	-0.063	-0.020
	(0.304)	(0.302)	(0.237)
Ordinary White Collar	0.240	0.312	-0.061
	(0.319)	(0.340)	(0.188)
Qualified Blue Collar	-0.142	-0.078	-0.082
	(0.212)	(0.179)	(0.137)
Observations	1962		

Notes:

Bootstrap standard errors in parentheses (50 replications).

References

Abramowitz, M. and I. A. Stegun 1964, *Handbook of Mathematical Functions*, National Bureau of Standards, Applied Mathematics Series Nr. 55, Washington, D.C.

Aitchison, J. and C.H. Ho 1989, "The multivariate Poisson-log normal distribution," *Biometrika* 76(4): 643-653.

Albert, J.H. and P.A. Pepple 1989, "A Bayesian Approach to Some Overdispersion Models," *Canadian Journal of Statistics* 17: 333-344.

Allison, P.D. 1984, *Event History Analysis: Regression for Longitudinal Event Data*, University Paper No. 46, Sage Publications: Beverly Hills.

Al-Osh, M.A. and A.A. Alzaid 1987, "First order integer valued autoregressive (INAR(1)) process," *Journal of Time Series Analysis* 8: 261-275.

Al-Osh, M.A. and A.A. Alzaid 1988, "Integer-valued moving average (INMA) process," *Statistical Papers* 29: 281-300.

Al-Qudsi, S. 1998a, "The Demand for Children in Arab Countries: Evidence from Panel and Count Data Models," *Journal of Population Economics* 11(3): 435-452.

Al-Qudsi, S. 1998b, "Labour participation of Arab women: Estimates of the fertility to labour supply link," *Applied Economics* 30: 931-941.

Alvarez, B. and M.A. Delgado 2002, "Goodness-of-Fit Techniques for Count Data Models: An Application to the Demand for Dental Care in Spain, " *Empirical Economics* 27: 543-567.

Andrews, D.W.K. 1988, "Chi-squared diagnostic tests for econometric models: introduction and applications," *Journal of Econometrics* 37: 135-156.

Angrist, J.D. 2001, "Estimations of Limited-Dependent Variable Models with Dummy Endogenous Regressors: Simple Strategies for Empirical Practice," *Journal of Business and Economic Statistics* 19(1): 2-16.

Aptech 1994, *GAUSS Command Reference Manual*, Aptech: Maple Valley, WA.

Arulampalam, W. and A.L. Booth 1997, "Who Gets Over the Training Hurdle? A Study of the Training Experiences of Young Men and Women in Britain," *Journal of Population Economics* 10(2): 197-217.

Atella, V. and F.C. Rosati 2000, "Uncertainty about children's survival and fertility: A test using Indian microdata," *Journal of Population Economics* 13(2): 263-278.

Barlow, R.E. and F. Proschan 1965, *Mathematical theory of reliability*, John Wiley: New York.

Barmby, T. and J. Doornik 1989, "Modelling trip frequency as a Poisson variable," *Journal of Transport Economics and Policy* 23(3): 309-315.

Barmby, T., M. Nolan and R. Winkelmann 2001, "Contracted Workdays and Absence," *Manchester School*, 69(3), 269-275.

Barron, D.N. 1992, "The analysis of count data: Overdispersion and autocorrelation," in P. Marsden (ed.) *Sociological Methodology 1992*, Blackwell: Cambridge, MA, 179-220.

Bates, G. and J. Neyman 1951, "Contributions to the theory of accident proneness. II: True or false contagion," *University of California Publications in Statistics*, 215-253.

Bauer, T., A. Million, R. Rotte, and K.F. Zimmermann 1998, "Immigrant Labor and Workplace Safety: A Bivariate Count Data Approach," mimeo.

Becker, G.S. 1962, "Investment in human capital: A theoretical analysis," *Journal of Political Economy* 70: 9-49.

Beckmann, M. 2002, "Wage compression and firm-sponsored training in Germany: Empirical evidence for the Acemoglu-Pischke model from a zero-inflated count data model," *Konjunkturpolitik* 48 (2/3).

Berglund, E. and K. Brännäs 1995, "Entry and Exit of Plants - A Study Based on Swedish Panel Count Data," Umea Economic Studies No. 374.

Berkhout, P. and E. Plug 1996, " A Bivariate Poisson Count Data Model Using Conditional Probabilities, " Tinbergen Institute Discussion Paper 96-88/7.

Blundell, R., R. Griffith and J. van Reenen 1995, "Dynamic count data models of technological innovation," *Economic Journal* 105: 333-344.

Blundell, R., R. Griffith and J. van Reenen 1999, "Market share, market value and innovation in a panel of British manufacturing firms," *Review of Economic Studies* 66: 529-554.

Blundell, R., R. Griffith and F. Windmeijer 2002, "Individual Effects and Dynamics in Count Data Models," *Journal of Econometrics* 108: 113-131.

Böckenholt, U. 1999, "Mixed INAR(1) Poisson Regression Models: Analyzing Heterogeneity and Serial Dependence in Longitudinal Count Data," *Journal of Econometrics* 89: 317-338.

Böhning, D., E. Dietz, P. Schlattmann, L. Mendonca, and U. Kirchner 1999, "The zero-inflated Poisson model and the decayed, missing and filled teeth index in dental epidemiology," *Journal of the Royal Statistical Society*, Ser. A., 162: 195-209.

Börsch-Supan, A. 1990, "Education and its double-edged impact on mobility," *Economics of Education Review* 9(1): 39-53.

Booth, A.L., W. Arulampalam, and P. Elias 1997, "Modelling Work Related Training and Training Effects Using Count Data Techniques," Centre for Economic Policy Research Discussion Paper No 1582.

Bortkiewicz, L. von 1898, *Das Gesetz der Kleinen Zahlen*, Teubner: Leipzig.

Bourlange, D. and C. Doz 1988, "Pseudo-maximum de vraisemblance: experiences de simulations dans le cadre d'un modele de Poisson," *Annales d'Economie et de Statistique* 10: 139-176.

Bowman, K.O., and Shenton, L.R. 1988, *Properties of Estimators for the Gamma Distribution*, New York: Marcel Dekker.

Brännäs, K. 1992a, "Finite Sample Properties of Estimators and Tests in Poisson Regression Models," *Journal of Statistical Simulation and Computation* 41: 229-241.

Brännäs, K. 1992b, "Limited dependent Poisson regression," *The Statistician* 41, 413-423.

Brännäs, K. 1994, "Estimating and Testing in Integer Valued AR(1) Models," Umea Economic Studies No. 335.

Brännäs, K. 1995a, "Explanatory Variables in the INAR(1) Model, " Umea Economic Studies No. 381.

Brännäs, K. 1995b, "Prediction and Control for a Time-Series Count Data Model," *International Journal of Forecasting* 11: 263-270.

Brännäs, K. and J. Hellström 2002, "Generalized Integer-Valued Autoregression," *Econometric Reviews* 20: 425-43.

Brännäs, K. and P. Johansson 1994, "Time Series Count Data Regression," *Communications in Statistics: Theory and Methods* 23: 2907-2925.

Brännäs, K. and P. Johansson 1996, "Panel data regression for counts," *Statistical Papers* 37(3): 191-213.

Brännäs, K. and G. Rosenqvist 1994, "Semiparametric Estimation of heterogenous count data models," *European Journal of Operational Research* 76: 247-258.

Breslow, N. 1990, "Tests of hypotheses in overdispersed Poisson regression and other quasi-likelihood models," *Journal of the American Statistical Association* 85: 565-571.

Brockett, P.L., L.L. Golden and H.H. Panjer 1996, "Flexible purchase frequency modeling," *Journal of Marketing Research*, 33: 94-107.

Broek, J. van den 1995, "A Score Test for Zero Inflation in a Poisson Distribution," *Biometrics* 51: 738-743.

Buck, A.J. 1984, "Modelling a Poisson process: strike frequency in Great Britain," *Atlantic Economic Journal* 12(1):60-64.

Butler, J.S. and R. Moffitt 1982, "A computationally efficient quadrature procedure for the one-factor multinomial probit model," *Econometrica* 50: 761-764.

Cameron, A. C. and P. Johansson 1997, "Count Data Regression Using Series Expansions: With Applications," *Journal of Applied Econometrics* 12(3): 203-223.

Cameron, A.C. and F.A.G. Windmeijer 1996, "R-squared measures for count data regression models with applications to health-care utilization," *Journal of Business and Economic Statistics* 14: 209-220.

Cameron, A.C., P.K. Trivedi, F. Milne, and J. Piggott 1988, "A Microeconometric Model of the Demand for Health Care and Health Insurance in Australia," *Review of Economic Studies* LV: 85-106.

Cameron, A.C. and P.K. Trivedi 1986, "Econometric models based on count data: comparisons and applications of some estimators and tests," *Journal of Applied Econometrics* 1: 29-53.

Cameron, A.C. and P.K. Trivedi 1990, "Regression-based tests for overdispersion in the Poisson model," *Journal of Econometrics* 46: 347-364.

Cameron, C. and P. Trivedi 1998, *Regression Analysis of Count Data*, Cambridge University Press.

Cameron, A.C. and F.A.G. Windmeijer 1996, "R-squared measures for count data regression models with applications to health-care utilization," *Journal of Business and Economic Statistics* 14: 209-220.

Caudill, S.B. and F.G. Mixon 1995, "Modeling household fertility decisions: Estimation and testing of censored regression models for count data," *Empirical Economics* 20: 183-196.

Chamberlain, G. 1992, "Comment: Sequential Moment Restrictions in Panel Data," *Journal of Business and Economic Statistics* 10: 20-26.

Chappell, W.F., M.S. Kimenyi and W.J. Mayer 1990, "A Poisson probability model of entry and market structure with an application to U.S. industries during 1972-77," *Southern Economic Journal* 56: 918-927.

Chatfield, C., A.S.C. Ehrenberg, and G.J. Goodhardt 1966, "Progress on a simplified model of stationary purchasing behaviour" (with discussion), *Journal of the Royal Statistical Society* A 129: 317-367.

Chesher, A. 1984, "Testing for neglected heterogeneity," *Econometrica* 52: 865-872.

Chernoff, H. 1954, "On the distribution of the likelihood ratio," *Annals of Mathematical Statistics* 25: 573-578.

Chib, S. and E. Greenberg 1995, "Understanding the Metropolis-Hastings Algorithm," *The American Statistician* 49: 327-335.

Chib, S. and E. Greenberg 1996, "Markov Chain Monte Carlo Simulation Methods in Econometrics," *Econometric Theory* 12: 409-431.

Chib, S., E. Greenberg, and R. Winkelmann 1998, "Posterior Simulation and Bayes Factors in Panel Count Data Models," *Journal of Econometrics* 86: 33-54.

Chib, S. and R. Winkelmann 2001, "Markov Chain Monte Carlo Analysis of Correlated Count Data," *Journal of Business and Economic Statistics* 19: 428-435.

Cincera, M. 1997, "Patents, R&D, and Technological Spillovers at the Firm Level: Some Evidence from Econometric Count Models for Panel Data," *Journal of Applied Econometrics* 12(3): 265-80.

Congdon, P. 1989, "Modelling migration flows between areas: An analysis for London using the Census and OPCS Longitudinal Study," *Regional Studies* 23(2): 87-103.

Consul, P.C. 1989, *Generalized Poisson distributions*, Marcel Dekker: New York.

Consul, P.C. and F. Famoye 1992, "Generalized Poisson regression model," *Communications in Statistics - Theory and Methods* 21(1): 89-109.

Cox, D.R. 1961, "Tests of separate families of hypotheses," *Proc. Fourth Berkeley Symposium on Mathematical Statistics and Probability* 1: 105-123.

Cox, D.R. 1962, *Renewal Theory*, John Wiley: New York.

Cox, D.R. 1972, "Regression models and life tables," *Journal of the Royal Statistical Society* B 34: 187-202.

Cramer, J.S. 1986, *Econometric Applications of Maximum Likelihood Methods*, Cambridge University Press: New York.

Creel, M.D. and J.B. Loomis 1990, "Theoretical and empirical advantages of truncated count data estimators for analysis of deer hunting in California," *American Journal of Agricultural Economics* 72(2): 434-441.

Crépon, B. and E. Duguet 1997a, "Estimating the Innovation Function from Patent Numbers: GMM on Count Panel Data," *Journal of Applied Econometrics* 12(3): 243-263.

Crépon, B. and E. Duguet 1997b, "Research and Development, Competition and Innovation Pseudo Maximum Likelihood and Simulated Maximum Likelihood Methods Applied to Count Data Models with Heterogeneity," *Journal of Econometrics* 79(2): 355-378.

Crouch, E.A.C. and D. Spiegelmann 1990, "The evaluation of integrals of the form $\int f(t) \exp(-t^2) dt$: application to logistic-normal models," *Journal of the American Statistical Association* 85: 464-469.

Davidson, R. and J. G. MacKinnon 1993, *Estimation and Inference in Econometrics*, Oxford University Press: Oxford.

Davutyan, N. 1989, "Bank failures as Poisson variates," *Economics Letters* 29(4): 333-338.

Dean, C. and J.F. Lawless 1989, "Tests for detecting overdispersion in Poisson regression models," *Journal of the American Statistical Association* 84: 467-472.

Dean, C., J.F. Lawless and G.E. Willmot 1989, "A mixed Poisson-inverse Gaussian regression model," *Canadian Journal of Statistics* 17(2): 171-181.

Deb, P. and P.K. Trivedi 1997, "Demand for Medical Care by the Elderly: A Finite Mixture Approach," *Journal of Applied Econometrics* 12(3): 313-336.

Deb, P. and P.K. Trivedi 2002, "The Structure of Demand for Health Care: Latent Class versus Two-Part Models," *Journal of Health Economics* 21: 601-625.

DeGroot, M.H. 1986, *Probability and Statistics*, Addison-Wesley: Reading, Mass.

Delgado, M.A. and T.J. Kniesner 1997, "Count Data Models with Variance of Unknown Forms: An Application to a Hedonic Model of Worker Absenteeism," *Review of Economics and Statistics* 79(1): 41-49.

Deutsches Institut für Wirtschaftsforschung (ed.) 1990, "Das Sozio-oekonomische Panel (SOEP), Benutzerhandbuch, Version 4 - Oktober/90," Berlin.

Dey, D.K. and Y. Chung 1992, "Compound Poisson Distributions: Properties and Estimation," *Communications in Statistics – Theory and Methods* 21: 3097-3121.

Diggle, P., K.-Y. Liang, and S. L. Zeger 1995, *Analysis of Longitudinal Data*, Oxford, Oxford University Press.

Dillon, W.R. and S. Gupta 1996, "A segment-level model of category volume and brand choice," *Marketing Science* 15: 38-59.

Dionne, G. et al. 1994, "Medical Conditions, Risk Exposure, and Truck Drivers' Accidents," Université de Cergy-Pontoise / CNRS-Thema Discussion Paper 9409.

Dionne, G. et al. 1995, "Medical Conditions and the Severity of Commercial Motor Vehicle Drivers' Road Accidents," University of Montreal Department of Economics Discussion Paper No. 9527.

Dionne, G., R. Gagné, F. Gagnon, and C. Vanasse 1997, "Debt, Moral Hazard and Airline Safety: An Empirical Evidence," *Journal of Econometrics* 79(2): 379-402.

Dionne, G. and C. Vanasse 1992, "Automobile insurance ratemaking in the presence of asymmetrical information," *Journal of Applied Econometrics* 7(2): 149-165.

Doorslaer, E. van, X. Koolman, and A.M. Jones 2002, "Explaining income-related inequalities in doctor utilisation in Europe: a decomposition approach," mimeo, Erasmus University, Rotterdam.

Duijn, M.A.J. van and U. Böckenholt 1995, "Mixture Models for the Analysis of Repeated Count Data, " *Applied Statistics* 44: 473-485.

Ebmer, R. 1990, "Placement service and offer arrival rates," *Economics Letters* 34: 289-294.

Faddy, M.J. 1997, "Extended Poisson process modelling and analysis of count data, " *Biometric Journal* 39: 431-440.

Famoye, F. 1993, "Restricted generalized Poisson regression," *Communications in Statistics – Theory and Methods* 22: 1335-1354.

Farrow, S. 1991, "Does analysis matter? Economics and planning in the department of interior," *Review of Economics and Statistics* 73: 172-176.

Feinstein, J.S. 1989, "The Safety Regulations of U.S. Nuclear Power Plants: Violations, Inspections, and Abnormal Occurrences," *Journal of Political Economy* 97: 115-154.

Feller, W. 1968, *An introduction to probability theory and its applications* Vol.1, 3rd ed., John Wiley: New York.

Feller, W. 1971, *An introduction to probability theory and its applications* Vol.2, 2nd ed., John Wiley: New York.

Firth, D. 1991, "Generalized Linear Models," in: Hinkley, D.V., N. Reid and E.J. Snell (eds.) *Statistical Theory and Modelling*, Chapman and Hall: London.

Firth, D. 1992, "Bias reduction, the Jeffreys prior, and GLIM," in: L. Fahrmeir et al. (eds.) *Advances in GLIM and statistical modelling: proceedings of the GLIM92 and the 7th International Workshop on Statistical Modelling, Munich, 13-17 June 1992.* Springer: New York, 91-100.

Flowerdew, R. and M. Aitkin 1982, "A method of fitting the gravity model based on the Poisson distribution," *Journal of Regional Science* 22(2): 191-202.

Freund, D.A., T.J. Kniesner and A.T. LoSasso 1996, "How Managed Care Affects Medicaid Utilization - A Synthetic Difference-in-Differences Zero-Inflated Count Data Model," CentER Discussion Paper No 9640.

Frome, E.L , M.H. Kutner and J.J. Beauchamp 1973, "Regression analysis of Poisson-distributed data," *Journal of the American Statistical Association* 68: 935-940.

Gameren, E. van and I. Woittiez 2002, "Determinants of the Demand for Home Care: The Effect of Supply Restrictions," mimeo. Central Planing Bureau.

Gamerman, D. 1997, *Markov Chain Monte Carlo*, Chapman & Hall: London.

Ganio, L.M. and D.W. Schafer 1992, "Diagnostics for overdispersion," *Journal of the American Statistical Association* 87: 795-804.

Geil, P., A. Million, R. Rotte and K.F. Zimmermann 1997, "Economic Incentives and Hospitalization in Germany," *Journal of Applied Econometrics* 12(3): 295-311.

Gerdtham, U.G. 1997, "Equity in Health Care Utilization: Further Tests Based on Hurdle Models and Swedish Micro Data," *Health Economics* 6(3): 303-319.

Gilbert, C.L. 1982, "Econometric models for discrete (integer valued) economic processes," in: E.G. Charatsis (ed.) *Selected Papers on Contemporary Econometric Problems*, The Athens School of Economics and Business science.

Goldberger, A.S. 1968, "The interpretation and estimation of Cobb-Douglas functions," *Econometrica* 35: 464-472.

Good, D.H., and M.A. Pirog-Good 1989, "Models for Bivariate Count Data with an Application to Teenage Delinquency and Paterny," *Sociological Methods & Research* 17: 409-431.

Gourieroux, C. 1989, *Econométrie des Variables Qualitatives*, 2nd ed., Economica: Paris.

Gourieroux, C., A. Monfort and A. Trognon 1984a, "Pseudo maximum likelihood methods: Theory," *Econometrica* 52: 681-700.

Gourieroux, C., A. Monfort and A. Trognon 1984b, "Pseudo Maximum likelihood methods: Applications to Poisson models," *Econometrica* 52: 701-721.

Gourieroux, C. and A. Monfort 1989, *Statistique et Modèles Econométriques*, Vol. 1 and 2, Economica: Paris.

Gourieroux, C. and A. Monfort 1993, "Simulation-Based Inference: A Survey with Special Reference to Panel Data Models," *Journal of Econometrics*, 59: 5-33.

Gourieroux, C. and M. Visser 1997, "A Count Data Model with Unobserved Heterogeneity," *Journal of Econometrics* 79(2): 247-268.

Greene, W.H. 1994, "Accounting for excess zeros and sample selection in Poisson and negative binomial regression models", Stern School of Business, New York University, Working Paper No. 94-10.

Greene, W.H. 1995, *LIMDEP 7.0 User's Manual*, Econometric Software Inc.: Bellport, NY.

Greene, W.H. 1997, "FIML Estimation of Sample Selection Models for Count Data," mimeo.

Greene, W.H. 1998, "Sample Selection in Credit-Scoring Models," *Japan and the World Economy* 10(3): 299-316.

Greene, W.H. 2000, *Econometric analysis*, 4th ed., Prentice Hall: New York.

Greenwood, M. and G.U. Yule 1920, "An inquiry into the nature of frequency distributions representative of multiple happenings with particular reference to the ocurrence of multiple attacks of disease or of repeated accidents," *Journal of the Royal Statistical Society* A 83: 255-279.

Grogger, J.T. 1990a, "The Deterrent Effect of Capital Punishment: An Analysis of Daily Homicide Counts," *Journal of the American Statistical Association* 85: 295-303.

Grogger, J.T. 1990b, "A Simple Test for Exogeneity in Probit, Logit, and Poisson Regression Models," *Economics Letters* 33: 329-332.

Grogger, J.T. and R.T. Carson 1991, "Models for truncated counts," *Journal of Applied Econometrics* 6: 225-238.

Grootendorst, P.V. 1995, "A comparison of alternative models of prescription drug utilization," *Health Economics* 4: 183-198.

Guldberg, A. 1931, "On discontinuous frequency-functions and statistical series," *Skandinauisk Aktuarietionskrift* 14: 167-197.

Guo, J.Q. and T. Li 2000, "Poisson Regression Models with Errors-in-Variables: Implication and Treatment," mimeo. Indiana University.

Guo, J.Q. and T. Li 2001, "Simulation-Based Estimation of the Structural Errors-in-Variables Negative Binomial Regression Model with an Application," *Annals of Economics and Finance* 2: 101-122.

Guo, J.Q. and M. Melkersson 2002, "A flexible count data model for the demand for dental care, " mimeo., Stockholm University.

Gupta, S. 1988, "Impact of Sales Promotions on When, What, and How Much to Buy," *Journal of Marketing Research* 25: 342-355.

Gurmu, S. 1991, "Tests for detecting overdispersion in the positive Poisson regression model," *Journal of the Business and Economics Statistics* 9: 215-222.

Gurmu, S. 1997, "Semi Parametric Estimation of Hurdle Regression Models with an Application to Medicaid Utilization," *Journal of Applied Econometrics* 12(3): 225-243.

Gurmu, S. 1998, "Generalized Hurdle Count Data Regression Models," *Economics Letters* 58(3): 263-268.

Gurmu, S. and J. Elder 1998, "Estimation of Multivariate Count Regression Models with Applications to Health Care Utilization," mimeo.

Gurmu, S. and J. Elder 2000, "Generalized bivariate count data regression models," *Economics Letters* 68: 31-36

Gurmu, S., P. Rilstone and S. Stern 1998, "Semiparametric Estimation of Count Regression Models," *Journal of Econometrics* 88(1): 123-150.

Gurmu, S. and P.K. Trivedi 1992, "Overdispersion tests for truncated Poisson regression models," *Journal of Econometrics* 54: 347-370.

Gurmu, S. and P.K. Trivedi 1996, "Excess Zeros in Count Models for Recreational Trips," *Journal of Business and Economic Statistics* 14(4): 469-477.

Haab, T.C. and K.E. McConnell 1996, "Count Data Models and the Problem of Zeros in Recreation Demand Analysis," *American Journal of Agricultural Economics* 78: 89-102.

Hall, R.E. 1982, "The importance of lifetime jobs in the U.S. Economy," *American Economic Review* 72: 716-724.

Hansen, L.P. 1982, "Large Sample Properties of Generalized Method of Moment Estimators," *Econometrica* 50: 1029-1054.

Hausman, J.A. 1978, "Specification tests in econometrics," *Econometrica* 46: 1251-1271.

Hausman, J.A. , B.H. Hall and Z. Griliches 1984, "Econometric models for count data with an application to the Patents-R&D relationship," *Econometrica* 52: 909-938.

Heckman, J.J. 1978, "Dummy endogenous variables in simultaneous equation systems," *Econometrics* 46: 931-959.

Heckman, J.J. 1979, "Sample selection bias as a specification error," *Econometrica* 47: 153-161.

Heckman, J.J. 1981, "Heterogeneity and state dependence," in: Rosen, S. (ed.) *Studies in Labor Markets* University of Chicago Press: Chicago.

Heckman, J.J. and G. Borjas 1980, "Does unemployment cause future unemployment? Definitions, questions and answers from a continuous time model for heterogeneity and state dependence," *Economica* 47: 247-283.

Heckman, J.J. and B. Singer 1984, "A method for minimizing the impact of distributional assumptions in econometric models for duration data," *Econometrica* 52: 271-320.

Hellström, J. 2002, "Count Data Modelling and Tourism Demand," Umea Economic Studies No. 584.

Hinde, J. 1982, "Compound Poisson regression models," in: R. Gilchrist (ed.) *GLIM 1982: Proceedings of the international conference on Generalized linear models*, Springer: Berlin.

Hinde, J. 1992, "Choosing between non-nested models: A simulation approach," in: L. Fahrmeir et al. (eds.) *Advances in GLIM and statistical modelling: proceedings of the GLIM92 and the 7th International Workshop on Statistical Modelling, Munich, 13-17 June 1992.* Springer: New York, 119-124.

Hinkley, D.V. and N. Reid 1991, "Statistical Theory," in: Hinkley, D.V., N. Reid and E.J. Snell (eds.) *Statistical Theory and Modelling*, Chapman and Hall: London.

Hogg, R.V. and A.T. Craig 1978, *Introduction to Mathematical Statistics*, 4th ed., Macmillan: New York.

Hsiao, C. 1986, *Analysis of Panel Data*, New York: Cambridge University Press.

Jaggia, S. and S. Thosar 1993, "Multiple bids as a consequence of target management resistance: a count data approach," *Review of Quantitative Finance and Accounting* 3: 447-457.

Jensen, E.J. 1987, "Research Expenditures and the Discovery of New Drugs," *Journal of Industrial Economics* 36(1): 83-95.

Jimenéz-Martin, S. , J.M. Labeaga and M. Martinez-Granado 2002, "Latent class versus two-part models in the demand for physician services across the European Union," *Health Economics* 11: 301-321.

Johansson, P. and K. Brännäs 1995, "Work absence within the household," in: C.S. Forbes, P. Kofman, and T.R.L. Fry (eds.) *Proceedings of the 1995 Econometrics Conference at Monash*, Department of Economics: Melbourne.

Johansson, P. and M. Palme 1996, "Do Economic Incentives Affect Work Absence? Empirical Evidence Using Swedish Micro Data," *Journal of Public Economics* 59(2): 195-218.

Johnson, N.L. and S. Kotz 1969, *Distributions in statistics: discrete distributions*, Wiley: New York.

Johnson, N.L. and S. Kotz 1970, *Distributions in statistics: continuous distributions*, Houghton-Mifflin: Boston.

Johnson, N.L., S. Kotz and N. Balakrishnan 1994, *Continuous Univariate Distributions*, Vol. 1, 2nd ed., Wiley: New York.

Jørgensen, B., Lundbye-Christensen, S., Xue-Kun Song, P. and Sun, L. 1999, "A State Space Models for Multivariate Longitudinal Count Data," *Biometrika* 86: 169-181.

Jorgenson, D.W. 1961, "Multiple regression analysis of a Poisson process," *Journal of the American Statistical Association* 56: 235-245.

Jovanovic, B. 1979a, "Job matching and the theory of turnover," *Journal of Political Economy* 87: 972-990.

Jovanovic, B. 1979b, "Firm specific capital and turnover," *Journal of Political Economy* 87: 1246-1260.

Jung, R. 1999, *Zeitreihenanalyse für Zähldaten: Eine Untersuchung ganzzahliger Autoregressiver-Moving-Average-Prozesse*, (Reihe: Quantitative Ökonomie, Vol. 100) Eul: Lohmar ; Köln.

Jung, R. and R. Liesenfeld 2001, "Estimating Time Series Models for Count Data Using Efficient Importance Sampling," *Allgemeines Statistisches Archiv* 85: 387-407.

Jung, R. and R. Winkelmann 1993, "Two aspects of labor mobility: A bivariate Poisson regression approach," *Empirical Economics* 18: 543-556.

Kalwij, A. 2000, "The Effects of Female Employment Status on the Presence and Number of Children," *Journal of Population Economics* 13: 221-240.

Kelly, M. 2000, "Inequality and Crime," *Review of Economics and Statistics* 82: 530-539.

Kenkel, D.S. and J.V. Terza 2001, "The Effect of Physician Advice on Alcohol Consumption: Count Regression with an Endogenous Treatment Effect," *Journal of Applied Econometrics* 16: 165-184.

Kennan, J. 1985, "The duration of contract strikes in U.S. manufacturing," *Journal of Econometrics* 28: 5-28.

Kennedy, P. 1981, "Estimation with correctly interpreted dummy variables in semilogarithmic equations," *American Economic Review* 71: 801.

Kennedy, W.J. and J.E. Gentle 1980, *Statistical Computation*, Marcel Dekker: New York.

King, G. 1988, "Statistical models for political science event counts: bias in conventional procedures and evidence for the exponential Poisson regression model," *American Journal of Political Science* 32: 838-862.

King, G. 1989a, "A seemingly unrelated Poisson regression model," *Sociological Methods & Research* 17: 235-255.

King, G. 1989b, "Variance specification in event count models: from restrictive assumptions to a generalized estimator," *American Journal of Political Science* 33: 762-784.

King, G. 1989c, *Unifying political methodology: the likelihood theory of statistical inference*, Cambridge University Press: Cambridge.

Knuth, D.E. 1969, *The Art of Computer Programing;* Vol.2 *Seminumerical Analysis.* Addison-Wesley: Reading, Mass.

Kocherlakota, S. and K. Kocherlakota 1992, *Bivariate discrete distributions*, Marcel Dekker: New York.

Koenker, R. and K. F. Hallok 2001, "Quantile regression," *Journal of Economic Perspectives* 15: 143-156.

Kostiuk, P.F. and D.A. Follmann 1989, "Learning curves, personal characteristics, and job performance," *Journal of Labor Economics* 7(2): 129-146.

Kozumi, H. 1999, "A Bayesian Analysis of Endogenous Switching Models for Count Data," mimeo.

Kulaserka, K.B. and D.W. Tonkyn 1992, "A new discrete distribution with applications to survival, dispersal and dispersion," *Communications in Statistics - Simulations and Computations* 21(2): 499-518.

Laird, N. 1978, "Nonparametric maximum likelihood estimation of a mixing distribution," *Journal of the American Statistical Association* 73: 805-811.

Lakshminarayana, J., S.N.N. Pandit and K.S. Rao 1999, "On a bivariate Poisson distribution," *Communications in Statistics – Theory and Methods* 28: 267-276.

Lambert, D. 1992,"Zero-inflated Poisson regression with an application to defects in manufacturing," *Technometrics* 34: 1-14.

Lancaster, T. 1990, *The econometric analysis of transition data.* Cambridge University Press: Cambridge, U.K.

Lawless, J.F. 1987a, "Regression methods for Poisson process data," *Journal of the American Statistical Association* 82: 808-815.

Lawless, J.F. 1987b, "Negative binomial and mixed Poisson regression," *The Canadian Journal of Statistics* 15(3): 209-225.

Lazear, E. 1990, "The job as a concept," The Hoover Institution Woking Paper E-90-24.

Lee, L.-F. 1983, "Generalized econometric models with selectivity, " *Econometrica* 51: 507-512.

Lee, L.-F. 1986, "Specification test for Poisson regression models," *International Economic Review* 27: 687-706.

Lee, L.-F. 1996, "Specification and Estimation of Count Data Regression and Sample Selection Models – A Counting Process and Waiting Time Approach," Hong Kong University of Sciece and Technology, Department of Economics Working Paper No. 96/19.

Leroux, B.G. 1992, "Consistent estimation of a mixing distribution," *Annals of Statistics* 20: 1350-1360.

Lindsay, B.G. 1983, "The geometry of mixture likelihoods," *Annals of Statistics* 11: 86-94.

List, J.A. 2002, "Determinants of securing academic interviews after tenure denial: Evidence from a zero inflated Poisson model," *Applied Economics* 33: 1423-1431.

Long and Bowyer 1953, "The influence of earnings on the mobility of labour," *Yorkshire Bulletin of Economic and Social Research* 5: 81-87.

Maasoumi, E. 1992, "Rules of thumb and pseudo science," *Journal of Econometrics* 53: 1-4.

Machado, J.A.F. and J.M.C. Santos Silva 2002, "Quantiles for Counts," Institute for Fiscal Studies, University College, London, cemmap working paper CWP22/02.

Maddala, G.S. 1983, *Limited-Dependent and Qualitative Variables in Econometrics*, Cambridge University Press: Cambridge.

Manski, C.F. and S.R. Lerman 1977, "The estimation of choice probabilities from choice based samples," *Econometrica* 45: 1977-1988.

Marshall, A.W. and I. Olkin 1990, "Multivariate distributions generated from mixtures of convolution and product families, " in: H.W. Block, A.R. Sampson, and T.H. Savits (eds.) *Topis in Statistical Dependence*, 371-393, IMS Lecture Notes Monograph Series, Volume 16.

Maul, A., A.H. El-Shaarawi and J.F. Ferard 1991, "Application of negative binomial regression models to the analysis of quantal bioassay data," *Environmetrics* 2: 253-261.

Mayer, J. and R. Riphahn 2000, "Fertility Assimilation of Immigrants: A Varying Coefficient Count Data Model," *Journal of Population Economics* 13: 241-262.

McCullagh, P. 1991, "Quasi-likelihood and estimating functions," in: Hinkley, D.V., N. Reid and E.J. Snell (eds.) *Statistical Theory and Modelling*, Chapman and Hall: London.

McCullagh, P. and J.A. Nelder 1989, *Generalized linear models*, 2nd ed., Chapman and Hall: London.

McIntosh, J. 1999, "An Analysis of Reproductive Behavior in Canada: Results from an Intertemporal Optimizing Model," *Journal of Population Economics* 12: 451-461.

McKelvey, R.D. and W. Zavoina 1975, "A statistical model for the analysis of ordinal level dependent variables," *Journal of Mathematical Sociology* 4: 103-120.

McKenzie E. 1986, "Autoregressive moving average provesses with negative binomial and geometric marginal distributions," *Advances in Applied Probability* 18: 679-705.

McKenzie E. 1988, "Some ARMA models for dependent sequences for Poisson counts," *Advances in Applied Probability* 20: 822-835.

Melkersson, M. and C. Olsson 1999, "Is Visiting the Dentist a Good Habit? Analyzing Count Data with Excess Zeros and Excess Ones," *Umea Economic Studies* No. 492.

Melkersson, M. and D. Roth 2000, "Modeling of Household Fertility Using Inflated Count Data Models," *Journal of Population Economics* 13: 189-204.

Merkle, L. and K.F. Zimmermann 1992, "The demographics of labor turnover: A comparison of ordinal probit and censored count data models," *Recherches Economiques de Louvain* 58: 283-306.

Michener, R. and C. Tighe 1992, "A Poisson regression model of highway fatalities," *American Economic Review* 82: 452-456.

Million, A. 1998, *Zähldatenmodelle für korrelierte abhängige Daten*, unpublished Ph.D. thesis, University of Munich.

Mincer, J. 1962, "On-the-job training: Costs, returns, and some implications," *Journal of Political Economy* 70: 50-79.

Mincer, J. and B. Jovanovic 1981, "Labor mobility and wages," in: S. Rosen (ed.) *Studies in Labor Markets*, The University of Chicago Press: Chicago, 21-62.

Moffatt, P.G. 1995, "Grouped Poisson Regression Models: Theory and an Application to Public House Visit Frequency," *Communications in Statistics: Theory and Methods* 24: 2779-2796.

Moffatt, P.G. 1997a, "Exploiting a matrix identity in the computation of the efficient score test for overdispersion in the Poisson regression model," *Statistics and Probability Letters* 32: 75-79.

Moffatt, P.G. 1997b, "Global log-concavity of the likelihood in models with grouped discrete data," *Australian Journal of Statistics* 39: 105-112.

Moffatt, P.G. and S.A. Peters 2000, "Grouped Zero-Inflated Count Data Models of Coital Frequency," *Journal of Population Economics* 13: 205-220.

Møller Danø, Anne 1999, "Unemployment and Health Conditions – A Count Data Approach," mimeo.

Montalvo, J.G. 1997, "GMM Estimation of Count Panel Data Models with Fixed Effects and Predetermined Instruments," *Journal of Business and Economic Statistics* 15(1): 82-89.

Mortensen, D.T. 1986, "Job search and labor market analysis," in: O. Ashenfelter and R. Layard (eds.) *Handbook of Labor Economics* Vol.II, Amsterdam, North-Holland.

Mukhopadhay, K. and P.K. Trivedi 1995, "Regression models for under-recorded count data," paper presented at the 7th World Congress of the Econometric Society.

Mullahy, J. 1986, "Specification and testing in some modified count data models," *Journal of Econometrics* 33: 341-365.

Mullahy, J. 1997a, "Instrumental Variable Estimation of Count Data Models: Applications to Models of Cigarette Smoking Behavior," *Review of Economics and Statistics* 79(4): 586-593.

Mullahy, J. 1997b, "Heterogeneity, Excess Zeros, and the Structure of Count Data Models," *Journal of Applied Econometrics* 12(3): 337-50.

Mullahy, J. 1999, "Interaction effects and difference-in-difference estimation in loglinear models," NBER Technical Working Paper No. 245.

Mullahy, J. and P.R. Portney 1990, "Air Pollution, Cigarette Smoking and the Production of Respiratory Health," *Journal of Health Economics* 9: 193-205.

Munkin, M.K. and P.K. Trivedi 1999, "Simulated maximum likelihood estimation of multivariate mixed-Poisson regression models, with application," *Econometrics Journal* 2: 29-48.

Nakamura, T. 1990, "Corrected score function for errors-in-variables models: methodology and application to generalized linear models," *Biometrika* 77: 127-137.

Nelder, J.A. and R.W. Wedderburn 1972, "Generalized Linear Models," *Journal of the Royal Statistical Society* A 135: 370-384.

Nguyen-Dinh, H. 1997, "A socioeconomic analysis of the determinants of fertility: the case of Vietnam," *Journal of Population Economics* 10: 251-272.

Okoruwa, A.A., J.V. Terza and H.O. Nourse 1988, "Estimating patronization shares for urban retail centers: An extension of the Poisson gravity model," *Journal of Urban Economics* 24(3): 241-259.

Ophem, H. van 1999, "A General Method to Estimate Correlated Discrete Random Variables," *Econometric Theory* 15(2): 228-237.

Ophem, H. van 2000, "Modeling Selectivity in Count Data Models," *Journal of Economic and Business Statistics* 18: 503-511.

Ozuna, T. and I.A. Gomez 1994, "Estimating a System of Recreation Demand Functions Using a Seemingly Unrelated Poisson Regression Approach," *Review of Economics and Statistics* 76(2): 356-360.

Ozuna, T. and I.A. Gomez 1995, "Specification and Testing of Count Data Recreation Demand Functions," *Empirical Economics* 20(3): 543-550.

Pesaran, M.H. 1974, "On the general problem of model selection," *Review of Economic Studies* 41: 153-171.

Pfeiffer, P.E. 1978, *Concepts of probability theory*, 2nd ed., Dover: New York.

Plassmann, F. and T.N. Tideman 2001, "Does the Right to Carry Concealed Handguns Deter Countable Crimes? Only a Count Analysis Can Say," *Journal of Law and Economics* 44: 771-798.

Pohlmeier, W. and V. Ulrich 1995, "An econometric model of the two-part decision making process in the demand for health care," *Journal of Human Resources* 30: 339-361.

Praag, B.M.S. van and E.M. Vermeulen 1993, "A count-amount model with endogenous recording of observations," *Journal of Applied Econometrics* 8: 383-395.

Prieger, J.E. 2002, "Regulation, innovation and the introduction of new telecommunications services," *Review of Economics and Statistics* 84: 704-715.

Ramaswamy, V., E.W. Anderson and W.S. DeSarbo 1996, "A disaggregate negative binomial regression procedure for count data analysis," *Management Science* 40: 405-417.

Rickard, J.M. 1988, "Factors influencing long distance rail passenger trip rates in Great Britain," *Journal of Transport Economics and Policy* 22(2): 209-233.

Riphahn, R.T., A. Wambach and A. Million 2003, "Incentive Effects in the Demand for Health Care: A Bivariate Panel Count Data Estimation," *Journal of Applied Econometrics*, forthcoming.

Rivers, D. and Q.H. Vuong 1988, "Limited information estimators and exogeneity tests for simultaneous probit models," *Journal of Econometrics* 39: 347-366.

Ronning, G. and R. Jung 1992, "Estimation of a first order autoregressive process with Poisson marginals for count data," in: L. Fahrmeir et al. (eds.) *Advances in GLIM and statistical modelling: proceedings of the GLIM92 and the 7th International Workshop on Statistical Modelling, Munich, 13-17 June 1992.* Springer: New York, 188-194.

Rose, N.L. 1990, "Profitability and product quality: Economic determinants of airline safety performance," *Journal of Political Economy* 98: 944-964.

Rosen, S. 1981, "Introduction," in: Rosen, S. (ed.) *Studies in Labor Markets*, University of Chicago Press: Chicago, 1-19.

Rosen, S. 1992, "Distinguished Fellow: Mincerian labor economics," *Journal of Economic Perspectives* 6(2): 157-170.

Ross, S.M. 1985, *Introduction to Probability Models*, Academic Press: New York.

Ruser, J. 1991, "Workers' Compensation and Occupational Injuries and Illnesses," *Journal of Labor Economics* 9(4): 325-350.

Ruser, J. 1993, "Workers' Compensation and the Distribution of Occupational Injuries," *Journal of Human Resources* 28(3): 593-617.

Saha, A. and D. Dong 1997, "Estimating Nested Count Data Models," *Oxford Bulletin of Economics and Statistics* 59(3): 423-430.

Sander, W. 1992, "The effects of women's schooling on fertility," *Economics Letters* 40: 229-233.

Santos Silva, J.M.C. 1997a, "Unobservables in Count Data Models for On Site Samples," *Economics Letters* 54(3): 217-220.

Santos Silva, J.M.C. 1997b, "Generalized Poisson regression for positive count data," *Communications in Statistics – Simulations and Computations* 26 (3).

Santos Silva, J.M.C. and F. Covas 2000, "A modified hurdle model for completed fertility," *Journal of Population Economics* 13: 173-188.

Santos Silva, J.M.C. 2001, "A Score Test for Non-Nested Hypotheses with Applications to Discrete Data Models," *Journal of Applied Econometrics* 16: 577-597.

Santos Silva, J.M.C. 2003, "A note on the estimation of mixture models under endogenous sampling," *Econometrics Journal* forthcoming.

Santos Silva, J.M.C. and F. Windmeijer 2001, "Two-Part Multiple Spell Models for Health Care Demand," *Journal of Econometrics* 104: 67-89.

Schellhorn, M. 2002, "A comparison of alternative methods to model endogeneity in count models. An application to the demand for health care and health insurance choice," *SEDAP Research Paper No. 40*

Schultz, T.P. 1990, "Testing the neoclassical model of family labor supply and fertility," *Journal of Human Resources* 25(4): 599-634.

Schwalbach, J. and K.F. Zimmermann 1991, "A Poisson Model of Patenting and Firm Structure in Germany," in: Z. Acs and D. Audretsch (Eds.) *Innovation and Technological Change*, Harvester Wheatsheaf, New York et al., 109-120.

Shaw, D. 1988, "On-site samples regression," *Journal of Econometrics* 37: 211-223.

Shonkwiler, J.S. and T.R. Harris 1996, "Rural retail business thresholds and interdependencies," *Journal of Regional Science* 36: 617-630.

Shonkwiler, J.S. and W.S. Shaw 1996, "Hurdle Count Data Models in Recreation Demand Analysis," *Journal of Agricultural and Resource Economics* 21(2): 210-219.

Silcock 1954, "The phenomenon of labour turnover," *Journal of the Royal Statistical Society* B 16: 429-440.

Simar, L. 1976, "Maximum likelihood estimation of a compound Poisson process," *Annals of Statistics* 4: 1200-1209.

Skvoretz, J. 1984, "Career mobility as a Poisson process," *Social Science Research* 13: 198-220.

Smith, T.E. 1987, "Poisson gravity models of spatial flows," *Journal of Regional Science* 27(3): 315-340.

Statistisches Bundesamt (ed.) 1985, "Statistisches Jahrbuch 1985 für die Bundesrepublik Deutschland," Wiesbaden.

StataCorp. (1997), "Stata 5.0 Reference Manual," StataPress: College Station.

StataCorp. (1999), *Stata Statistical Software: Release 6.0.* Stata Corporation: College Station, TX.

Terza, J.V. 1985, "A Tobit type estimator for the censored Poisson regression model," *Economics Letters* 18: 361-365.

Terza, J.V. 1998, "Estimating Count Data Models with Endogenous Switching: Sample Selection and Endogenous Treatment Effects," *Journal of Econometrics* 84(1): 129-154.

Tomlin, K.M. 2000, "The Effects of Model Specification on Foreign Direct Investment Models: An Application of Count Data Models," *Southern Economic Journal* 67: 460-468.

Topel, R.H. and M.P. Ward 1992, "Job mobility and the careers of young men," *Quarterly Journal of Economics* 106(2): 439-479.

Veall, M.R. and K.F. Zimmermann 1992, "Performance measures from prediction-realization tables," *Economics Letters* 39: 129-134.

Vistnes, J.P. 1997, "Gender Differences in Days Lost from Work Due to Illness," *Industrial and Labor Relations Review* 50(2): 304-323.

Vuong, Q.H. 1989, "Likelihood ratio tests for model selection and non-nested hypotheses," *Econometrica* 57(2): 307-333.

Wagner, G.G., R.V. Burkhauser and F. Behringer 1993, "The English language public use file of the German Socio-Economic Panel," *Journal of Human Resources* 28: 429-433.

Wang, P., M.L. Puterman, I.M. Cockburn, and N. Lee 1997, "Mixed Poisson Regression Model With Covariate Dependent Rates," *Biometrics* 52: 381-400.

Wang, P., I.M. Cockburn, and M.L. Puterman 1998, "Analysis of Patent Data: A Mixed Poisson Regression Model Approach," *Journal of Business and Economic Statistics* 16(1): 27-41.

Wang, W. and F. Famoye 1997, "Modeling Household Fertility Decisions with Generalized Poisson Regression," *Journal of Population Economics* 10(3): 273-283.

Wedel, M., W.S. Desarbo, J.R. Bult and V. Ramaswamy 1993, "A latent class Poisson regression model for heterogeneous count data," *Journal of Applied Econometrics* 8: 397-411.

Weiss, A. 1999, "A Simultaneous Binary Choice / Poisson Regression Model With an Application to Credit Card Approvals," mimeo.

White, H. 1982, "Maximum Likelihood Estimation of Misspecified Models," *Econometrica* 50(1): 1-25.

Williams, D.A. 1970, "Discriminating between regression models to determine the pattern of enzyme synthesis in sychronous cell cultures," *Biometrics* 28: 23-32.

Wilson, P.W. 1992, "Count data models without mean-variance restrictions," presented at the European Meeting of the Econometric Society, Brussels

Windmeijer, F.A.G. 2000, "Moment conditions for fixed effects count data models with endogenous regressors," *Economics Letters* 68: 21-24.

Windmeijer, F.A.G. and J.M.C. Santos Silva 1997, "Endogeneity in Count Data Models: An Application to Demand for Health Care," *Journal of Applied Econometrics* 12(3): 281-294.

Winkelmann, R. 1995, "Duration Dependence and Dispersion in Count Data Models," *Journal of Business and Economic Statistics* 13: 467-474.

Winkelmann, R. 1996a, "A Count Data Model for Gamma Waiting Times," *Statistical Papers* 37: 177-187.

Winkelmann, R. 1996b, "Markov chain Monte Carlo analysis of underreported count data with an application to worker absenteeism," *Empirical Economics* 21: 575-587.

Winkelmann, R. 1997, "Poisson Regression with Endogenous Reporting," in: P. Bardsley and V.L. Martin (eds.), Proceedings of the Econometric Society Australasian Meeting 2nd-4th July 1997 volume 4: Microeconometrics, 385-398.

Winkelmann, R. 1998, "Count Data Models with Selectivity," *Econometric Reviews* 17: 339-359.

Winkelmann, R. 1999, "Wages, Firm Size, and Absenteeism," *Applied Economics Letters* 6: 337-341.

Winkelmann, R. 2000, "Seemingly Unrelated Negative Binomial Regression," *Oxford Bulletin of Economics and Statistics*, 62(4), 553-560.

Winkelmann, R. 2001, "Correctly Interpreting the Results from a Log-Linear Regression under Heteroskedasticty - Methods and an Application to the Relative Wages of Immigrants," *Jahrbücher für Nationalökonomie und Statistik* 221: 418-431.

Winkelmann, R. 2003, "Health Care Reform and the Number of Doctor Visits - An Econometric Analysis," *Journal of Applied Econometrics* forthcoming.

Winkelmann, R. and K.F. Zimmermann 1991, "A new approach for modeling economic count data," *Economics Letters* 37: 139-143.

Winkelmann, R. and K.F. Zimmermann 1992a, "Robust Poisson regression," in: L. Fahrmeir et al. (eds.) *Advances in GLIM and statistical modelling: proceedings of the GLIM92 and the 7th International Workshop on Statistical Modelling, Munich, 13-17 June 1992.* Springer: New York, 201-206.

Winkelmann, R. and K.F. Zimmermann 1992b, "Recursive probability estimators for count data," in: Haag, G., U. Mueller and K.G. Troitzsch (eds.)

Economic Evolution and Demographic Change. Springer: New York, 321-329.

Winkelmann, R. and K.F. Zimmermann 1993a, "Job separations in an efficient turnover model," in: Bunzel, H., P. Jensen and N. Westergard-Nielsen, *Panel Data and Labour Market Dynamics* North-Holland: Amsterdam, 107-122.

Winkelmann, R. and K.F. Zimmermann 1993b, "Ageing, migration and labour mobility," in: P. Johnson and K.F. Zimmermann (eds.) *Labour Markets in an Ageing Europe*, Cambridge University Press: Cambridge U.K., 255-283.

Winkelmann, R. and K.F. Zimmermann 1993c, "Poisson-Logistic Regression," Department of Economics, University of Munich, Working Paper No. 93-18.

Winkelmann, R. and K.F. Zimmermann 1994, "Count data models for demographic data," *Mathematical Population Studies* 4: 205-221.

Winkelmann, R. and K.F. Zimmermann 1995, "Recent developments in count data modeling: Theory and applications," *Journal of Economic Surveys* 9: 1-24.

Winkelmann, R. and K.F. Zimmermann 1998, "Is Job Stability Declining in Germany? Evidence From Count Data Models," *Applied Economics* 30: 1413-1420.

Winkelmann, R. and K.F. Zimmermann 2000, "Editorial: Fertility Studies Using Count Data Models," *Journal of Population Economics* 13: 171-172.

Winkelmann, R., C. Signorino, and G. King 1995, "A Correction for an Underdispersed Event Count Probability Distribution," *Political Analysis* 5: 215-228.

Wooldridge, J.M 1997a, "Multiplicative Panel Data Models Without the Strict Exogeneity Assumption," *Econometric Theory* 13: 667-679.

Wooldridge, J.M 1997b, "Quasi-likelihood methods for count data," in: M.H. Pesaran amd P. Schmidt (eds.) *Handbook of Applied Econometrics* Vol. II, Microeconometrics, Blackwell.

Wooldridge, J.M 1999, "Distribution-free estimation of some nonlinear panel data models," *Journal of Econometrics* 90: 77-97.

Wun, L.-M. 1991, "Regression analysis for autocorrelated Poisson distributed data," *Communications in Statistics - Theory and Methods* 20(10): 3083-3091.

Yen, S.T. 1999, "Gaussian versus count-data hurdle models: cigarette consumption by women in the US," *Applied Economics Letters* 6: 73-76.

Yousry, M.A. and R.C. Srivastava 1987, "The hyper negative binomial distribution," *Biometric Journal* 29: 875-884.

Zeger, S.L. 1988, "A regression model for time series of counts," *Biometrika* 75(4): 621-629.

Zellner, A. 1962, "An Efficient Method of Estimating Seemingly Unrelated Regressions and Tests of Aggregation Bias," *Journal of the American Statistical Association* 57: 348-368.

Zellner, A. 1971, *An Introduction to Bayesian Inference in Econometrics*, John Wiley: New York.

Zimmermann, K.F. and J. Schwalbach 1991, "Determinanten der Patentaktivität," *Ifo-Studien* 37: 201-227.

Author's Index

Subject Index